Some Topics
in
Two-Person Games

Modern Analytic *and* Computational Methods *in* Science *and* Mathematics

A GROUP OF MONOGRAPHS
AND ADVANCED TEXTBOOKS

Richard Bellman, EDITOR
University of Southern California

Some Topics in Two-Person Games

T. Parthasarathy
Case Western Reserve University
Cleveland, Ohio

T.E.S. Raghavan
University of Essex
England

University of Illinois
Chicago, Illinois

American Elsevier
Publishing Company, Inc.
NEW YORK · 1971

AMERICAN ELSEVIER PUBLISHING COMPANY, INC.
52 Vanderbilt Avenue, New York, N.Y. 10017

ELSEVIER PUBLISHING COMPANY, LTD.
Barking, Essex, England

ELSEVIER PUBLISHING COMPANY
335 Jan Van Galenstraat, P.O. Box 211
Amsterdam, The Netherlands

International Standard Book Number 0-444-00059-3

Library of Congress Card Number 76-75524

AMS 1970 Subject Classifications
90D05, 90D10, 90D12, 46A05, 52XX02

Printed in the United States of America

CONTENTS

PREFACE

This book deals with the mathematical aspects of the theory of two-person games. We have discussed only certain features of two-person games. We have given references at the end of each chapter, but we have not attempted to give an exhaustive list.

In Chapter One, basic requirements in set topology and linear spaces (necessary for understanding general minimax theorems) have been developed. Also theorems on convex sets (such as the separation theorem and the fixed-point theorem) and convex functions have been proved. In Chapter Two, notion of games in extensive and normal form is given, and the fundamental minimax theorem due to von Neumann is established.

Chapter Three contains results on the properties of optimal mixed strategies and their relationship to the form of the payoff matrix and the value of the game. Some of the theorems are concerned with positive matrices and the relationship between their spectral properties and their values. A section contains results on saddle-point theorems. Chapter Four describes the simple algorithm for finding a pair of optimal strategies. Since this is closely linked to the problem of linear programming, some aspects of the general problems and the algorithm have been discussed. The Kuhn–Tucker theorem on non-linear programming is also included.

In Chapter Five, we prove certain general minimax theorems that are extensions of Wald and Ky Fan. In Chapter Six we have indicated that game theory can be successfully used as a tool to prove theorems in other branches of sciences.

Chapter Seven deals with non-zero-sum noncooperative two-person games. A theorem due to Nash on the existence of equilibrium pair in mixed strategies for finite non-zero-sum two-person games is proved. The relationship between equilibrium points and solutions of a certain nonlinear programming problem is also discussed.

Chapter Eight gives an introduction to differential games, and Chapter Nine gives an introduction to n-person solution theory due to von Neumann and Morgenstern. In Chapter Nine, the example due to Lucas—which tells us that every n-person game need not possess a solution—is given. In Chapter Ten, we prove that the stochastic game has a value if and only if certain dummy games have a value, using the results on dynamic programming when the state space is uncountable. We have also established some partial results on limiting average effective payoff on a stochastic game. In the last chapter a few open problems are mentioned.

We take great pleasure in thanking Professors C. R. Rao, G. A. Barnard, and Richard Bellman for their interest in this work. We are extremely indebted to Dr. A. Maitra and Dr. L. S. Shapley for their helpful comments. We are grateful to the following institutions: Indian Statistical Institute, Calcutta, University of California, Berkeley, and University of Essex, Colchester, for the research facilities that they have given to us while writing this book. Our thanks are also due to Miss Jean McMillan for her efficient typing of the manuscript.

T. PARTHASARATHY
T. E. S. RAGHAVAN

Chapter One

PRELIMINARIES

In this chapter we shall develop our basic requirements in set topology and linear spaces necessary for understanding general minimax theorems. We shall also prove theorems on convex sets and convex functions that will be needed in the sequel. Occasionally we may explain without proof certain results in normed linear spaces and in the theory of measures on abstract spaces. For a proof of them the reader is referred to the book by Kolmogorov and Fomin [4].

1.1 SETS AND ORDER

Sets

A set is a collection of elements. If an element x belongs to a set X, we say that $x \in X$. If it does not, then we say that $x \notin X$. A set consisting of the single element x is denoted as $\{x\}$. The *empty set* is a set with no elements, and we denote it by ø. We say that X is a *subset* of Y if every element of X is an element of Y. This is written as $X \subset Y$. We say that $X = Y$ if $X \subset Y$ and $Y \subset X$. The set $A_1 \cup A_2$ denotes the new set consisting of elements that are in A_1 or in A_2. The set $\bigcup_{n=1}^{\infty} A_n$ is the set whose elements belong to at least one A_n. In general, $\bigcup_{\alpha \in p} A_\alpha$ denotes the set whose elements belong to at least one A_α, where α is just some index. This symbol $\cup$ is called *union*. The set $\bigcap_{\alpha \in p} A_\alpha$ denotes the set of elements where each element belongs to every one of the A_α's. This symbol $\cap$ is called *intersection*. We say that A and B are disjoint whenever $A \cap B = $ ø. For any set A, the set of all points $x \notin A$ is denoted as A', the complement of A. Sometimes we write $X \ni x$ for $x \in X$.

Functions

Let X and Y be two sets. Then f is a *function* on X into Y if, for every $x \in X$, we associate exactly one element $y \in Y$ denoted by $f(x)$. Then X is called the *domain* of f, and the set of such $f(x)$ for x ranging over X denoted by $\{f(x): x \in X\}$ is called the *range* of f. Sometimes it is also written as $f(X)$. The set of points or elements in X whose functional value lies in a set $M \subset Y$ is denoted by $\{x: f(x) \in M\}$. It is also written as $f^{-1}(M)$, the *inverse image* of M.

Symbolically we write $f: X \to Y$. If $f(X) = Y$, then f is called *onto*. If f is onto and if, for every $y \in Y$, $f^{-1}(y)$ is a single-element set, then f is called a *one-to-one* mapping. If $f: X \to Y$ and $g: Y \to Z$, then $g \circ f: X \to Z$. Here $g \circ f$ denotes the function that takes $x \in X$ to $g(f(x))$ in Z. This is called a *composite function*.

The following is a well-known axiom in set theory.

Axiom of Choice

Let $\{U_\alpha, \alpha \in A\}$ be a family of sets indexed by $\alpha \in A$. Then it is possible to form a new set by choosing exactly one element from each U_α.

Product Set

Let (a, b) be all possible ordered pairs of elements with $a \in X$ and $b \in Y$. This new set is called the *Cartesian product* of X and Y, denoted by $X \times Y$. In fact, this could be thought of as the set of all functions on the set $\{1, 2\}$ consisting of numbers 1 and 2 with $f(1) \in X$ and $f(2) \in Y$. A typical such function with $f(1) = a$ and $f(2) = b$ is our ordered pair (a, b). The *product space* $\prod_{\alpha \in A} X_\alpha$ consists of the space of all functions f on the set A with

$$f(\alpha) \in X_\alpha \text{ for } \alpha \in A.$$

Partial Order

A set X is said to be *partially ordered* with respect to the binary relation $\leq$ among certain pairs of X if, for any x, y, $z \in X$, (1) $x \leq x$; (2) if $x \leq y$, $y \leq x$, then $x = y$; (3) if $x \leq y$, $y \leq z$, then $x \leq z$.

Every subset of a partially ordered set is also partially ordered with respect to the same binary relation. The set X with partial order $\leq$ is denoted by $(X, \leq)$. Thus $x \in (X, \leq)$ means that the element x belongs to the set X, partially ordered by $\leq$.

We call c an *upperbound* for the set $M \subset (X, \leq)$ if $x \leq c$ for all $x \in M$. Then c is the *supremum* for M if c is an upperbound for M and if, for any other upperbound d of M, $c \leq d$. An element $x_0 \in (X, \leq)$ is a *maximal* element if there exists no $x \neq x_0$, with $x_0 \leq x$. Equivalently, x_0 is maximal for $(X, \leq)$ if $x_0 \leq y$ for some $y \in X$ implies $x_0 = y$. A partially ordered set $(X, \leq)$ is called a *directed set* if for every pair of points there is an upperbound.

Linear Order

A set X is said to be linearly ordered with respect to the relation $\leq$ if, for any pair x, $y \in X$, either $x \leq y$ or $y \leq x$. If every pair in X can be compared by $\leq$, then it becomes a linear order. Thus certain subsets of a partially ordered set $(X, \leq)$ are linearly ordered. We also call a linearly ordered set a *chain*.

Zorn's Lemma

If in a partially ordered set $(X, \leq)$ every linearly ordered subset of X has an upperbound, then $(X, \leq)$ has a maximal element. This could be proved by the axiom of choice. For a proof see [3].

1.2 TOPOLOGICAL SPACES

Open and Closed Sets

A set X with a collection T of subsets of X is called *a topological space* if, (1) for any $A, B \in T$ implies $A \cap B \in T$; (2) for any $A_\alpha \in T$, $\alpha \in I$ implies $\underset{\alpha \in I}{\cup} A_\alpha \in T$; (3) $\emptyset$ and $X \in T$. Elements of the collection T or the sets belonging to the collection T are called *open*.

A set C has x as a *limit point* if, for any open set G containing x, $G \cap C$ contains a point other than x. A set C is *closed* if and only if (abbreviated as iff) it contains all its limit points. One can easily prove from the definition that C is closed iff C' is open. We say that G is a *neighborhood* of a point x whenever G contains an open set containing x.

A topological space (X, T) is Hausdorff if, for any arbitrary x, $y \in X$, $x \neq y$, there exist disjoint neighborhoods N_x and N_y for x and y, respectively.

The *closure* of a set M, denoted by $\overline{M}$, consists of the set of all points of M and their limit points. One can check that $(\overline{\overline{M}}) = \overline{M}$ for any set $M \subset (X, T)$.

A point x is interior to M if M contains some neighborhood N_x of x. A point y is a boundary point of M if every neighborhood G_y of y contains points of M and M'.

For any set $Y \subset (X, T)$, the relative topology for Y consists of all sets of the form $Y \cap G$, where $G \in T$. (Check that it is a topology for Y.)

If (X, T) and (Y, T_0) are two topological spaces with $f: X \rightarrow Y$, then f is *continuous* if, for every $G \in T_0$, $f^{-1}(G) \in T$. Call f *continuous at a point* $x_0 \in X$ if, for any $V \in T_0$, and containing $f(x_0)$, there exists a $U \in T$ with $x_0 \in U$ and $f(U) \subset V$. (Check the fact that, if f is continuous at each $x \in X$, then f is continuous.)

Compact Sets

A family $\{G_\alpha\}$ of sets is an *open cover* for (X, T) if $\underset{\alpha}{\cup} G_\alpha = X$ and $G_\alpha \in T$ for all α. Call (X, T) a *compact* topological space if, for any open cover $\{G_\alpha\}$ of X, there exists a finite of them—say, G_{α_1}, G_{α_2}, $\cdots$, G_{α_k}, with

$\overset{k}{\underset{=1}{\cup}} G_{\alpha_j} = X$. A set $Y \subset (X, T)$ is *relatively compact* or just compact if it is compact in its relative topology.

Theorem 1.2.1

Compact subset of a Hausdorff space is closed.

PROOF: Let $Y \subset X$ be a compact subset of the Hausdorff X. Clearly Y is Hausdorff in its relative topology. Let y_0 be a limit point of Y, with $y_0 \notin Y$. There exists for each $y \in Y$ an open set V_y containing y disjoint from an open set $G_{y_0}^{(y)}$ containing y_0. The V_y's with $y \in Y$ are evidently an open cover of Y. Since Y is compact, we have a finite of them—say, $V_{y_1}, V_{y_2}, \cdots, V_{y_s}$, which themselves cover Y. Now $G_{y_0}^{(y_j)} \cap V_{y_j} = \emptyset$ for $j = 1, 2, \cdots, s$. Thus $\overset{s}{\underset{j=1}{\cap}} G_{y_0}^{(y_j)}$ is an open set containing y_0, and it is disjoint with every V_{y_j} and therefore with Y. This contradicts our assumption that y_0 is a limit point of Y. Thus $y_0 \in Y$. Since y_0 is arbitrary, Y contains all its limit points and hence it is closed.

Theorem 1.2.2

Closed subset of a compact set is compact.

PROOF: Let $C \subset (X, T)$ be closed, and let (X, T) be compact. The set $X \cap C'$, henceforth denoted by $(X - C)$, is open. Let $\{G_\alpha\}$ be any open cover of C. Then $G_\alpha = H_\alpha \cap C$, where $H_\alpha \in T$. Since $C = \cup G_\alpha$, $(\underset{\alpha}{\cup} G_\alpha) \cup (X - C) = X$. By assumption, X is compact, and hence there exists a finite of the H_α's which with $X - C$ cover X. Say, for example, that H_{α_1}, $H_{\alpha_2}, \cdots, H_{\alpha_s}$, $X - C$ cover X. Clearly $H_{\alpha_1}, H_{\alpha_2}, \cdots, H_{\alpha_s}$ cover C. Thus $H_{\alpha_1} \cap C = G_{\alpha_1}, H_{\alpha_2} \cap C = G_{\alpha_2}, \cdots, H_{\alpha_s} \cap C = G_{\alpha_s}$ cover C. The theorem is proved.

Theorem 1.2.3

If (X, T) is a compact topological space and if $f: X \to Y$ is a continuous map of X onto (Y, T_0), then (Y, T_0) is compact.

PROOF: Let $\{G_\alpha\}$ be an open cover of Y. Since f is continuous, $H_\alpha = f^{-1}(G_\alpha) \in T$. Further, $f^{-1}(\underset{\alpha}{\cup} G_\alpha) = f^{-1}(Y) = X$. But $f^{-1}(\cup G_\alpha) = \cup f^{-1}(G_\alpha) = \underset{\alpha}{\cup} H_\alpha = X$. Thus for the open cover $\{H_\alpha\}$ of X, we have by assumption a finite subcover—say, $H_{\alpha_1}, H_{\alpha_2}, \cdots, H_{\alpha_s}$. Thus $f(\overset{s}{\underset{j=1}{\cup}} H_{\alpha_j}) = f(X) = Y$. But $f(\overset{s}{\underset{j=1}{\cup}} H_{\alpha_j}) = \overset{s}{\underset{j=1}{\cup}} f(H_{\alpha_j}) = \overset{s}{\underset{j=1}{\cup}} G_{\alpha_j}$. Hence the theorem is proved.

Product Topology

Let (X_α, T_α) be a family of topological spaces. Then a topology T on ΠX_α can be introduced as follows. Any set $\Pi\, G_\alpha \subset \Pi X_\alpha$ is called open if G_α is an open subset of (X_α, T_α) for each α, and $G_\alpha = X_\alpha$ for all but a finite of the indices α. This topology is called the *product topology*. Let $\{x_\alpha\}$ be an element of ΠX_α; then the map $P_{\alpha_0}: \{x_\alpha\} \to x_{\alpha_0}$, which sends the set ΠX_α to X_{α_0}, is called the *projection* of the product space onto the α_0th coordinate space. In this topology T, the projections $\{P_\alpha\}$ are continuous functions from $\Pi X_\alpha \to X_\alpha$. Further, the map P_α is *open* (that is, the P_α image of an open set in ΠX_α is open in X_α).

Generalized Sequence

The set of integers is evidently a directed set, and we have the following generalization of sequences. Since any sequence is just a function on integers, any *generalized sequence* is defined as a function on a directed set $(D, \le)$ with range in a topological space X. We generally denote it by $\{x_d\}$, $d \in (D, \le)$, $x_d \in X$. A generalized sequence $\{x_d\}$ in X is said to converge to x_0 if, for any neighborhood N_{x_0} of x_0, $x_d \in N_{x_0}$ for all $d \ge d_0(x_0)$, for some d_0. Then $\{x_{d_j}\} \subset \{x_d\}$ is called a *generalized subsequence* of the generalized sequence $\{x_d\}$ if, for any $d \in D$, there exists a $d_k \ge d$ such that $x_{d_k} \in \{x_{d_j}\}$.

Theorem 1.2.4

A set $Y \subset (X, T)$ has y_0 in its closure $\overline{Y}$ if there exists a generalized sequence in Y converging to y_0.

PROOF: Let $\{y_d\}$ be a generalized sequence in Y with $d \in (D, \le)$ a directed set. Let $y_d \to y_0$. Since each neighborhood N_{y_0} of y_0 contains some $y_d \in N_{y_0}$ for $d \ge d(y_0)$, and since there exists at least one $d \ge d(y_0)$, we are through; that is, y_0 belongs to the set $\overline{Y}$. Conversely, let y_0 be a limit point of $Y \subset (X, T)$. Consider all neighborhoods of y_0, and define $N_{y_0} \le M_{y_0}$ for any two of the neighborhoods N_{y_0}, M_{y_0} when $N_{y_0} \supseteq M_{y_0}$. Then the collection $\{N_{y_0}^\alpha\}$ of the neighborhoods of y_0 is directed by $\le$. Choose one element $y_\alpha \in N_{y_0}^\alpha$ for each α. This is possible by the axiom of choice. Clearly $y_\alpha \to y_0$, and it is a generalized sequence.

Theorem 1.2.5

A set (X, T) is compact if every generalized sequence in X has a convergent generalized subsequence.

Let (X, T) be compact, and let $\{x_d\}$ be any generalized sequence in X with the d's in the directed set $(D, \le)$. We shall construct a generalized convergent subsequence. For that we need first of all the following lemma.

LEMMA: A set (X, T) is compact if, for any arbitrary family $\{C_\alpha\}$ of closed sets with the property that every finite subfamily of them has a non-null intersection, their total intersection is non-null.

PROOF: Let $\cap_\alpha C_\alpha = \emptyset$; that is, $(\cap_\alpha C_\alpha)' = \cup_\alpha C_\alpha' = (\emptyset)' = X$.

By the compactness of X there exists a finite subfamily of the open sets C_α' which cover X, say $C_{\alpha_1}', C_{\alpha_2}', \cdots, C_{\alpha_s}'$. Then $\overset{s}{\underset{j=1}{\cup}} C_{\alpha_j}' = X$ implies $\overset{s}{\underset{j=1}{\cap}} C_{\alpha_j} = \emptyset$. This contradicts the assumption that the finite subfamily has a non-null intersection. Thus when X is compact, $\underset{\alpha}{\cap} C_\alpha \neq \emptyset$. (The C's are said to possess *finite intersection property*.)

Conversely, let (X, T) be a topological space where for any family $\{C_\alpha\}$ of closed sets with finite intersection property their total intersection is non-null. Let $\{G_\tau\}$ be any open cover of X. Suppose that $\overset{s}{\underset{j=1}{\cup}} G_{\tau_j} \neq X$ for any finite indices $\tau_1, \tau_2, \cdots, \tau_s$. We shall arrive at a contradiction. We have $\overset{s}{\underset{j=1}{\cap}} G_{\tau_j}' \neq \emptyset$. Thus the family $\{G_\tau'\}$ is a family of closed sets with finite intersection property. By assumption we have $\underset{\tau}{\cap} G_\tau' \neq \emptyset$. Let $x \in \underset{\tau}{\cap} G_\tau'$; that is, $x \in (\underset{\tau}{\cup} G_\tau)'$, $x \notin (\underset{\tau}{\cup} G_\tau)$. This is a contradiction to the assumption that $\{G_\tau\}$ is a cover of X.

PROOF OF THE THEOREM: Let $C_\alpha = \{x_d : \alpha \leq d\}$. Clearly $\{\overline{C_\alpha}\}$ is a family of closed sets with finite intersection property. By assumption, (X, T) is compact. The lemma above shows that $\underset{\alpha}{\cap} \overline{C}_\alpha \neq \emptyset$. Let $x_0 \in \underset{\alpha}{\cap} \overline{C}_\alpha$. Clearly for any neighborhood N_{x_0} of x_0, $N_{x_0} \cap \overline{C}_\alpha$ contains some x_d in $\overline{C}_\alpha$ for each α. Let it be x_{d_α}. Clearly by the axiom of choice we can choose one such x_{d_α} from each C_α, and we are through.

Conversely, let every generalized sequence have a convergent generalized subsequence. We shall prove that (X, T) is compact. Let $\{C_\alpha\}$ be an arbitrary family of closed sets with finite intersection property. Take the family $\mathscr{F}$ of their finite intersections; $F \in \mathscr{F}$ implies that $F = \overset{s}{\underset{j=1}{\cap}} C_{\alpha_j} \neq \emptyset$ for some finite indices $\alpha_1, \alpha_2, \cdots, \alpha_s$ and that $F \neq \emptyset$. Partial-order $\mathscr{F}$ by $\supset$; that is, $F \leq G$ if $F \supset G$. Clearly $(\mathscr{F}, \leq)$ is a directed set, for you can go on intersecting. Choose an $x_F \in F$ for each $F \in \mathscr{F}$ (as $F \neq \emptyset$, this is possible). Being a generalized sequence, $\{x_F\}$ has a convergent generalized subsequence $x_{F_\tau} \to x_0$. Since every neighborhood N_{x_0} contains some x_{F_τ}, and further since for every $F \in \mathscr{F}$ we have an $F_\tau \geq F$ or equivalently $F_\tau \subset F$ with $x_{F_\tau} \in F_\tau$, the set $F = C_\alpha$ has the property that $N_{x_0} \cap F \supset N_{x_0} \cap F_\tau \ni x_{F_\tau}$, and that $N_{x_0} \cap F \neq \emptyset$ for all $F = C_\alpha$. Since N_{x_0} is any arbitrary neighborhood of x_0, x_0 is a limit point of C_α, or $x_0 \in C_\alpha$ for each α. Since C_α's are closed, $x_0 \in C_\alpha$ for each α; that is, $\underset{\alpha}{\cap} C_\alpha \neq \emptyset$. Hence X is compact.

1.3 FILTERS AND PRODUCT TOPOLOGY

Filter

A collection $\mathscr{F}$ of subsets of a set X is said to form a *filter* if (1) $\emptyset \notin \mathscr{F}$; (2) $A, B \in \mathscr{F}$ implies $A \cap B \in \mathscr{F}$; (3) $A \supset B$, $B \in \mathscr{F}$ implies $A \in \mathscr{F}$. A filter $\mathscr{F}_0$ is called an ultrafilter in X if there exists no other filter $\mathscr{F}$ in X that strictly contains $\mathscr{F}_0$ as a subcollection.

Theorem 1.3.1

For any ultrafilter $\mathscr{F}_0$ in X and for any $A \subset X$, exactly one of A or $A' \in \mathscr{F}_0$.

PROOF: If $A \cap B = \emptyset$ for some $B \in \mathscr{F}_0$, then $B \subset A'$ and, by the definition of $\mathscr{F}_0$ as a filter, $A' \in \mathscr{F}_0$. Further, $A \notin \mathscr{F}_0$, for otherwise $A \cap A' = \emptyset \in \mathscr{F}_0$, which is not true. On the other hand, if $A \cap B \neq \emptyset$ for every $B \in \mathscr{F}_0$, then the collection $\mathscr{F}_*$ consisting of

$$\mathscr{F}_* = \{D: D \in \mathscr{F}_0\} \cup \{A \cap D: D \in \mathscr{F}_0\} \cup \{E: E \supset A \cap D\}$$

is itself a filter containing properly $\mathscr{F}_0$ if $A \neq D$ for any $D \in \mathscr{F}_0$. This contradicts the maximality of $\mathscr{F}_0$. In this case, $A \in \mathscr{F}_0$.

Filter Convergence

A filter $\mathscr{F}$ is said to converge to a point x_0 ($\mathscr{F} \to x_0$) if every neighborhood of x_0 is in $\mathscr{F}$.

Theorem 1.3.2

(X, T) is compact if and only if every ultrafilter in X converges.

PROOF: Let X be compact, and let $\mathscr{F}_0$ be an ultrafilter in X. Suppose that $\mathscr{F}_0$ does not converge to any point x. Then for each $x \in X$, we have an open neighborhood G_x of x such that $G_x \notin \mathscr{F}_0$. By theorem 1.3.1, $G_x' \in \mathscr{F}_0$. Now $\underset{x \in X}{\cup} G_x$ is a cover of X and hence has a finite subcover, say $\overset{k}{\underset{i=1}{\cup}} G_{x_i} = X$. Clearly $G_{x_i} \notin \mathscr{F}_0$ for $i = 1, 2, \cdots, k$ implies simply by induction $\overset{k}{\underset{i=1}{\cap}} G_{x_i}' = (X)' = \emptyset \in \mathscr{F}_0$. This contradicts the filter property of $\mathscr{F}_0$. Conversely, let every ultrafilter $\mathscr{F}_0$ in X converge to some point. Further, let $\{C_\alpha\}$ be any arbitrary family of closed sets in X with finite intersection property. Form all finite intersections of such C_α's, and take supersets of such sets. Let $\mathscr{F} = \{F: F \supseteq \bigcap_I C_\alpha,$ I's being finite subsets of the index set $\Lambda \ni \alpha\}$. Then $\mathscr{F}$ is evidently a filter, for $F \in \mathscr{F}$ implies $F \supset \bigcap_I C_\alpha$ for some finite indices $I = (\alpha_1, \alpha_2, \cdots, \alpha_k)$. By assumption, $\bigcap_I C_\alpha \neq \varnothing$, and hence $F \neq \varnothing$. Trivially the other requirements of a filter are satisfied. Now consider the family of all filters containing it. They are partially ordered by collection inclusion $(\supset)$, and for any chain $\{\mathscr{F}_\tau\} \supset \mathscr{F}$, $\bigcup_\tau \mathscr{F}_\tau$ is itself an upperbound. By Zorn's lemma there exists a maximal element in the collection of filters containing $\mathscr{F}$. Call it $\mathscr{F}_0$, an ultrafilter containing $\mathscr{F}$. By assumption, $\mathscr{F}_0 \to x_0$ for some x_0; that is, every neighborhood N_{x_0} of x_0 is an element of $\mathscr{F}_0$. Since $C_\alpha \in \mathscr{F}_0$, $N_{x_0} \cap C_\alpha \neq \varnothing$ for any α. In this case N_{x_0} is an arbitrary neighborhood. This shows that $x_0 \in C_\alpha$, or x_0 is a limit point of C_α for each α. Since C_α's are closed, $x_0 \in C_\alpha$ for all α in either case. Thus $\bigcap_\alpha C_\alpha \neq \varnothing$. By the lemma to theorem 1.2.5, X is compact.

Theorem 1.3.3

Let (X_α, T_α) be compact spaces for $\alpha \in \Delta$. Then $\prod_{\alpha \in \Delta} X_\alpha$ is compact in the product topology.

PROOF: We shall show that any ultrafilter $\mathscr{F}_0$ in $\prod_{\alpha \in \Delta} X_\alpha$ converges; this would by the previous theorem prove our statement. Consider any $V \in \mathscr{F}_0$. Project V to X_α; that is, consider $P_\alpha V = \{\alpha\text{th coordinate of } x \in V\}$. Since $\mathscr{F}_0$ is an ultrafilter, $P_\alpha \mathscr{F}_0 = \{G_\alpha: P_\alpha G = G_\alpha, G \in \mathscr{F}_0\}$ is also an ultrafilter. As $\{G_\alpha, G \in \mathscr{F}_0\}$ is an ultrafilter in X_α for a fixed α, $P_\alpha \mathscr{F}_0 \to x_\alpha^\circ$ by the compactness of X_α. We shall show that the point $\{x_\alpha^\circ\}$ is the one to which $\mathscr{F}_0$ converges in the product topology. Let H be an open set of $\prod_{\alpha \in \Delta} X_\alpha$. By definition $H = \prod_{\alpha \in \Delta} H_\alpha$, H_α's open for all α, and $H_\alpha = X_\alpha$ except for a finite $\alpha \in \Delta$. We have $P_\alpha \mathscr{F}_0 \to x_\alpha^\circ$ for each α, and $P_{\alpha_1, \alpha_2, \cdots, \alpha_k} \mathscr{F}_0 \to \{x_{\alpha_1}^\circ, x_{\alpha_2}^\circ, \cdots, x_{\alpha_k}^\circ\}$ for any finite $\alpha_1, \alpha_2, \cdots, \alpha_k$. (Here $P_{\alpha_1, \cdots, \alpha_k}$ is the projection map of $\prod_{\alpha \in \Delta} X_\alpha$ into $\prod_{i \leq j \leq k} X_{\alpha_j}$.) In fact this is what we want, as any neighborhood of $(x_{\alpha_1}^\circ, x_{\alpha_2}^\circ, \cdots, x_{\alpha_k}^\circ)$ in $\prod_{i \leq j \leq k} X_{\alpha_j}$ belongs to $P_{\alpha_1, \alpha_2, \cdots, \alpha_k} \mathscr{F}_0$. The fact that $H_\alpha = X_\alpha$ for the rest of the indices, coupled with the above statement, shows that $\mathscr{F}_0 \to \{x_\alpha^\circ\}$ in the product topology.

1.4 CONVEX SETS AND LINEAR SPACES

Convex Sets

A set K in a real or complex linear space E is said to be *convex* if $x, y \in K$, $0 \leq \lambda \leq 1$ implies $\lambda x + (1 - \lambda) y \in K$.

Locally Convex Space

A linear space E over the field Λ with a topology is said to constitute a linear topological space if (1) $(x, y) \to x - y$ is continuous from $E \times E$ to E, and (2) $(\alpha, x) \to \alpha x$ is continuous from $\Lambda \times E$ to E.

A Hausdorff linear topological space over the real or complex number field is a *locally convex space* if every neighborhood of a point contains a convex symmetric neighborhood U of the point (that is, $y \in U$ implies $-y \in U$).

Theorem 1.4.1

Let $\{K_\alpha\}$ be a family of convex sets in a real or complex linear space E. Then $\underset{\alpha}{\cap}\, K_\alpha$ is convex.

PROOF: If $x, y \in \underset{\alpha}{\cap}\, K_\alpha$, then $x, y \in K_\alpha$ for each α, and by the convexity of K_α, $\lambda x + (1 - \lambda) y \in K_\alpha$ for each α for any $0 \le \lambda \le 1$. Thus $\lambda x + (1 - \lambda) y \in \underset{\alpha}{\cap}\, K_\alpha$. Hence the theorem.

Theorem 1.4.2

The closure of a convex set K is a convex set in a real or complex linear topological space E.

Let $x_0, y_0 \in \overline{K}$, the closure of K. For any λ in $0 \le \lambda \le 1$, $f(x, y) = \lambda x + (1 - \lambda) y$ is a continuous map of $E \times E$ into E. Let G be any neighborhood of $\lambda x_0 + (1 - \lambda) y_0$. If we show that $G \cap K \neq \emptyset$, then $\lambda x_0 + (1 - \lambda) y_0 \in \overline{K}$, which would prove the result. Since $f(x, y)$ is continuous, there exist V and W, neighborhoods of x_0 and y_0 such that $f(x, y) \in G$ when $x \in V$ and $y \in W$. Since $x_0 \in \overline{K}$, there exists an $x_1 \in V$, and $x_1 \in K$. Similarly, $y_1 \in W$, and $y_1 \in K$. By the convexity of K, $\lambda x_1 + (1 - \lambda) y_1 \in K \subset \overline{K}$. Also $f(x_1, y_1) \in G$. Hence the result.

Theorem 1.4.3

Let K be a convex set with non-null interior in a real or complex linear topological space. Let x be a boundary point of K, and y an interior point of K. Then for any $0 < \lambda < 1$, $\lambda x + (1 - \lambda) y$ is an interior point of K. For this we need the following lemma.

LEMMA: If U is open and $A \subset E$, then $U + A = \{z : z = x + a, x \in U, a \in A\}$ is open. Also $\alpha U = \{z : z = \alpha x, x \in U\}$ for $\alpha \neq 0$ is open.

PROOF: The map $x \to x + a$ is a homeomorphism of E onto itself. Thus $U + a$ is open for any a, and therefore $\underset{a \in A}{\cup}\, U + a = U + A$ is open. The homeomorphism $x \to \alpha x$ proves the second assertion.

PROOF OF THE THEOREM: Let U be an open set containing 0 with $y + U \subset K$. Since y is an interior point of K, such a U exists. Thus $\lambda x + (1 - \lambda)\{y + U\} \subset K$; that is, $\lambda x + (1 - \lambda)y + (1 - \lambda)U \subset K$. Since $0 < \lambda < 1$ and $x \to \mu x$ is a homeomorphism of E onto itself for $\mu \neq 0$, $V = (1 - \lambda)U$ is an open set containing the origin. Thus $\lambda x + (1 - \lambda)y + V \subset K$ proves the fact that $\lambda x + (1 - \lambda)y$ is an interior point of K.

Remark: We have the same result when both x and y belong to the interior of K. This also proves the fact that the interior points of a convex set constitute a convex set.

Theorem 1.4.4

Let K be a compact convex set and S a closed convex set in a real or complex linear topological space E. Then $K + S = \{z : z = x + y; x \in K, y \in S\}$ is closed in E.

Let z_0 be a limit point of $K + S$. Then by theorem 1.2.4 we have a generalized sequence $z_d = x_d + y_d$, $x_d \in K$, $y_d \in S$, $d \in (D, \leq)$ with $z_d \to z_0$. Since K is compact, there exists a generalized subsequence $\{x_{d_j}\}$ of $\{x_d\}$ with $x_{d_j} \to x_0$ for some $x_0 \in K$. Trivially $z_{d_j} \to z_0$. By the continuity of $(z, x) \to z - x$ we have $z_{d_j} - x_{d_j} = y_{d_j} \to (z_0 - x_0)$. Let $(z_0 - x_0) = y_0$. As S is closed, $y_0 \in S$. We already have $x_0 \in K$. Thus $x_0 + y_0 \in K + S$. But $x_0 + y_0 = z_0 \in K + S$. This shows that $K + S$ contains all its limit points. Hence the theorem.

1.5 HAHN-BANACH THEOREM

Linear Functional

A function f from a real linear space E to the set of reals is linear if $f(\alpha x + \beta y) = \alpha f(x) + \beta f(y)$ for any $x, y \in E$, and for any α, β scalars.

Sublinear Functionals

A real valued function $p(x)$ on E is a sublinear functional if (1) $p(x + y) \leq p(x) + p(y)$ for $x, y \in E$; (2) $p(\alpha x) = \alpha p(x)$ if $\alpha \geq 0$.

Theorem 1.5.1 (Hahn–Banach)

Let E be a real linear space with a sublinear functional $p(x)$ on E. Let M be a linear manifold in E. If f is a linear functional on M with $f(x) \leq p(x)$ for all $x \in M$, then f can be extended linearly to f^* with $f^*(x) \leq p(x)$ for all $x \in E$, and $f^*(x) = f(x)$ for $x \in M$.

PROOF: Consider the family $\mathscr{F}$ of all linear extensions g of f with $g(x) \leq p(x)$ over the domain of g. Partial-order $\mathscr{F}$ by saying $g_1 \leq g_2$ if the domain of g_1 is a subset of the domain of g_2. Clearly Zorn's lemma is applicable, and we have an f^* that is maximal in $\mathscr{F}$. Let the domain of f^* be E_0. We shall show that $E_0 = E$. Let $E_0 \neq E$. Then there exists a $y \in E$, $y \notin E_0$. Consider the linear manifold L spanned by E_0 and y. Any typical element is $x + \alpha y$, $x \in E_0$, α real. Define on L a linear functional g with $g(x + \alpha y) = f^*(x) + \alpha c$ for some constant c.

We shall choose c so that $g \in \mathscr{F}$. This would contradict the maximality of f^*. That is, we want $g(x + \alpha y) \leq p(x + \alpha y)$ for all $x \in E_0$. Let $z, x \in E_0$. Then $g(z) - g(x) = g(z - x) \leq p(z - x) \leq p(z + y) + p(-y - x)$; that is, $-p(-y - x) - g(x) \leq p(z + y) - g(z)$. Here the left-hand side is independent of z, and the right-hand side is independent of x. Therefore

$$-p(-y - x) - g(x) \leq c \leq p(z + y) - g(z)$$

because we could take supremum over x on the left and infimum over z on the right, which would yield two constants.

Now let $x_0 + \alpha y \in L$, $\alpha > 0$, $x_0 \in E_0$. In the inequality above, putting $z = x_0/\alpha$, we have

$$c \leq p\left(\frac{x_0}{\alpha} + y\right) - g\left(\frac{x_0}{\alpha}\right)$$

That is,

$$c \leq \left(\frac{1}{\alpha}\right) p(x_0 + \alpha y) - \left(\frac{1}{\alpha}\right) g(x_0)$$

Thus $g(x_0) + \alpha c \leq p(x_0 + \alpha y)$; that is, $g(x_0 + \alpha y) \leq p(x_0 + \alpha y)$. If $\alpha < 0$ we can take the other inequality and prove the same; $\alpha = 0$ is the trivial case. Thus we have a proper extension g of f^*.

Remark: An extension to complex linear spaces is also true of this theorem with a suitable formulation.

1.6 SEPARATION THEOREMS

Minkowski Functional

Let E be a real linear topological space. Let K be a convex set with the origin as an interior point. The functional

$$p_K(x) = \inf \left\{t: t > 0, \frac{x}{t} \in K\right\}$$

is called the Minkowski functional. It is a sublinear functional. To see that $p_K(x)$ exists, let $t \to \infty$. Then $x/t \to O$, and since O is an interior point of K,

$x/t \in K$ for some $t > 0$. This shows that $p_k(x)$ exists. To prove subadditivity, let $x, y \in E$, and $p_K(x) = \alpha$, $p_K(y) = \beta$. That is, for any $\varepsilon > 0$, there exists $0 < \delta < \varepsilon$ with

$$\frac{x\cdot}{\alpha + \delta} \in K, \frac{y}{\beta + \delta} \in K$$

Further,

$$\frac{x + y}{\alpha + \beta + 2\delta} = \frac{\alpha + \delta}{\alpha + \beta + 2\delta}\left(\frac{x}{\alpha + \delta}\right) + \frac{\beta + \delta}{\alpha + \beta + 2\delta}\left(\frac{y}{\beta + \delta}\right)$$

and by the convexity of K, $\dfrac{x + y}{\alpha + \beta + 2\delta} \in K$. That is, $p_K(x + y) \le$ $\alpha + \beta + 2\delta$. Since ε is arbitrary with $\delta < \varepsilon$, we have $p_K(x + y) \le \alpha + \beta = p_K(x) + p_K(y)$. Lastly, let $\lambda > 0$. Consider $p_K(\lambda x)$. Let $p_K(x) = \alpha$. For any $\varepsilon > 0$ there exists $0 < \delta < \varepsilon$ with $\dfrac{x}{\alpha + \delta} \in K$ or $\dfrac{\lambda x}{\lambda\alpha + \lambda\delta} \in K$. Thus $p_K(\lambda x) \le \lambda\alpha + \lambda\delta$. Since ε is arbitrary and λ is fixed, $p_K(\lambda x) \le \lambda\alpha = \lambda p_K(x)$. Similarly, one can prove the other side of the inequality. Clearly, $p_K(0) = 0$.

Remark: For $x \in K$, $p_K(x) \le 1$, and $\{x : p_K(x) < 1\}$ is the interior of K. The boundary of $K = \{x : p_K(x) = 1\}$.

We shall now prove the so-called weak separation theorem, which is very useful in the proof of minimax theorem.

Theorem 1.6.1

Let S and T be convex sets in a real locally convex space E. Let T have a nonempty interior, and let S be disjoint with the interior of T. Then there exists a continuous linear functional f and a real constant c such that $f(x) \le c$ for $x \in T$ and $f(x) \ge c$ for $x \in S$.

PROOF: We can assume without loss of generality that the origin is an interior point of T, for otherwise we can consider $T - x_0$, $S - x_0$ as our set, where x_0 is interior to T. Since there exist neighborhoods N of O containing a vector u and $-u$, choose some such u with $-u \in T - S$. Consider $K = T - S + u$. $T - S$ does not contain the origin as an interior point, for otherwise there would be some x interior to T and belonging to S. This shows that u is not interior to $T - S + u = K$. If $p_K(x)$ is the Minkowski functional on K, then $p_K(u) \ge 1$. Let $g(\alpha u) = \alpha p_K(u)$. Then g is a linear functional on the one-dimentional subspace spanned by u. Further, $g(\alpha u) = p_K(\alpha u)$ for all $\alpha > 0$; and for $\alpha < 0$, $g(\alpha u) = \alpha g(u) < 0 \le p_K(\alpha u)$.

Thus $g(\alpha u) \le p(\alpha u)$ for all α. By the Hahn–Banach theorem (theorem 1.5.1), g could be extended linearly as f to the whole of E with $f(x) \le p_K(x)$ for all $x \in E$.

Thus $f(x) \leq 1$ for $x \in K$, and $f(u) \geq 1$. That is, when $x = y - z + u$, $y \in T$, $z \in S$, $f(x) = f(y) - f(z) + f(u) \leq 1$. Or $f(y) - f(z) \leq 1 - f(u)$. That is, $f(y) \leq f(z) + 1 - f(u)$.

Choosing $\sup_{y \in T} f(y) \leq c \leq \inf_{z \in S} f(z) + 1 - f(u)$, we have $f(x) \geq c$ when $x \in S$, and $f(x) \leq c$ when $x \in T$. Lastly we shall show that f is continuous. Let v be interior to T, and let U be a neighborhood of O with $v + U \subset T$. Then $f(U) \subset f(T) - f(v)$. Without loss of generality we can assume that U is convex. (This is where we need the local convexity of the space.) Then $f(U)$ is contained in an interval $(-\infty, a]$ or $[-a, \infty)$ for some $a > 0$.

Let $U - U = G$; then $G = -G$, and $f(G) \subset [-a\,a]$. Thus $f(\varepsilon a^{-1}G) \subset [-\varepsilon, \varepsilon]$. Since $\varepsilon a^{-1}G$ is a neighborhood of the origin, f is continuous at O. Further, open sets containing a point x are translates of open sets containing O. Thus the linearity of f with its continuity at O implies that f is everywhere continuous.

The following is the strong separation theorem.

Theorem 1.6.2

Let E be a real locally convex space. Let S be a closed convex set disjoint with a compact convex set K. Then there exists a continuous linear f and a constant c such that

$$f(x) \geq c, \qquad x \in K$$

$$f(x) < c, \qquad x \in S$$

PROOF: By theorem 1.4.4, $K - S$ is a closed set. It is trivially convex, and $O \notin K - S$. Let U be a convex neighborhood of O disjoint with $K - S$. By the previous theorem we have a non-null continuous linear f and constant α with $f(K - S) \geq \alpha$ and $f(U) \leq \alpha$. Since $f \not\equiv 0$, $f(x) = 1$ for some $x \in E$. For a suitable λ, $\lambda x \in U$. Thus $f(\lambda x) = \lambda$. Hence $f(U)$ contains every λ of modulus less than ε for some ε. That is, $f(K - S) \geq \alpha \geq \varepsilon$, and

$$f(x) - f(y) \geq \alpha \geq \varepsilon > 0 \text{ for } x \in K \text{ and } y \in S$$

If $c = \inf_K f(x) = \min_K f(x)$

$$f(x) \geq c \qquad \text{for } x \in K$$

$$f(x) \leq c - \varepsilon < c \text{ for } x \in S$$

1.7 WEAK AND WEAK* TOPOLOGIES

Let E be a real locally convex space. Let E^* be the collection of continuous linear functionals on E. Then E^* itself becomes a linear space. We can give a new topology to E called the weak topology. It consists of open sets of the form

$$N(x_0, A, \varepsilon) = \{x : |f_i(x) - f_i(x_0)| < \varepsilon, f_i \in A, \varepsilon > 0,$$

$$A \text{ a finite subset of } E^*\}$$

We want the topology to be the smallest one with $x_d \rightarrow x_0$ iff $f(x_d) \rightarrow f(x_0)$ for every generalized sequence $\{x_d\}$ and for each $f \in E^*$. Similarly, we could give a topology for E^* called the w^* topology. It is the smallest topology in which $f_d \rightarrow f$ iff $f_d(x) \rightarrow f(x)$ for each $x \in E$ and for every generalized sequence $\{f_d\}$.

Remark: Actually the w^* topology of E^* is the relative topology of $E^* \subset R^E$ with its product topology. Further, the w^* and w topologies are locally convex.

Theorem 1.7.1

Any set K of the form $K = \{f : f \in E^*, |f(x)| \leq c_x\} \subset E^*$ is w^* compact.

PROOF: Clearly $K \subset \prod_{x \in E} [-c_x\, c_x]$. Further, the weak* topology of K is the relative topology in the product space $\prod_{x \in E} [c_x, c_x]$ with its product topology. By theorem 1.3.3, K is compact in w^* topology.

1.8 KREIN-MILMAN THEOREM

Extreme Point

Let K be a convex set in a real locally convex space E. A point $x_0 \in K$ is an *extreme point* of K if $K - \{x_0\}$ as a set is convex. Equivalently we can say that there exists no $x, y \in K$, $x \neq y$, with $\lambda x + (1 - \lambda) y = x_0$ for some $0 < \lambda < 1$.

The following is one of the deepest results on compact convex sets.

Theorem 1.8.1 (Krein-Milman)

Let K be a nonempty compact convex subset of a real locally convex space E. Then K has at least one extreme point. Further, if K_e is the set of extreme points of K, then the closure of the smallest convex set containing K_e coincides with K.

PROOF: Consider the family $\mathscr{A}$ of nonempty closed subsets of K with the following property: $A \in \mathscr{A}$ if $x, y \in K$ with $\lambda x + (1 - \lambda) y \in A$ for $0 < \lambda < 1$ implies $x, y \in A$. Then $\mathscr{A}$ is a nonempty collection, and it is ordered by inclusion. We say that $A_1 \leq A_2$ if $A_1 \supset A_2$, and $A_1, A_2 \in \mathscr{A}$.

Clearly $\mathscr{A}$ is a partially ordered set. Let $\mathscr{A}_1$ be a chain in $\mathscr{A}$. Clearly $\underset{A \in \mathscr{A}_1}{\cap} A$ has the same property, and it is an upperbound for $\mathscr{A}_1$. Zorn's lemma is applicable, and we have a maximal element. We shall show that every maximal element A_1 of $\mathscr{A}_1$ has exactly one element. Suppose that $x, y \in A_1 \in \mathscr{A}_1$, and $x \neq y$. By the strong separation theorem (theorem 1.6.2) we have f continuous and linear on E with $f(x) \leq c$ and $f(y) > c$ for some c, and $f(x) \neq f(y)$.

Let $A_0 = \{x : x \in A_1, f(x) = \inf_{y \in A_1} f(y)\}$. Then A_0 is a proper subset of A_1. On the other hand, if $u, v \in K$ with $\lambda u + (1 - \lambda) v \in A_0$, $0 < \lambda < 1$, then by definition of A_0, $u, v \in A_0$. Thus $A_0 \in \mathscr{A}$ and contradicts the maximality of A_1. Being nonempty, these single-element sets or our maximal elements constitute the extreme points of K.

To prove the second part of the theorem, let $z \in K$ and $z \notin \overline{\text{con}} \{K_e\}$, the smallest closed convex set containing all extreme points of K. By the strong separation theorem we have a continuous linear ψ with $\psi(z) < c$ and $\psi(u) \geq c$ for some c and for all $u \in \overline{\text{con}} \{K_e\}$.

Let $K_1 = \{x : \psi(x) = \inf_{y \in K} \psi(y)\}$. Then $\lambda x_1 + (1 - \lambda) x_2 \in K_1$ if and only if $x_1, x_2 \in K_1$ for any $0 < \lambda < 1$; that is, $K_1 \in \mathscr{A}$. Further, K_1 as a closed convex subset of K has extreme points. Further, any extreme point of K_1 is also an extreme point of K. But $K_1 \cap K_e = \emptyset$ by assumption. Hence we have a contradiction.

1.9 CONVEX SETS IN R^n

In this section we shall study many properties of convex sets in R^n which often depend on the dimension n.

Dimension of a Convex Set

A convex set $K \subset R^n$ has dimension k if, for some $x \in K$, $K - x$ generates a linear manifold of dimension k.

Remark: Since $K - x$ and $K - y$ for any $y \in K$ generates the same linear manifold, the number k is unique, and the dimension is well-defined. Thus we say that all translates of K have the same dimension.

Relative Interiors: Since $x \rightarrow x - y$ is a homeomorphism, $K - x$ and $K - y$ are homeomorphic. If $K - x$, for $x \in K$, has an interior point relative to the linear manifold it spans, then we say that K has relative interior.

Theorem 1.9.1

Every convex set K in R^n has relative interior.

PROOF: Assuming that $O \in K$, we shall show that K has relative interior. This would prove the theorem by the above definition. Let $\mathscr{L}$ be the linear manifold spanned by K. If K has dimension k, $\mathscr{L}$ is k-dimensional and therefore $\mathscr{L}$ is homeomorphic to R^k. It is clear that the set

$$K' = \{x: x = \lambda_1 e_1 + \lambda_2 e_2 + \cdots + \lambda_k e_k, \lambda_1, \lambda_2, \cdots, \lambda_k > 0,$$

$$\sum_1^k \lambda_k < 1, e_1, e_2, \cdots, e_k, \text{ a basis in } K\}$$

is an open set in R^k and further $K' \subseteq K$. Hence the result.

Theorem 1.9.2

Any two compact convex sets of the same dimension are homeomorphic.

PROOF: Let A and B be the compact convex sets. Without loss of generality, let O be an interior point to both of them. Consider the Minkowski function $p(x)$ for A and $q(y)$ for B where $x \in A$ and $y \in B$. Since A and B are compact, they are closed and bounded. Further, $x/p(x) \in A$, $y/q(y) \in B$, as A and B are closed.

If $\|u\| \le \alpha$ for $u \in A$ or $u \in B$, then $\|x/p(x)\| \le \alpha$ for any x; that is, $p(x) \ge \|x\|/\alpha$. (Here $\| \cdot \|$ denotes the norm.)

Let $\{x: \|x\| < \beta\} \subseteq A \cap B$. Since $x/p(x)$ is a boundary point of A, for any x, $\|x\|/p(x) \ge \beta$; that is, $p(x) \le \|x\|/\beta$.

Thus by symmetry $\|x\|/\alpha \le p(x)$, $q(x) \le \|x\|/\beta$.

Further, by the triangular inequality of p and q we have

$$|p(x) - p(x^1)| \le \max(p(x - x^1), p(x^1 - x)) \le \frac{\|x - x^1\|}{\beta}$$

with a similar inequality for $q(x)$. The map $\tau: x \to \dfrac{p(x)}{q(x)} x$ is a one-to-one onto map of A onto B. For if $x \in A$, then $\dfrac{x}{q(x)} \in B$. Since $0 \le p(x) \le 1$ for $x \in A$, $p(x)\dfrac{x}{q(x)} \in B$. Given any $y \in B$, clearly $\tau^{-1} = \dfrac{q(y)}{p(y)} y$. We shall show that τ is continuous, and by symmetry τ^{-1} would be continuous. Let $x, x^1 \ne 0$. Then

$$\|\tau x - \tau x^1\| = \left\| \frac{p(x)}{q(x)} x - \frac{p(x^1)}{q(x^1)} x^1 \right\|$$

$$= \left\| \frac{p(x)}{q(x)} (x - x^1) + \left(\frac{p(x)}{q(x)} - \frac{p(x^1)}{q(x^1)} \right) x^1 \right\|$$

$$\leq \frac{\alpha}{\beta} \|x - x^1\| + \left\| \left(\frac{p(x)}{q(x)} - \frac{p(x^1)}{q(x^1)} \right) x^1 \right\|$$

$$\leq \frac{\alpha}{\beta} \|x - x^1\| + \left\| \frac{p(x)}{q(x)} (q(x^1) - q(x)) \frac{x^1}{q(x^1)} + \right.$$
$$\left. + (p(x) - p(x^1)) \frac{x^1}{q(x^1)} \right\|$$

$$\leq \frac{\alpha}{\beta} \|x - x^1\| + \frac{\alpha}{\beta} \frac{\|x - x^1\|}{\beta} \alpha + \frac{\|x - x^1\|}{\beta} \alpha$$

$$\left(\text{Here } \frac{x^1}{q(x^1)} \in B, \text{ and } \frac{\|x^1\|}{q(x^1)} \leq \alpha \text{ is used.} \right)$$

$$\leq \left(2 \frac{\alpha}{\beta} + \frac{\alpha^2}{\beta^2} \right) \|x - x^1\|$$

Thus τ is continuous. Hence the result.

Theorem 1.9.3

If S is any subset of R^n, then every point of con $[S]$ (smallest convex set containing S) is at most a convex linear combination of $(n + 1)$ points of S.

Let us prove that, for any

$$x = \sum_1^m \lambda_i x_i, \quad 0 \leq \lambda_i \leq 1, \quad \sum_1^m \lambda_i = 1, \quad x_1, x_2, \cdots, x_m \in S,$$

x has a representation with $n + 1$ elements of S.

If $m \leq n + 1$, we have nothing to prove; so we shall assume the proposition for $m = k > n$ and prove for $m = k + 1$.

Now $x_1 - x_{k+1}, x_2 - x_{k+1}, \cdots, x_k - x_{k+1}$ are k vectors, with $k > n$ in R^n. Thus by their linear dependence

$$\alpha_1(x_1 - x_{k+1}) + \alpha_2(x_2 - x_{k+1}) + \cdots + \alpha_k(x_k - x_{k+1}) = 0$$

and $(\alpha_1, \alpha_2, \cdots, \alpha_k) \neq 0$. If $\alpha_{k+1} = -\sum_1^k \alpha_i$, then

$$\sum_{i=1}^{k+1} \alpha_i x_i = 0$$

In our $x = \sum_{i=1}^m \lambda_i x_i$ we shall assume that all λ_i's are positive; otherwise we have a convex combination with less than m elements. Consider $\lambda_i - \mu \alpha_i$

for a general μ. Choose μ with $\theta_i = \lambda_i - \mu\alpha_i \geq 0$ for all $i = 1, 2, \cdots, k + 1$, and $\lambda_i - \mu\alpha_i = 0$ for some i. That is,

$$\frac{1}{\mu} = \max_{1 \leq i \leq k+1} \frac{\alpha_i}{\lambda_i}$$

gives us the required condition. Now

$$\sum_{i=1}^{k+1} \theta_i x_i = \sum_{1}^{k+1} \lambda_i x_i - \mu \sum_{1}^{k+1} \alpha_i x_i = \sum_{1}^{k+1} \lambda_i x_i$$

Further

$$\sum_{1}^{k+1} \theta_i = \sum (\lambda_i - \mu\alpha_i) = \sum \lambda_i = 1$$

Thus $\sum_{i=1}^{k+1} \theta_i x_i = x$ with some $\theta_i = 0$, so that x is a combination of, at most, k elements. By induction we have proved the result.

Remark: If x is a boundary point of con $[S]$, then it is a combination of, at most, n elements from S! (Hint: Use theorem 1.6.1.)

Theorem 1.9.4

If S is any bounded subset of R^n, then the extreme points of $\overline{\text{con}}\,[S]$ denoted by S_e have the property that con $[S_e] = \overline{\text{con}}\,[S_e] = \overline{\text{con}}\,[S]$.

PROOF: By the Krein-Milman theorem, the closed bounded set $\overline{\text{con}}\,[S]$ has extreme points, and further $\overline{\text{con}}\,[S] = \overline{\text{con}}\,[S_e]$. To show that con $[S_e] = \overline{\text{con}}\,[S_e]$, let $x \in$ boundary of con $[S_e]$. (Here they could differ only by the boundary points of con $[S_e]$!) We shall assume without loss of generality that con $[S_e]$ has interior. Since x is not an interior point, the weak separation theorem applies to the convex sets $\{x\}$ and con $[S_e]$. Thus we have $f(x) = c$, $f(u) \leq c$, $u \in$ con $[S_e]$. Take $T = \{u : f(u) = c\} \cap \overline{\text{con}}\,[S_e]$. This is a compact convex set. Since x is an element of this set, it is nonempty and it has extreme points. If x_0 is an extreme point of T, so it is for $\overline{\text{con}}\,[S_e]$, for if $x_0 = (x_1 + x_2)/2$, x_1, $x_2 \in$ con $[S_e]$, then $f(x_1) \leq c$, $f(x_2) \leq c$, and $f(x_0) \leq c$. But $f(x_0) = c$. This implies that $f(x_1) = c$, $f(x_2) = c$; that is, $x_1, x_2 \in T$. This contradicts the fact that x_0 is an extreme point of T. Further, T is at least one dimension less than that of $\overline{\text{con}}\,[S_e]$. If we have proved the theorem for sets of dimension $\leq k$, then we have by the above arguments proved it for the case of dimension $\leq k + 1$. For $k = 1$ it is trivial. Thus our mathematical induction proves the theorem.

The following theorems concern intersections of a family of convex sets.

Theorem 1.9.5 (Berge)

Let $K_1, K_2, \cdots, K_p$ $(p > 2)$ be compact convex sets in R^n with $\bigcup\limits_{j=1}^{p} K_j$ convex. If every $(p - 1)$ of them has a point in common, then all of them have a common point.

PROOF: Let us prove this by induction.

Case 1: $p = 2$. Let $K_1, K_2 \neq \emptyset$, but $K_1 \cap K_2 = \emptyset$. Then by strong separation we have $f(x) < c$ for $x \in K_1$ and $f(x) > c$ for $x \in K_2$ for some c and linear f. If $P = \{x: f(x) = c\}$, then for some $x \in K_1$, $y \in K_2$, and $0 < \lambda < 1, f(\lambda x + (1 - \lambda) y) = \lambda f(x) + (1 - \lambda) f(y) = c$. As $x, y \in K_1 \cup K_2$, and since $K_1 \cup K_2$ is convex, $\lambda x + (1 - \lambda) y \in K_1 \cup K_2$, and suppose it belongs to, say, K_1. But $f(x) < c$ for all $x \in K_1$. This contradiction proves that $K_1 \cap K_2 \neq \emptyset$.

Case 2: p is general. By induction we shall assume the result for $p = r$ and prove for $p = r + 1$. Let $K = \bigcap\limits_{1}^{r} K_j \neq \emptyset$ and $K_{r+1} \neq \emptyset$, with $\bigcap\limits_{1}^{r+1} K_j = \emptyset$. By the strong separation theorem we have a linear g and constant η with

$$g(x) < \eta \text{ for } x \in K$$
$$g(x) > \eta \text{ for } x \in K_{r+1}$$

Let
$$K_j^{\circ} = \{x: x \in K_j, g(x) = \eta\}$$

Then
$$\bigcup\limits_{i=1}^{r} K_j^{\circ} = \{x: x \in \bigcup\limits_{j=1}^{r} K_j, g(x) = \eta\} = \{x: x \in \bigcup\limits_{j=1}^{r+1} K_j, g(x) = \eta\}$$

(Here K_{r+1} does not contribute anything additional to $\bigcup\limits_{1}^{r+1} K_j^{\circ}$.) Since $\bigcup\limits_{j=1}^{r+1} K_j$ is compact convex and $\{x: g(x) = \eta\}$ is closed convex, $\bigcup\limits_{1}^{r} K_j^{\circ}$ is compact convex.

Further, the intersection of any $(r - 1)$ of $K_1, K_2, \cdots, K_r$ contains K and intersects K_{r+1} in a non-null set, and hence the intersection of any $(r - 1)$ of $K_1, K_2, \cdots, K_r$ intersects $\{x: g(x) = \eta\}$. That is, $\bigcap\limits_{\substack{j \neq i \\ 1 \leq j \leq r}} K_j^{\circ} \neq \emptyset$.

$\bigcup\limits_{j=1}^{r} K_j^{\circ}$ being convex with the induction assumption proves that $\bigcap\limits_{1}^{r} K_j^{\circ} \neq \emptyset$. This contradicts $K \cap \{x: g(x) = \eta\} = \emptyset$. Hence the theorem.

Remark: Instead of $K_1, K_2, \cdots, K_p$ $(p > 2)$ being compact convex, it is enough that they are closed convex. For if $x_i \in \bigcap\limits_{j \neq i} K_j$, then for the smallest convex set X containing $x_1, x_2. \cdots, x_p$, we have $K_i' = X \cap K_i$ are compact convex sets with $\cup K_i' = \cup X \cap K_i$ convex, and the fact that $\overset{p}{\underset{i=1}{\cap}} K_i' \neq \emptyset$

implies $\overset{p}{\underset{i=1}{\cap}} K_i \neq \emptyset$. Further, $\underset{j \neq i}{\cap} K_j' = X \cap \underset{j \neq i}{\cap} K_j$ contains x_i.

Theorem 1.9.6 (Helley)

Let $K_1, K_2, \cdots, K_p$ $(p > n + 1)$ be convex sets in R^n. every $(n + 1)$ of them have a common point, then all of them have a point common.

PROOF: Let us assume that it is true for $p = r(n + 1 < r)$ and prove for $p = r + 1$. Let $K_1, K_2, \cdots, K_r, K_{r+1}$ be convex sets with any r of them having a common point. We shall show that $\overset{r+1}{\underset{j=1}{\cap}} K_j \neq \emptyset$.

Let $x_i \in \underset{\substack{j \neq i \\ 1 \leq j \leq r+1}}{\cap} K_j$ and $Q_i =$ the smallest convex set containing $\{x_1, x_2, \cdots, x_{i-1}, x_{i+1}, \cdots, x_{r+1}\}$. Let $K_i^\circ = K_i \cap Q_i$. Clearly, $K_i \cap Q_i \neq \emptyset$, $i = 1, 2, \cdots, r + 1$. Further, if $K_i^* = \underset{\substack{j \neq i \\ 1 \leq j \leq r+1}}{\cap} K_j^\circ$, then

$K_i^* \neq \emptyset$ for $x_i \in K_1, K_2, \cdots, K_{i-1}, K_{i+1}, \cdots, K_{r+1}$, and in $Q_1, Q_2, \cdots, Q_{i-1}, Q_{i+1}, \cdots, Q_{r+1}$.

Further, $\underset{i}{\cup} K_i^\circ = \underset{i}{\cup} Q_i$. For $K_i^\circ \subset Q_i$ for all i implies $\underset{i}{\cup} K_i^\circ \subset \underset{i}{\cup} Q_i$. Conversely, let $x \in Q_i$. Then $x = \underset{j \neq i}{\sum} \lambda_j x_j$, $\lambda_j \geq 0$ for all j, and $\underset{j \neq i}{\sum} \lambda_j = 1$. But $x_1, x_2, \cdots, x_{i-1}, x_{i+1}, \cdots, x_{r+1} \in K_i$ by definition. Thus by the convexity of K_i, $x \in K_i$, and $x \in K_i \cap Q_i = K_i^\circ$. Thus $\overset{r+1}{\underset{i=1}{\cup}} K_i^\circ = \overset{r+1}{\underset{i=1}{\cup}} Q_i$. Further,

$\overset{r+1}{\underset{i=1}{\cup}} Q_i$ is the convex set generated by $\{x_1, x_2, \cdots, x_{r+1}\}$, for, by theorem 1.9.3, any point in the smallest convex set containing $\overset{r+1}{\underset{i=1}{\cup}} Q_i$ is a combination of at most $(n + 1)$ of the points from $\{x_1, x_2, \cdots, x_{r+1}\}$. As $r > n + 1$, it is in some Q_i, and hence $\overset{r+1}{\underset{i=1}{\cup}} Q_i = \overset{r+1}{\underset{i=1}{\cup}} K_i^\circ$ is convex. By Berge's theorem (1.9.5), $\overset{r+1}{\underset{1}{\cap}} K_i^\circ \neq \emptyset$; that is, $\overset{r+1}{\underset{1}{\cap}} K_i \neq \emptyset$. This completes the proof of the theorem.

1.10 CONVEX FUNCTIONS

Let K be a convex set and f a real valued function on K. Call f a *convex function*, if $f(\lambda x + (1 - \lambda) y) \leq \lambda f(x) + (1 - \lambda) f(y)$ for any

$0 \leq \lambda \leq 1$ and $x, y \in K$. Thus for every pair $x, y \in K$, if $\varphi_{x,y}(\lambda) = f(\lambda x + (1 - \lambda)y)$, then f is convex on K if and only if $\varphi_{x,y}(\lambda)$ is a convex function on $0 \leq \lambda \leq 1$. The negative of a convex function is called a *concave function*.

Theorem 1.10.1

Let K be a convex set in a real linear space E. A real valued function f on K is convex if and only if the set $C_f = \{(x, r) : x \in K, f(x) \leq r\}$ is convex.

PROOF: Let (x, r), $(y, s) \in C_f$, that is, $f(x) \leq r$, $f(y) \leq s$. Thus $\lambda f(x) + (1 - \lambda)f(y) \leq \lambda r + (1 - \lambda)s$. But, as f is convex, by assumption $f(\lambda x + (1 - \lambda)y) \leq \lambda f(x) + (1 - \lambda)f(y) \leq \lambda r + (1 - \lambda)s$. Thus $(\lambda x + (1 - \lambda)y, \lambda r + (1 - \lambda)s) \in C_f$. That is, C_f is convex. Conversely, let C_f be convex. Then $(x, f(x))$, $(y, f(y)) \in C_f$ implies that $(\lambda x + (1 - \lambda)y, \lambda f(x) + (1 - \lambda)f(y)) \in C_f$, which shows that $f(\lambda x + (1 - \lambda)y) \leq \lambda f(x) + (1 - \lambda)f(y)$, and that f is convex.

Theorem 1.10.2

Let K be a convex set and f a convex function on K. If $x_1, x_2, \cdots, x_k \in K$ and $0 \leq \lambda_1, \lambda_2, \cdots \lambda_k \leq 1$ with $\sum_{i=1}^{k} \lambda_i = 1$, then

$$f\left(\sum_{1}^{k} \lambda_i x_i\right) \leq \sum_{i=1}^{k} \lambda_i f(x_i)$$

PROOF: Induction.

Theorem 1.10.3

Let $f(x)$ be a twice-differentiable function on $a < x < b$. Then f is convex on (a, b) if and only if the second derivative $f''(x) \geq 0$ on (a, b).

PROOF: Let $f''(u) \geq 0$ for all u in (a, b), and let $x < y$. Consider $\lambda x + (1 - \lambda)y$, $0 < \lambda < 1$. Thus $x < \lambda x + (1 - \lambda)y < y$. Now

$$f(\lambda x + (1 - \lambda)y) - f(x) = (1 - \lambda)(y - x)f'(\xi)$$

by the mean value theorem for some $x < \xi < \lambda x + (1 - \lambda)y$. Similarly,

$$f(\lambda x + (1 - \lambda)y) - f(y) = \lambda(x - y)f'(\zeta), \quad \lambda x + (1 - \lambda)y < \zeta < y$$

Thus, multiplying the first equation by λ and the second one by $(1 - \lambda)$, and adding, we have

$$f(\lambda x + (1 - \lambda)y) - \lambda f(x) - (1 - \lambda)f(y) = \lambda(1 - \lambda)(y - x)(f'(\xi) - f'(\zeta))$$

Since $f'' \geq 0$, f' is nondecreasing, and $f'(\xi) - f'(\zeta) \leq 0$ for $\xi < \zeta$. Here $x < \xi < \lambda x + (1 - \lambda)y < \zeta < y$. Thus $f(\lambda x + (1 - \lambda)y) \leq \lambda f(x) + (1 - \lambda)f(y)$.

This proves the first part of the theorem. Conversely, let a convex function f be twice-differentiable in (a, b). Let $x < y$, and for any $\lambda x + (1 - \lambda)y$, $0 < \lambda < 1$, by the previous argument,

$$f(\lambda x + (1 - \lambda)y) - \lambda f(x) - (1 - \lambda)f(y)$$
$$= \lambda(1 - \lambda)(y - x)(f'(\xi) - f'(\zeta)), \quad x < \xi < \zeta < y$$

By the convexity of f we have $f'(\xi) - f'(\zeta) \leq 0$. Since x and y càn be chosen as near as we please, we have $(f'(\xi) - f'(\zeta))/(\xi - \zeta) \geq 0$, which implies $f''(x) \geq 0$. (This is assured by the existence of f''.) As x is arbitrary, we are through.

The following theorems are useful in the theory of nonlinear programming.

Theorem 1.10.4 (Farkas–Minkowski)

Let $K \subset R^n$ be convex, and let $g_1, g_2, \cdots, g_p (p > n + 1)$ be convex functions on K. If $\sup_i g_i(x) \geq 0$ for $x \in K$, then there exists $\lambda_1, \lambda_2, \cdots$,

$$\lambda_{n+1} \geq 0, \sum_1^{n+1} \lambda_j = 1, \text{ with } \inf_{x \in K} \sum_{i=1}^{n+1} \lambda_i g_{k_i}(x) \geq 0, \text{ for some } k_i\text{'s}.$$

PROOF: Let

$$G = \{(y_1, y_2, \cdots, y_p): (y_1, y_2, \cdots, y_p) \in R^p, g_i(x) < y_i, i = 1, 2, \cdots, p, x \in K\}$$

By assumption, $O \notin G$, and G is a convex set by the convexity of g_i's. By the weak separation theorem, there exists $(l_1, l_2, \cdots, l_p) \neq 0$ with $\sum_{i=1}^{p} l_i y_i \geq 0$, $y = (y_1, \cdots, y_p) \in G$. If $\varepsilon > 0$ is added to the ith coordinate of $y = (y_1, \cdots, y_p) \in G$, the new vector $y_i^* = (y_1, y_2, y_{i-1}, y_i + \varepsilon, y_{i+1}, \cdots, y_p)$ also belongs to G, and $\sum_{i=1}^{p} l_i y_i + \varepsilon l_i \geq 0$ for all $\varepsilon > 0$. Thus $l_i \geq 0$ for any i.

Thus in the limit $\sum_{i=1}^{p} l_i g_i(x) \geq 0$, with $l_i \geq 0$, $i = 1, 2, \cdots, p$, for all $x \in K$. As $l \neq 0$, we can assume that $\Sigma l_i = 1$. Let $p > n + 1$. Take

$$K_i = \{x: g_i(x) < 0, x \in K\}, i = 1, 2, \cdots, p$$

Now $\bigcap_1^p K_i = \emptyset$, by assumption, and $p > n + 1$. By Helley's theorem, $\bigcap_1^{n+1} K_i = \emptyset$ for some $n + 1$ of them. Let them be $K_{k_1}, \cdots, K_{k_{n+1}}$. Then

$$\{x: g_{k_i}(x) < 0, i = 1, 2, \cdots, n + 1, x \in K\} = \emptyset$$

That is, $\sup\limits_{1\leq i\leq n+1} g_{k_i}(x) \geq 0$, $x \in K$. From the argument used above, we have

$$\sum_{i=1}^{n+1} \lambda_i g_{k_i}(x) \geq 0, \quad \lambda_i \geq 0, \quad i = 1, 2, \cdots, n+1; \quad \sum_{1}^{n+1} \lambda_i = 1.$$ Hence the

theorem. This theorem leads to theorem 1.10.5.

Theorem 1.10.5 (Bohnenblust, Karlin, and Shapley)

Let K be a compact convex set in R^n, and let $\{g_\alpha\}$ be an arbitrary family of convex functions with $\{x : g_\alpha(x) \leq r\}$ closed in R^n for each α and r. If $\sup\limits_{\alpha} g_\alpha(x) > 0$ for each $x \in K$, there exist finite indices $\alpha_1, \alpha_2, \cdots, \alpha_{n+1}$ and

constants $\lambda_1, \lambda_2, \cdots, \lambda_{n+1}$ in $0 \leq \lambda_1, \lambda_2, \cdots, \lambda_{n+1} \leq 1$, $\sum\limits_{i} \lambda_i = 1$, with

$$\inf_{x \in K} \sum_{i=1}^{n+1} \lambda_i g_{\alpha_i}(x) > 0.$$

PROOF: Consider $K_{\alpha,m} = \{x : x \in K, g_\alpha(x) \leq 1/m\}$ for each α and positive integer m. By assumption $K_{\alpha,m}$ are closed, and they are compact, as K is compact. Further, $\bigcap\limits_{\alpha,m} K_{\alpha,m} = \varnothing$. Thus there exists a finite of them, say $K_{\beta_1,m_1}, K_{\beta_2,m_2}, \cdots, K_{\beta_k,m_k}$ with empty intersection (see the lemma to theorem 1.2.5). That is, $g_{\beta_i}(x) - 1/m_i \leq 0$, $i = 1, 2, \cdots, k$, has no solution in K. By the previous theorem (1.10.4) there exist indices $\alpha_1, \alpha_2, \cdots, \alpha_{n+1} \subset \{\beta_1, \beta_2, \cdots, \beta_k\}$ with

$$\sum_{j=1}^{n=1} \lambda_j g_{\alpha_j}(x) - \sum_{j=1}^{n+1} \lambda_j \cdot \frac{1}{m_{i_j}} \geq 0 \text{ for all } x \in K$$

Hence

$$\inf_{x \in K} \sum_{i=1}^{n+1} \lambda_j g_{\alpha_j}(x) > 0$$

Theorem 1.10.6 (Farkas–Minkowski Theorem)

Let $f(x), g_1(x), g_2(x), \cdots, g_p(x)$ be concave functions, and let $q \leq p$ be such that $g_{q+1}(x) - g_{q+1}(o), \cdots, g_p(x) - g_p(o)$ are linear.

If $\left. \begin{array}{l} g_i(x) \geq 0, i = 1, 2, \cdots, p \\ f(x) > 0 \end{array} \right\}$ has no x in R^n

but $\left. \begin{array}{l} g_i(x) > 0, i = 1, 2, \cdots, q \\ \geq 0, i = q + 1, \cdots, p \end{array} \right\}$ has an $x \in R^n$

then there exists $\lambda_1, \lambda_2, \cdots, \lambda_p \geq 0$ with

$$f(x) + \sum_{1}^{p} \lambda_i g_i(x) \leq 0 \text{ for all } x$$

PROOF: By theorem 1.10.4, we have

$$\mu_0 f(x) + \sum_{i=1}^{p} \mu_i g_i(x) \leq 0 \text{ for all } x$$

$$\mu_0, \mu_1, \cdots, \mu_p \geq 0, \sum_{0}^{p} \mu_i = 1$$

If $\mu_0 \neq 0$, then by taking $\mu_i/\mu_0 = \lambda_i$, $i = 1, 2, \cdots, p$, we get the required result. In fact, we shall show that $\mu_0 \neq 0$. Suppose that $\mu_0 = 0$. Then $\sum_{i=1}^{p} \mu_i g_i(x) \leq 0$ for all x. By assumption we have a y with

$$g_i(y) > 0, i = 1, 2, \cdots, q$$

$$g_i(y) \geq 0, i = q + 1, \cdots, p$$

For this y, $\sum_{1}^{p} \mu_i g_i(y) \geq 0$; that is, $\sum_{1}^{p} \mu_i g_i(y) = 0$. Since $g_i(y) > 0$, $i = 1, 2, \cdots, q$, we have $\mu_1 = \mu_2 = \cdots = \mu_q = 0$. That is, $\sum_{q+1}^{p} \mu_i g_i(y) = 0$.

Since $g_i(x) - g_i(0)$ is linear in x, for $i = q + 1, \cdots, p$, $\sum_{i=q+1}^{p} \mu_i(g_i(x) - g_i(y))$ ≤ 0 implies that $\sum_{i=q+1}^{p} \mu_i g_i(x) = 0$.

Without loss of generality, let $g_i(x) \geq 0$, $f(x) > 0$ have a solution if i ranges over any strict subset of $\{1, 2, \cdots, p\}$. Thus the intersection of the $(p - q)$ closed convex sets

$$C_i = \{x: g_i(x) \geq 0\}, i = q + 1, q + 2, \cdots, p$$

has no point in common with

$$C = \{x: f(x) > 0, g_i(x) \geq 0, i = 1, 2, \cdots, q\}$$

But the intersection of any $p - q - 1$ of C_i's meets C. Thus $\cup_i C_i \not\supset C$ is a consequence of Berge's theorem; that is, there exists an $x_0 \in C$ such that $g_i(x_0) < 0$, $i = q + 1, \cdots, p$. Thus $\sum_{i=q+1}^{p} \mu_i g_i(x_0) < 0$ is a contradiction.

1.11 FIXED-POINT THEOREMS

Simplex

Let $\{x_1, x_2, \cdots, x_{k+1}\}$ be $k + 1$ points in R^n whose convex hull K has dimension k. Then K is called a k-simplex with vertices $x_1, x_2, \cdots, x_{k+1}$. The x_i's are the only extreme points of K. The point $(1/k)(x_1 + x_2 + \cdots + x_k)$ is called the *centroid* of the simplex. Every subset of the vertices with r elements $(r = 1, 2, \cdots, k + 1)$ generates an $(r - 1)$ simplex.

Theorem 1.11.1

Let $T = \{x_1, x_2, \cdots, x_k\}$ be any finite set in R^n generating a $(k - 1)$ simplex with $\max\limits_{1 \le i,j \le k} \|x_i - x_j\| = d$. Let T' be any subset of T. If b and b' are the centroids of the simplexes generated by T and T', then

$$\|b - b'\| \le \frac{k - 1}{k} \cdot d.$$

PROOF: Let us assume, without loss of generality, $T' = \{x_1, x_2, \cdots, x_{k_1}\}$, for otherwise we can re-order them. Then

$$b - b' = \frac{1}{k} \sum_1^k x_j - b' = \frac{1}{k} \sum_1^k (x_j - b')$$

But

$$x_j - b' = x_j - \frac{1}{k_1} \sum_1^{k_1} x_i = \frac{1}{k_1} \sum_1^{k_1} (x_j - x_i)$$

Thus

$$b - b' = \frac{1}{kk_1} \sum_{j=1}^k \sum_{i=1}^{k_1} (x_j - x_i)$$

But $x_j - x_i = 0$ for $i = j$. Thus we have in effect only $kk_1 - k_1 = (k - 1)k_1$ nonzero vectors in the sum. Thus

$$\|b - b'\| \le \frac{1}{kk_1} \sum_{j=1}^k \sum_{i=1}^{k_1} \|x_j - x_i\| \le \frac{(k - 1)k_1}{kk_1} d = \frac{k - 1}{k} d$$

Face of a Simplex

Let K be a simplex generated by $\{x_1, x_2, \cdots, x_{k+1}\}$. The convex set generated by any subset of $\{x_1, x_2, \cdots, x_{k+1}\}$ is called a *face* of K. If x is a point of K with $x = \sum\limits_{j=1}^r \lambda_j x_j, \lambda_1, \lambda_2, \cdots, \lambda_r > 0, \sum\limits_1^r \lambda_j = 1$, $r < k + 1$, then x is contained in the face generated by $\{x_1, x_2, \cdots, x_r\}$. This unique face is called the *carrier* of x. Each boundary point is contained in a proper face.

Simplicial Partition

Let K be partitioned into a finite number of open simplexes $K_1, K_2, \cdots, K_p$ with

$(1)\ \overset{p}{\underset{i=1}{\cup}}\ K_i = K$ $(2)\ K_i \cap K_j = \emptyset$ when $i \neq j$ and

(3) Every face of K_i belongs to the partition. Then such a partition is called a simplicial partition.

Theorem 1.11.2 (Sperner)

Let π be a simplicial partition of a k-simplex K generated by $\{x_1, x_2, \cdots, x_{k+1}\}$. Let φ be a mapping of the vertices of every subsimplex in π into the vertices of the carrier of them. Then there exists a subsimplex of K which is a k-simplex with vertices $\{x_1', x_2', \cdots, x_{k+1}'\}$ with $\varphi(x_1') = x_1$, $\varphi(x_2') = x_2, \cdots, \varphi(x_{k+1}') = x_{k+1}$. In fact, the number of such simplexes is odd.

PROOF: The theorem will be proved by induction. Let $k = 0$. The truth is apparent. Assume the truth of the proposition for all $(k - 1)$-dimensional simplexes. We shall prove it for k-dimensional simplexes. Let $K_1, K_2, \cdots, K_\alpha$ be the k-dimensional subsimplexes of K in π. We say that a $(k - 1)$-dimensional face is *special* if its vertices say $\{x_1', x_2', \cdots, x_k'\}$ are mapped onto $\{x_1, x_2, \cdots, x_k\}$ under φ.

If any K_j satisfies the conclusion of the theorem, then it has only one special face, by the one-to-one correspondence imposed in our theorem. On the contrary, K_i for some i do not satisfy the conclusion of the theorem. If $\{x_1', x_2', \cdots, x_{k+1}'\}$ are its vertices, then for any of its subsets with k elements either the φ image is $\{x_1, \cdots, x_k\}$ or not. Suppose that $\{x_{n_1}', x_{n_2}', \cdots, x_{n_k}'\} \subset \{x_1', x_2', \cdots, x_k'\}$ has the property $\varphi(x_{n_1}') = x_1, \varphi(x_{n_2}') = x_2, \cdots, \varphi(x_{n_k}') = x_k$. Since K_i does not satisfy the conditions of the theorem, its vertex $x_{n_{k+1}}'$ different from $x_{n_1}', x_{n_2}', \cdots, x_{n_k}'$ under φ goes to $\varphi(x_{n_{k+1}}') \neq x_{k+1}$. Thus $\varphi(x_{n_{k+1}}')$ is again one of $x_1, x_2, \cdots, x_k$, say x_1. Thus $\{x_{n_1}', x_{n_2}', \cdots, x_{n_k}'\}$ and $\{x_{n_{k+1}}', x_2', \cdots, x_{n_k}'\}$ are two *special* $(k - 1)$-dimensional faces of K_i. As the last case, no $(k - 1)$-dimensional face of a K_j has its vertices mapped to $\{x_1, x_2, \cdots, x_k\}$ under φ. Thus the number of special faces of K_j is either 0 or 2 when the K_j does not satisfy the conclusion of the theorem. When it does, it has exactly one such face. If we prove that the total number of special faces for all $K_1, \cdots, K_\alpha$ is odd, then the theorem is proved.

Let V be a $(k - 1)$-dimensional special face in π. Then V either is a subset of the boundary of K or contains some interior points of K. If V lies on the boundary of K, then it could be the face of only one k-dimensional subsimplex in π. In this case for each vertex x of V, $\varphi(x)$ must be a vertex of the foregoing $(k - 1)$-dimensional face of K which could be spanned by

$\{x_1, x_2, \cdots, x_k\}$. If V possesses an interior point of K, then it is a special face of 2, k-dimensional simplexes in π contiguous to V. Thus V is counted twice in the sum. Thus the total number of special faces is congruent (mod) 2 to the number of $(k-1)$-dimensional simplexes V of π lying on the $(k-1)$-dimensional face of K spanned by $\{x_1, x_2, \cdots, x_k\}$ whose vertices are mapped under φ onto the vertices $\{x_1, \cdots, x_k\}$. By our induction assumption it is odd. Hence our required sum is odd. This proves the theorem.

Theorem 1.11.3 (Kuratowski–Knaster–Mazurkiewicz)

If the $k+1$ closed subsets $C_1, C_2, \cdots, C_{k+1}$ of the k-dimensional simplex K with vertices $\{x_1, x_2, \cdots, x_{k+1}\}$ satisfy the condition that for each index set $\{i, j, \cdots, l\} \subset \{1, 2, \cdots, k+1\}$ the face of K spanned by $\{x_i, x_j, \cdots, x_l\}$ is contained in $C_i \cup C_j \cup \cdots \cup C_l$, then $\overset{k+1}{\underset{1}{\cap}} C_j \neq \varnothing$.

PROOF: Let π_μ be a sequence of simplicial partitions of K. By applying theorem 1.11.1, we could have the diameter of all subsimplexes of π_μ be $< \delta_\mu \to 0$ as $\mu \to \infty$. (Here the diameter of a simplex is the maximum distance between the vertices of the simplex.) Consider a sequence $\{\varphi_\mu\}$ of functions, φ_μ corresponding to the partition π_μ defined as in the previous theorem. Further, let φ_μ be determined for the partition π_μ as follows. Suppose that x is a vertex of a subsimplex in π_μ with $\{x_{j_1}, x_{j_2}, \cdots, x_{j_s}\}$ as its carrier. Then $x \in \overset{s}{\underset{r=1}{\cup}} C_{j_r}$. Define $\varphi_\mu(x) = x_{j_t}$ if j_t is the minimum subscript for which $x \in \overset{t}{\underset{r=1}{\cup}} C_{j_r}$. By the previous theorem we have a k-dimensional simplex in π_μ for whose vertices $\{x_1^{(\mu)}, x_2^{(\mu)}, \cdots, x_{k+1}^{(\mu)}\}$ one has $x_j^{(\mu)} \in C_j, j = 1, 2, \cdots, k+1$. Going to a subsequence we can assume that $x_j^{(\mu)}$ converges to a limit as $\mu \to \infty$ for $j = 1, 2, \cdots, k+1$. Since $\delta_\mu \to 0$, these $x_j^{(\mu)}$ converge to the same point x^* for all j. Since C_j's are closed, $\underset{\mu}{\lim} x_j^{(\mu)} = x^* \in C_j, j = 1, 2, \cdots, k+1$; that is, $\overset{k+1}{\underset{j=1}{\cap}} C_j \neq \varnothing$. Now we shall prove the main theorems of this section.

Theorem 1.11.4 (Brouwer)

Let S be a compact convex set with interior in R^k. If φ is a continuous map of S into itself, then there exists an $x^* \in S$ with $\varphi(x^*) = x^*$. This is called *a fixed point*.

PROOF: Let K be a k-simplex in R^k. By theorem 1.9.2, S is homeomorphic to K. Let $\psi : S \to K$ be a homeomorphism. The map $\psi \circ \varphi \circ \psi^{-1}$ maps $y \in K$ to $\psi(\varphi(\psi^{-1}(y)))$, which is a point of K. Since ψ, φ, ψ^{-1} are continuous, this composite map is continuous. If we have proved the theorem for K, then $\psi \circ \varphi \circ \psi^{-1}(y) = y$ for some $y \in K$. Let $\psi^{-1}(y) = z$. Then $\psi(\varphi(z)) = y = \psi(z)$.

Since ψ is one-to-one, $\varphi(z) = z$. Thus φ has a fixed point. So, without loss of generality we shall assume φ to be a continuous map of the k-simplex K into itself. Let $\{x_1, x_2, \cdots, x_{k+1}\}$ be its vertices. Let $x = \sum_{j=1}^{k+1} \lambda_j(x)x_j$ for any $x \in K$. Let $C_j = \{x : x \in K, \lambda_j(\varphi(x)) \leq \lambda_j(x)\}$. Since φ is continuous, C_j's are closed. Further, the C_j's satisfy the conditions of the previous theorem (1.12.3) for, if x is a point of the face of K whose vertices are $x_{j_1}, x_{j_2}, \cdots, x_{j_s}$, then

$$\lambda_j(x) = 0 \text{ for all } j \neq j_1, j_2, \cdots, j_s, \text{ and } \sum_{r=1}^{s} \lambda_{j_r}(x) = 1 \geq \sum_{r=1}^{s} \lambda_{j_r}(\varphi(x)). \text{ Then}$$

there exists $\lambda_{j_r}(x) \geq \lambda_{j_r}(\varphi(x))$ and $x \in C_{j_r}$. Thus by the previous theorem $\bigcap_{j=1}^{k+1} C_j \neq \varnothing$. That is, there exists an x^* with $\lambda_j(\varphi(x^*)) \leq \lambda_j(x^*), j = 1, 2, \cdots,$ $k + 1$. Since we know that $\Sigma\lambda_j(\varphi(x^*)) = \Sigma\lambda_j(x^*) = 1, \lambda_j(\varphi(x^*)) = \lambda_j(x^*)$. The λ_j's are unique for any $x \in K$, and therefore $\varphi(x^*) = x^*$. The proof of the theorem is complete.

Theorem 1.11.5 (Kakutani)

Let S be a compact convex set in R^k. Let ψ be a mapping of points of S into compact convex subsets of S satisfying the following condition: If $x_n \to x$, $y_n \in \psi(x_n)$, and $y_n \to y$ implies $y \in \psi(x)$ in S. Then there exists an $x^* \in S$ with $x^* \in \psi(x^*)$.

PROOF: Let $\tau : K \to S$ be a homeomorphism of a k-simplex K onto S. Further, let π_μ be a sequence of simplicial partitions of K with diameter δ_μ for each subsimplex in π_μ and $\delta_\mu \to 0$. Let $\psi^{(\mu)}$ be the following continuous map of K to S. For each vertex x of π_μ, choose $\psi^{(\mu)}(x) \in \psi(\tau(x)) \subset S$, and then express $\psi^{(\mu)}$ linearly in terms of the simplexes of π_μ. That is, if $x = \sum_{j=1}^{k+1} \lambda_j x_j$ (λ_j's $\geq 0, \Sigma\lambda_j = 1$) is a point of a subsimplex in π_μ with vertices $\{x_1, x_2, \cdots, x_{k+1}\}$, then $\psi^{(\mu)}(x) = \sum_{j=1}^{k+1} \lambda_j\psi^{(\mu)}(x_j)$. This is obviously unique and defines a continuous map $\psi^{(\mu)}$ of K into S. Then the map $\varphi = \tau^{-1}\psi^{(\mu)}$ defines a continuous map of K into K and has a fixed point $x^{(\mu)}$. If

$$x^{(\mu)} = \sum_{1}^{k+1} \lambda_j^{(\mu)} x_j^{(\mu)}, \ \sum_{1}^{k+1} \lambda_j^{(\mu)} = 1 \ (\lambda_j^{(\mu)} \geq 0, j = 1, 2, \cdots, k + 1)$$

without loss of generality $x_j^{(\mu)}$'s, $\lambda_j^{(\mu)}$'s $\psi^{(\mu)}(x_j^{(\mu)})$'s converge as $\mu \to \infty$ for $j = 1, 2, \cdots, k + 1$. The sequence $x_j^{(\mu)}, j = 1, 2, \cdots, k + 1$, as well as $x^{(\mu)}$, must all have the same limit $x^* \in K$. Let $\lim_\mu \lambda_j^{(\mu)} = \lambda_j^*$, and $\lim_\mu \psi^{(\mu)}(x_j^{(\mu)}) = \eta_j^*$.

Now

$$\tau(x^\mu) = \psi^{(\mu)}(x^{(\mu)}) = \Sigma\lambda_j^{(\mu)}\psi^{(\mu)}(x_j^\mu)$$

and $\psi^{(\mu)}(x_j^{(\mu)}) \in \psi(\tau(x_j^\mu))$, $\lim_\mu \tau(x_j^\mu) = \tau(x^*)$, and $\lim_\mu \tau(x^\mu) = \tau(x^*)$ by the continuity of τ. Thus $\tau(x^*) = \Sigma\lambda_j^*\eta_j^*$, and, by the nature of ψ, $\eta_j^* \in \psi(\tau(x^*))$, for all j. Then $\tau(x^*) \in \psi(\tau(x^*))$ by the convexity of $\psi(\tau(x^*))$. Thus $\tau(x^*)$ is the required element in K.

Theorem 1.11.6 (Tychonoff)

Let S be a compact convex set in R^k. Let $\{\varphi_t\}$ be an arbitrary family of continuous maps of S into itself with the following properties: (1) $\varphi_t o\varphi_s = \varphi_s o\varphi_t$ for any two indices t, s; (2) $\varphi_t(\lambda x + (1 - \lambda)y) = \lambda\varphi_t(x) + (1 - \lambda)\varphi_t(y)$ for any $x, y \in S$, and $0 \le \lambda \le 1$. Then there exists an $x^* \in S$ with $\varphi_t(x^*) = x^*$ for all t.

PROOF: By Brouwer's theorem, for any φ_t we have a fixed point. Consider any finite $\varphi_{t_1}, \varphi_{t_2}, \cdots, \varphi_{t_k}$. Let $H_{t_1} = \{x: \varphi_{t_1}(x) = x\}$. Clearly H_{t_1} is a convex set by condition 2. By continuity of φ_{t_1}, H_{t_1} is closed. Thus H_{t_1} is a compact convex set. For any $x \in H_{t_1}$, $\varphi_{t_2}(x) = \varphi_{t_2}o\varphi_{t_1}(x) = \varphi_{t_1}o\varphi_{t_2}(x)$; that is, $\varphi_{t_1}(\varphi_{t_2}(x)) = \varphi_{t_2}(x)$; thus $\varphi_{t_2}(x)$ is a fixed point under φ_{t_1}. So if $x \in H_{t_1}$, $\varphi_{t_2}(x) \in H_{t_1}$. Thus φ_{t_2} maps continuously H_{t_1} into itself. Once again by Brouwer's theorem we have $\varphi_{t_2}(y) = y$ for some $y \in H_{t_1}$; that is, $\varphi_{t_1}(y) = y$, $\varphi_{t_2}(y) = y$. In this way we could extend this to any finite indices, and $H_{t_1} \cap H_{t_2} \cap \cdots \cap H_{t_k} \ne \emptyset$ for any finite $t_1, t_2, \cdots, t_k$. Since K is compact, $\cap_t H_t \ne \emptyset$; that is, $\varphi_t(x^*) = x^*$ for some x^* and for all t.

Theorem 1.11.7 (Schauder)

Let S be a compact convex set in a real locally convex space E. Let φ be a continuous map of S into itself. Then φ has a fixed point (that is, $\varphi(x^*) = x^*$ for some $x^* \in S$).

PROOF: Since E is locally convex, any neighborhood of O contains a closed symmetric convex neighborhood of O (prove this!). Let G be any closed symmetric convex neighborhood of O. Consider the set $x + G$ for each $x \in S$. Then $\cup_{x \in S} x + G \supset S$. Since $x + G$ is a neighborhood of x and S is compact, there exists a finite $x_1, x_2, \cdots, x_{k+1}$ with $\cup_{i=1}^{k+1} x_i + G \supseteq S$.

Let K be the simplex generated by $\{x_1, x_2, \cdots, x_{k+1}\}$. Define the map

$$K \to K$$

$$x \to (\varphi(x) + G) \cap K$$

We shall show that ψ is a map of the type assumed in theorem 1.11.5. By theorem 1.4.4 $(\varphi(x) + G) \cap K$ is evidently closed convex. Further, if $x_n \to x_0$ in K, then $\varphi(x_n) \to \varphi(x_0)$. If $y_n \in \varphi(x_n) + G$ and $y_n \to y_0$, then $y_n - \varphi(x_n) \in G$ and $y_n - \varphi(x_n) \to y_0 - \varphi(x_0)$. Since G is closed, $y_0 - \varphi(x_0) \in G$, and therefore $y_0 \in \varphi(x_0) + G$. Further, K is finite-dimensional, and theorem 1.12.5 is valid. Let $x_0 \in K$ with $x_0 \in \psi(x_0)$. That is, $x_0 \in (\varphi(x_0) + G) \cap K$.

Let $N_G = \{x : x \in (\varphi(x) + G) \cap S\}$. We have proved that $N_G \neq \emptyset$. Further, N_G is closed. If G_1, G_2 are two symmetric neighborhoods, then $N_{G_1 \cap G_2} \subseteq N_{G_1} \cap N_{G_2}$. But $N_{G_1 \cap G_2} \neq \emptyset$, and $N_{G_1} \cap N_{G_2} \neq \emptyset$. Thus N_G's for every finite number of G's have non-null intersection. By the compactness of S, we have $\underset{G}{\cap} N_G \neq \emptyset$. If x^* is a point of this, then $x^* - \varphi(x^*) \in G$ for all closed neighborhoods of O. If $x^* - \varphi(x^*) \neq 0$, by the Hausdorff property of the space we have a G disjoint with $x^* - \varphi(x^*)$. Thus $x^* - \varphi(x^*) = 0$; that is, $x^* = \varphi(x^*)$. Hence the theorem.

REFERENCES

[1] Berge, C. (1963). "Topological Spaces, Vector Spaces and Convexity" (English translation), Oliver & Boyd Co., Edinburgh and London.
[2] Dunford, N., and Schwartz, J. T. (1964). "Linear Operators," Part I, Interscience Publishers, New York.
[3] Kelley, J. L. (1955). "General Topology," D. Van Nostrand Company, Princeton, New Jersey.
[4] Kolmogorov, A. N., and Fomin, F. V. (1957). "Functional Analysis," Vols. 1 and 2, Grelock, New York.

Chapter Two

GAMES IN EXTENSIVE AND NORMAL FORM

2.1 GRAPH

Let X be a finite set with its elements called vertices. Any ordered pair (x, y) with $x, y \in X$ is called an *arc*. Let $\mathcal{U}$ be a certain collection of ordered pairs in X. Then we say that $(X, \mathcal{U})$ constitutes a *graph*. We say that two arcs are *adjacent* if they have a vertex in common. Two vertices are called *adjacent vertices* when there is an arc u joining them. If $(x, y) \in \mathcal{U}$, we say that x is the predecessor of y and that y is a *successor* of x. We say that an arc (a, b) is *incident from A* if $a \in A$ and $b \notin A$. A sequence of arcs in a graph where the terminal vertex of each arc coincides with the initial vertex of the succeeding arc is called a *path*. A graph is *connected* if for any distinct pair of vertices a, b we have a finite sequence of arcs $(u_1, u_2, \cdots, u_r, u_{r+1}, \cdots, u_s)$ such that each of its intermediate arcs u_r is attached to u_{r-1} at one of its extremities and to u_{r+1} at the other (this is called a *chain*) with a vertex a of u_1 and a vertex b of u_s.

A *cycle* in a graph is a sequence of arcs $(u_1, u_2, \cdots, u_s)$ such that (1) it forms a chain, (2) the sequence does not use the same arc twice, and (3) the initial and terminal vertices of the chain coincide. A *tree* is a connected

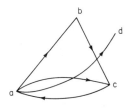

Fig. 1.

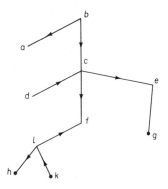

Fig. 2.

graph with at least two vertices and no cycles. An *edge* is an unordered pair of adjacent vertices.

One can draw the picture of a graph on the plane. For example, Fig. 1 represents a graph with four vertices, and the directions → and ← indicate that (a, b), (a, c), (c, a), (b, c), and (a, d) are the arcs for this graph. Here $\{(a, b), (b, c), (c, a)\}$ constitute a chain.

Figure 2 is a picture of a tree. For example, given the vertices h and c, we have the chain $\{(l, h), (l, f), (c, f), (c, e)\}$. This graph has no cycles. In any tree we have at least one vertex x_0 with no arc incident at x_0. For if it were not true, we have for any vertex $x \in X$ an arc emanating from x. Let y_1 be the terminal vertex of this arc. If one does not use the same edge twice, one cannot go to the same vertex twice (since a tree has no cycles). Thus, by assumption we have an arc emanating from y_1 with its end vertex $y_2 \neq y_1$ and $y_2 \neq x_0$. Proceeding inductively we can get an infinite sequence $x_0, y_1, y_2, \cdots$ of vertices. This would contradict the assumption that S is finite.

2.2 GAMES AND GAME THEORY

Game

If a group of people agree to obey certain rules and to act individually or in coalition without violating the specified rules, the results of their joint action lead to certain situations called *outcomes*. The members of the group are called the *players*, and the rules they agree to obey constitute a *game*. The rules in general specify a sequence of moves in which each move is allotted to specific players in the group or to a random experiment to be conducted at that move (this statement will become clear soon). The rules specify an initial move and the person entitled to that move. If a move is allotted to a player, it is called a *player move*. If it is allotted to a chance experiment, it is called a *chance move*. In a chance move the rules specify the selection of an alternative from a set of alternatives according to a specific probability distribution. In a player move, the player concerned is informed of the alternatives available to him at that move, and he is asked to select one among the alternatives as he wishes. Any particular alternative chosen in that move is called a *choice*. The rules specify what amount of information about all the previous moves and the alternatives chosen can be given to him before he makes his specific choice in his move. The rules specify the allotted player for the next move if it is a player move. The rules also specify a termination when no more moves are allowed. The termination occurs when some specific sequences of choices are made. There will be an outcome when the termination occurs, and we assume that the final outcome will be in the form of a score or a payment order to the players. We also assume that the players have a preference pattern over the outcomes reflected in the scores or in the payment orders. Our aim in this section will be to give a mathematical formulation to these notions.

Game Tree

Let us start with a specific game and try to formulate it mathematically. (We shall in fact try to draw a graph for that game, and this will be the mathematical description of the game.)

Example 2.2.1: Two players (say, player I and player II) play the following game. A referee distributes a card to each of the players selected randomly from the cards numbered 1, 2, and 3. Each one of the players knows only the number of his card. The game begins with player I. He is allowed to say "low" or "high." If both say low or if both say high or if player I says high and player II says low, the game terminates. If player I says low and player II says high, player I has a second chance to say low or high. A card with a greater number is considered more valuable than one with a smaller number. The players receive payment as follows. If both say high, then the player with the more valuable card gets 2 units of money from his opponent. If both say low, the player with the more valuable card gets 1 unit from his opponent. If player I says high and player II says low, player I gets 1 unit from his opponent (player II). If player I says low first and player II says high with player I saying low once again, player II gets 1 unit from player I. Lastly, if player I says low first and player II says high, and player I says high in his second chance, then the player with the more valuable card gets 2 units from the opponent.

This game has at most four moves. The first move is the chance move, which decides the deal the players get. The possible deals are $(1, 2)$, $(1, 3)$, $(2, 1)$, $(2, 3)$, $(3, 1)$, and $(3, 2)$, where each pair represents a deal, with the first number for player I and the second number for player II. Here each deal is equally likely; we have six alternatives available in the first move, and one of them is to be chosen at random. This move may be denoted by a point in a plane with six arcs emanating from it. Since the next move is the move for player I, the other vertex of these arcs is labeled 1. The player is informed of only his card number. This is represented geometrically by encompassing the end vertices of the pairs $\{(1, 2), (1, 3)\}$, $\{(2, 1), (2, 3)\}$, and $\{(3, 1), (3, 2)\}$ together. When for $(1, 2)$, $(1, 3)$ the end vertices are encompassed as in Fig. 3, this means that in that move the player corresponding to those vertices knows that his position is in one of them, or equivalently that the game has approached one of those cases; that is, the deals are either $(1, 2)$ or $(1, 3)$ when his card is 1. Since the first move is a chance move, we

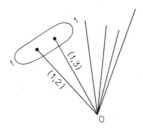

Fig. 3.

label it 0. If the information for all the other moves was also represented, then the representation of the game in the form of a graph would be like the one in Fig. 4. The vertices are labeled 0 or 1 or 2, the arcs are labeled H, L, etc. The vertices with no arcs emanating from them indicate the termination of the game. From each vertex labeled 1 or 2 we have two arcs emanating with labels H and L. They indicate that the player concerned has the choice of saying "high" or "low." With these intuitive notions, let us rigorously define a game in general.

Definition 2.2.1: A game tree is defined by a tree with a distinguished vertex ∗. The edges that are incident at any vertex are called *alternatives*. They are indexed by the integers 1, 2, · · ·, j if there are exactly j of them. Those vertices with at least one alternative are called *moves*. The rest of the vertices are called *plays*. Sometimes we also refer to the unique path beginning at the special vertex ∗ and ending at a vertex corresponding to a play as a play. By a partition of the vertices we mean an exhaustive decomposition of the vertices into disjoint sets. The partition of the moves into sets A_j, $j = 1, 2, · · ·$, where A_j consists of all moves with exactly j alternatives, is known as the *alternative partition*. The rank of any move x is defined as the number of moves used in the unique noncyclic path from the vertex ∗ to the move x.

Definition 2.2.2: A game G with n players is a game tree T with the following specifications.

1. A partition of the moves into $n + 1$ indexed sets $S_0, S_1, · · ·, S_n$, called the player partition. The moves in S_0 are chance moves. The moves in S_j are for player j, for $j = 1, 2, · · ·, n$.

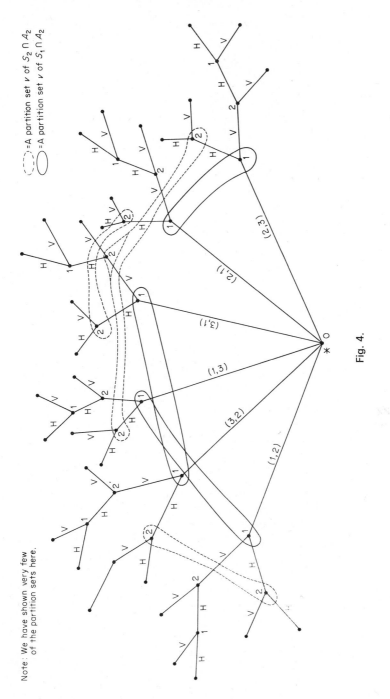

Fig. 4.

$\{ \ \ \}$ = A partition set ν of $S_2 \cap A_2$

$\bigcirc$ = A partition set ν of $S_1 \cap A_2$

Note: We have shown very few of the partition sets here.

2. The $\{S_j \cap A_k\}$, with $j \neq 0$, are further partitioned into sets U where no U contains two moves of the same play. This partition is called the *information partition*, and the sets U are called information sets.

3. For each $U \subset S_0 \cap A_k$, there is a probability distribution on the integers $1, 2, \cdots, k$ which assigns positive probability to $1, 2, \cdots, k$. These U's are assumed to be one-element sets.

4. Any play (the unique noncyclic path from $*$ to a terminal position) contains at most one vertex of any information set U.

5. An outcome function $k(w) = (k_1(w), k_2(w), \cdots, k_n(w))$, a vector in R^n, is given for each play w. The function h will be called the *payoff function*.

A play begins at the vertex $*$. Suppose that it has progressed to the move x. Let $x \in S_j \cap A_k$. This means that it is a move with k alternatives, and it is for the jth player (if $j \geq 1$) to have his choice in the move. Further, $x \in U \subset S_j \cap A_k$ for some information set. From this information set, the jth player gets the information about the progress of the play until then. He then chooses a positive integer $\leq j$, knowing only that he is choosing an alternative at one of the moves in his information set. If $x \in S_0 \cap A_k$ for some k, then an alternative is chosen according to the probability distribution specified in rule 3. Thus a path with $*$ as its initial point is constructed. With T being a finite tree, we reach a unique play w by this. At this point player j receives the amount $k_j(w)$, $j = 1, 2, \cdots, n$.

The same game may have two trees as its game tree. We say that two trees, T_1, T_2, induce the same game if (1) T_1 and T_2 have the same number of vertices; (2) there is a one-to-one map σ between the vertices of T_1 and T_2 such that, if $(G_1, \mathcal{U}_1)$, $(G_2, \mathcal{U}_2)$ are the graphs of T_1, T_2, then for any (x, y) with $(x, y) \in \mathcal{U}_1$, we have $(\sigma x, \sigma y) \in \mathcal{U}_2$, and conversely; (3) the homomorphism σ preserves the distinguished vertex and the properties 1, 2, 3, and 4 stated in definition 2.2.2.

Remark 2.2.1: In example 2.2.1, if we had split up the chance move into two chance moves, which corresponds to choosing one card first randomly and then choosing another card randomly, we would have represented this by a different tree. This evidently changes the number of vertices.

2.3 PURE STRATEGY AND NORMAL FORM

Suppose that in our card game (example 2.2.1) the players decide beforehand what they would do in any possible situation to which they may be led. For example, player II may decide as follows. If he gets card number 1 and if he is confronted with a "low" from player I, he will say low. If he is confronted with a "high" from player I, he will say low. He makes similar decisions beforehand for cards numbered 2 and 3. Thus, for example, whatever card he receives, he can say low whatever his opponent says. He has many more such actions. This means that he has decided beforehand what alternative he has to take in each situation that is possible. If both players have decided beforehand what action they would take if they were faced with a particular situation, then any representative could carry out a play, keeping the players' interest by following their specification without their presence. Any such plan for a player is called a *pure strategy*.

Definition 2.3.1: Let $\mathcal{U}_j = \{U : U \subset A_k \cap S_j$ for some $k\}$ be the collection of all information sets available to player $j(j \geq 1)$. A pure strategy for player j is a function τ_j mapping $\mathcal{U}_j$ into the set of positive integers such that $U \subseteq A_k$ implies $\tau_j(U) \leq k$. We shall say that τ_j chooses the alternative x incident at $x \in U$ if $\tau_j(U) = v(x)$. (Here $v(x)$ is actually the index of the alternative x, and $v(x) \leq k$.)

In our card game with only two alternatives, high ($= 1$) and low ($= 2$) for the players in any move, $\tau_2(U) \equiv 2$ denotes that player II chooses the call "low" always.

Suppose that $\lambda\tau = (\tau_1, \tau_2, \cdots, \tau_n)$ are the pure strategies chosen by the players 1, 2, $\cdots$, n. This induces a probability distribution on the alternatives in each information set of the tree. Suppose that x is the alternative chosen at a personal move of player j by τ_j in the information set U. Then we have the distribution

$$p_{\tau_j}(x) = 1 \text{ if } \tau_j(U) = v(x)$$

$$= 0 \text{ otherwise}$$

If x is an alternative in a chance move, we already have a probability distribution. Let $p_x(w)$ on the terminal vertices (plays) be defined by

$$p_\tau(w) = \prod_{x \in S} p_\tau(x)$$

where S is the set of all vertices x through which we can reach w starting at $*$ through the unique path.

Though the players have a preference over the terminal outcomes, they cannot always reach any play they like by the rules of the game. The presence of chance moves assigns a probability of their reaching a terminal vertex. This, coupled with their own pure strategies, leads to the $p_\tau(w)$ on the terminal w's. Thus when their preference pattern recognizes the presence of chance moves that have a say in the final outcome, a new preference pattern will be introduced that also takes this into account. We define this new preference pattern measure as the expected payoff $H_j(\tau) = \sum_w p_\tau(w) k_j(w)$

for player j, when $\tau = (\tau_1, \tau_2, \cdots, \tau_n)$ were the pure strategies chosen by the respective players. A question arises as to the logic behind accepting the expectation as a plausible measure of their preferences.

(For a discussion, see the chapter on utility in [2] and the Appendix in [15].)

We shall call τ_j and τ_j^* *equivalent* for player j if

$$\tau = (\tau_1 \tau_2, \cdots, \tau_{j-1}, \tau_j, \tau_{j+1}, \cdots, \tau_n)$$

and

$$\tau^* = (\tau_1, \tau_2, \cdots, \tau_{j-1}, \tau_j^*, \tau_{j+1}, \cdots, \tau_n)$$

have $p_\tau(w) = p_{\tau^*}(w)$ for all τ, τ^* containing τ_j, τ_j^* as the pure strategies of the jth player and for all terminal vertices w. One can see that the pure strategies τ_j and τ_j^* are equivalent if and only if they define the same relevant information sets and coincide on these sets (see theorem 1 [9], p. 202).

Any game tree representing a game among n players can be brought to the following form. Each of the players 1, 2, 3, $\cdots$, n chooses a pure strategy from the sets of their available pure strategies. This means that they all decide beforehand what choices they would make in their moves, for various situations that may arise by others' choices in their moves. Suppose that τ_1, τ_2, $\cdots$, τ_n are their pure strategies. Then this defines the probability distribution $p_\tau(w)$ on the terminal w's. The expected payoff to player j is $H_j(\tau) = \sum_w p_\tau(w)k_j(w)$. Since the tree is finite, the available number of pure strategies for any player is finite. If the number of players is 2, this reduces to the following new game.

Normal Form: Two players play a game as follows. Player I chooses an integer i from the integers $\{1, 2, \cdots, m\}$. Player II chooses an integer j from $\{1, 2, \cdots, n\}$. Both players do not know their opponent's choice. An amount a_{ij} is paid to player I, and an amount b_{ij} is paid to player II. Then such a game is called a two-person game in *normal form*.

If the τ's are enumerated for each player, then choosing a τ can be identified with a positive integer for the players. If player I has m pure strategies and player II has n pure strategies, then $H_1(\tau_1, \tau_2)$, $H_2(\tau_1, \tau_2)$ can be identified with the numbers a_{ij}, b_{ij}, where τ_1 is the ith pure strategy for player I and τ_2 is the jth pure strategy for player II, and $H_1(\tau_1, \tau_2) = a_{ij}$, $H_2(\tau_1, \tau_2) = b_{ij}$. Thus we can reduce any n-person game to its normal form with n payoff functions in n variables where the jth variable is at the hand of the jth player and is to be chosen from a finite set. The payoff for the jth player is the value of the jth function at the n-dimensional point chosen by all the players. From now onward we shall assume that someone has simplified our job and has given to us the normal form of the game. Moreover, we shall assume in general that the number of players is only two.

A two-person game in normal form with payoff matrices (a_{ij}, b_{ij}) is called *zero-sum* if, for any strategy i of player I and j of player II, $a_{ij} + b_{ij} \equiv 0$. Thus, physically we find these games with opposite interest where a gain to player I (if $a_{ij} > 0$) is a loss to player II ($b_{ij} < 0$), and conversely.

Unless otherwise stated, by a game we mean a zero-sum, two-person game.

Even for very simple games, with a few moves the number of pure strategies becomes very large and sometimes astronomical. Even to enumerate all pure strategies for a player in a game needs skill. Example 2.2.1 (our card game) illustrates how large a number it could be, even for such simple games. Let us in fact enumerate the pure strategies available to player II.

Let H denote calling high and L denote calling low. Since any pure strategy should specify what action to take when card 1 or 2 or 3 occurs (player II obtains), and since it should also instruct the action for any specific call made by player I in his first move, any pure strategy can be written by three groups of pairs of letters H or L formally written as, say, (LL, LH, HH), whose meaning is as follows.

The first pair contains the complete instruction for the case when he gets the card numbered 1, the second pair for card numbered 2, and the third pair for card numbered 3. The first instruction is taken when he hears low from player I, and the second instruction is taken when he hears high from player I. Thus (LL, LH, HH) would instruct player II with card 2 in his hand to say low when player I says low and to say high when player I says high. Similarly, (HH, HL, LH) would instruct player II with card 3 in his hand to say low when player I says low and to say high when player I says high. Thus, there are $4^3 = 64$ possible pure strategies for player II. One can show that player I has $3^3 = 27$ pure strategies.

Suppose that we do not restrict the game tree to possess only a finite number of vertices and a finite number of arcs connecting them, and suppose that such a tree could be reduced to the normal form with a not necessarily finite number of pure strategies. This new game could be played as follows. Players I and II choose an element $x \in X$ (an infinite set) and $y \in Y$ (an infinite set), respectively, each not knowing the choice of the opponent. Player I is paid $K_1(x, y)$, and player II is paid $K_2(x, y)$. The game is called an infinite two-person game. If $K_1(x, y) + K_2(x, y) = 0$ on $X \times Y$, then the game is zero-sum, and in that case the game is defined by the *payoff* $K_1(x, y)$ to player I. The moment we step into infinite pure strategies, then things are not that safe and the analysis of such games involves the topological structure of the spaces X and Y. This will be discussed in the chapter on minimax theorems.

Often we are faced with redundant pure strategies for a player, and these could be eliminated without affecting the strategic behavior of the players. Consider the following game.

Example 2.3.1: Players I and II choose a positive integer from the set of positive integers, each without the knowledge of the other. Player I receives 1 unit from player II if the sum is even and pays 1 unit to player II if it is odd.

In this game the pure strategies are the set of positive integers for the two players; and for any pure strategy i of player I and j of player II, if $K_1(i,j) = 1$, then $K_2(i,j) = -1$, and conversely, and thus the game is zero-sum. Let a_{ij} be the payoff $K_1(i,j)$ to player I. Then $a_{1j} = a_{3j} = \cdots = a_{2n+1,j} = \cdots$ for any j and $a_{2j} = a_{4j} = \cdots$ for any j, and player I can as well restrict himself just to the choice of $i = 1$ and $i = 2$. So it is for player II also. This drastically cuts the game to a payoff matrix with two rows and two columns. More generally, we have the following definition.

Definition 2.3.2: Let $G_1 = (T_1, T_2, H)$ and $G_2 = (T_1{}^*, T_2{}^*, K)$ be two games with H, K as their payoffs. (We shall not assume the pure strategy spaces $T_1, T_2, T_1{}^*, T_2{}^*$ to be finite.) We say that G_1 is a *reduction* of G_2 if either (1) $T_2 = T_2{}^*$ and $H(\tau_1, \tau_2) = K(f(\tau_1), \tau_2)$ for some function f on T_1 with range $T_1{}^*$ for all $(\tau_1, \tau_2) \in T_1 \times T_2$, or (2) $T_1 = T_1{}^*$ and $H(\tau_1, \tau_2) = K(\tau_1, g(\tau_2))$ for some function g on T_1 with range $T_2{}^*$ for all $(\tau_1, \tau_2) \in T_1 \times T_2$.

Definition 2.3.3: Two games G and G^* are equivalent if, through a finite number of games, G can be reduced to G^* or G^* to G.

2.4 SADDLE POINTS

Consider a game $G = (T_1, T_2, K)$, where T_1 is the set of positive integers $1, 2, \cdots, m$ and $T_2 = 1, 2, \cdots, n$ and $K(i,j) = a_{ij}$. Any finite game in normal form is of this type, and the pure strategies are the rows and columns of the matrix (a_{ij}) for player I and player II, respectively. Let $r_1, r_2, \cdots, r_m$ be the minimum of the first, second, $\cdots$, mth row. Clearly player I gets at least r_k units from player II by always choosing the kth row. Thus, independent of player II's choice, he can get at least the max $(r_1, r_2, \cdots, r_m)$ by always choosing the i_0th row where the max $(r_1, r_2, \cdots, r_m) = r_{i_0}$. Here

$$r_{i_0} = \max_i \min_j a_{ij}.$$

Similarly, for player II if c_j denotes the maximum of the jth column, then his choice of the jth column would result in a loss, which at is most c_j. Thus the choice of the column j_0 where $c_{j_0} = \min(c_1, c_2, \cdots, c_n)$ for player II would result in a loss to him which is at most c_{j_0}. This is independent of what player I does. Here $c_{j_0} = \min_j \max_i a_{ij}$.

We have the following elementary proposition for any matrix $A = (a_{ij})$.

Theorem 2.4.1

For any matrix $A = (a_{ij})$,

$$\max_i \min_j a_{ij} \leq \min_j \max_i a_{ij}$$

where, as usual, the minimum is taken over all columns j and the maximum is taken over all rows i.

PROOF: Fix any j, say, $j = 1$. Then

$$a_{i1} \leq \max a_{i1} \text{ for any } i$$

This is true for every j, and therefore

$$a_{ij} \leq \max_i a_{ij} \text{ for all } i \text{ and for any } j$$

Taking the minimum does not alter the inequality, and hence

$$\min_j a_{ij} \leq \min_i \max_j a_{ij} \text{ for all } i$$

Since the right-hand side is a constant and since the left-hand side is bounded by a constant, for all i we have

$$\max_i \min_j a_{ij} \leq \min_j \max_i a_{ij}$$

Hence the theorem.

This theorem tells us that $r_{i_0} \leq c_{j_0}$ in our game, which means that, with all his effort, player I cannot raise his income from player II beyond the limit of c_{j_0} if player II adopts the strategy of choosing the j_0th column. Suppose that for a matrix

$$r_{i_0} = \max_i \min_j a_{ij} = \min_j \max_i a_{ij} = c_{j_0}$$

Then player II has to give player I at least r_{i_0}, and player I receives at most c_{j_0} from player II, which are equal. Thus it is better for player I to restrict himself to the choice of the i_0th row, and for player II to restrict himself to the j_0th column. As an example, consider the payoff matrix

$$\begin{bmatrix} 2 & 3 \\ 1 & 0 \end{bmatrix}$$

Here $\min(2, 3) = 2$ and $\min(1, 0) = 0$ and $r_{i_0} = \max(2, 0) = 2$, which occurs at $i = 1$. Similarly $\max \begin{pmatrix} 2 \\ 1 \end{pmatrix} = 2$ and $\max \begin{pmatrix} 3 \\ 0 \end{pmatrix} = 3$, with $\min(2, 3) = 2$ occurring at $j = 1$. Thus it is always preferable for player I to choose row 1 and for player II to choose column 1. Here the first row dominates the second row in the sense that each entry in the first row is bigger than the corresponding entry in the second row. But there is no domination among the columns. If

$$\max_i \min_j a_{ij} = \min_j \max_i a_{ij} = a_{i^0 j^0}$$

for some i^0 and j^0, then we say that the matrix has a *saddle point* in the pure strategies i^0, j^0. In the infinite case, one is not always safe with minimum and maximum, and in such cases we can replace them by infimum and supremum. The problem then is to see whether

$$\sup_{x \in X} \inf_{y \in Y} K(x, y) = \inf_{y \in Y} \sup_{x \in X} K(x, y)$$

where x and y range over arbitrary sets X and Y, and $K(x, y)$ is a real valued function on $X \times Y$. In fact, this will be the main discussion in Chapter Five.

2.5 PERFECT INFORMATION GAMES

Suppose that in a zero-sum, two-person game in its extensive form the players are informed at any move of the choices of all the previous moves in a play; then one feels intuitively the optimality of a specific pure strategy. As for the game tree, such games correspond to game trees with information sets constituting single point sets. Such a game is called a game of *perfect information*. For example, consider the following game, known as tick-tack-toe.

Example 2.5.1: Players I and II choose alternately a small square with a number as shown in Fig. 5. The game terminates as soon as all the small squares are exhausted by their choices. The one who first occupies a complete

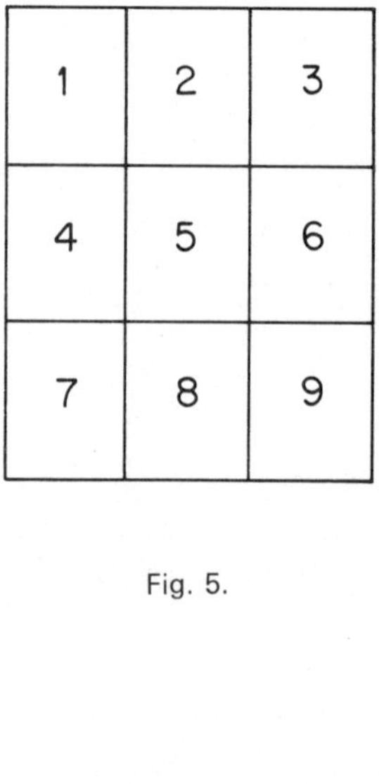

1	2	3
4	5	6
7	8	9

Fig. 5.

row or column or main diagonal wins the game. (Equivalently, the one who first gets three numbers from 1, 2, $\cdots$, 9 that could form an arithmetic progression wins the game.) Otherwise it is a draw.

This is an example of a zero-sum, two-person game with perfect information. In fact, one could find out an optimal pure strategy for a player by playing this game several times. Such a strategy for a player either yields a victory if the opponent is not playing correctly, or yields a draw. In fact, we shall prove the following more general proposition:

Theorem 2.5.1

All finite two-person games (not necessarily zero-sum) with perfect information have an equilibrium pair; that is,

$$K_1(\tau_1^0, \tau_2) \geq K_1(\tau_1, \tau_2) \qquad \text{and} \qquad K_2(\tau_1, \tau_2^0) \geq K_2(\tau_1, \tau_2)$$

for all τ_1, τ_2, where K_1, K_2 are the payoffs for the normal form of the matrix with τ_1, τ_2 as the generic pure strategies.

Remark 2.5.1: When the game is zero-sum with $K_1(\tau_1, \tau_2) = K(\tau_1, \tau_2)$, then the above statement reduces to

$$K_1(\tau_1^0, \tau_2) = K(\tau_1^0, \tau_2) \geq K(\tau_1, \tau_2) \text{ for all } \tau_1, \tau_2$$

and

$$+ K_2(\tau_1, \tau_2^0) = - K(\tau_1, \tau_2^0) \geq - K(\tau_1, \tau_2) \text{ for all } \tau_1, \tau_2$$

That is,

$$K(\tau_1, \tau_2^0) \leq K(\tau_1, \tau_2) \text{ for all } \tau_1, \tau_2$$

Thus, if τ_1, τ_2 range over finite sets, we have

$$\max_{\tau_1} K(\tau_1, \tau_2) \leq K(\tau_1^0, \tau_2) \text{ for all } \tau_2$$

That is,

$$\min_{\tau_2} \max_{\tau_1} K(\tau_1, \tau_2) \leq \min_{\tau_2} K(\tau_1^0, \tau_2)$$

Similarly

$$K(\tau_1, \tau_2^0) \leq \min_{\tau_2} K(\tau_1, \tau_2)$$

for all τ_1 and hence

$$\max_{\tau_1} K(\tau_1, \tau_2^0) \leq \max_{\tau_1} \min_{\tau_2} K(\tau_1, \tau_2)$$

That is,

$$\min_{\tau_2} \max_{\tau_1} K(\tau_1, \tau_2) \leq \min_{\tau_2} K(\tau_1^0, \tau_2) \leq K(\tau_1^0, \tau_2^0)$$

$$\leq \max_{\tau_1} K(\tau_1, \tau_2^0) \leq \max_{\tau_1} \min_{\tau_2} K(\tau_1, \tau_2)$$

But by theorem 2.4.1, the reverse inequality is true:

$$\max_{\tau_1} \min_{\tau_2} K(\tau_1, \tau_2) \leq \min_{\tau_2} \max_{\tau_1} K(\tau_1, \tau_2)$$

Thus

$$\max_{\tau_2} \min_{\tau_1} K(\tau_1, \tau_2) = \min_{\tau_1} \max_{\tau_2} K(\tau_1, \tau_2)$$

and we have a saddle point. In fact, one could show that τ_1^0, τ_2^0 is an equilibrium pair if $K(\tau_1^0, \tau_2^0)$ is a saddle point for the function $K(\tau_1, \tau_2)$

The proof of the theorem for the finite game tree with perfect information will be given by induction. The proof needs the knowledge of the truncation of games which will first be explained below.

Suppose that the rules of the game of tick-tack-toe are modified so that player I begins, but his choice should always be the principle top subsquare (the square with number 1). Then this is a new game with fewer moves for player I (in the sense that the first move's choice was thrust upon him), and player II has the same number of moves. Once again, it is a game of perfect information. The trees of such games have the following property, illustrated in Figs. 6, 7, 8, and 9.

If the game tree of a perfect information game looks like Fig. 6, we have three subgames with perfect information whose trees are as shown in Figs. 7, 8, and 9 when the choice in the first move is conditioned by the left or the middle or the right edge emanating from the special vertex in Fig. 6. Thus, by deleting the first move we still obtain such game trees because of the fact that each information set consists of a single vertex. Since a pure strategy for a player of a game of perfect information is a function that picks out an alternative at each of the players' moves, we can also consider the *truncations of a given strategy* corresponding to the various *truncations* of the game. For any pure strategy of the original game, the truncated strategy is defined only over the branch points of the corresponding truncation of the game, and it picks out the same alternatives at those branch points as does the original strategy.

PROOF OF THEOREM 2.5.1: Let T be the tree corresponding to the perfect information game with special vertex $*$. (We ask the reader to refer to the notions of tree partitions, truncations, pure strategy, etc.) Let the other end of the alternatives at the first move $*$ be $a_1, a_2, \cdots, a_r$. Thus we have the special vertex $* \in A_r$. Let $T_1, T_2, \cdots, T_r$ be the truncated games with $a_1, a_2, \cdots, a_r$ as their special vertices. Since each information set has a single element, we have these games unambiguously defined, and furthermore these truncated games are also games of perfect information.

Let the longest play in T be of length n (the number of vertices in the longest play). We shall prove the theorem by an induction argument on n. Clearly the games $T_1, T_2, \cdots, T_r$ have length at most $(n - 1)$. Let p_j, q_j be any generic pure strategy for the game corresponding to T_j for players I and II. Let τ_1, τ_2 be any generic pure strategy for the two players for the

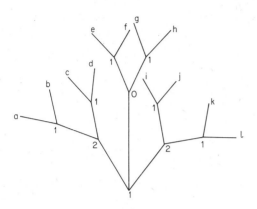

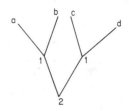

Fig. 6.

Fig. 7.

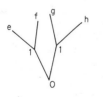
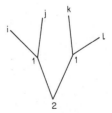

Fig. 8.

Fig. 9.

game T. Let $K_1^j(p_j, q_j)$ and $K_2^j(p_j, q_j)$ be the payoffs for players I and II for T_j. Let $K_1(\tau_1, \tau_2)$ and $K_2(\tau_1, \tau_2)$ be payoffs for players I and II for T.

For games of length 0, the theorem is trivial. So we shall assume the existence of equilibrium pairs for games of length $(n - 1)$, in particular for $T_1, T_2, \cdots, T_r$. Let (p_j^0, q_j^0) be such pairs for $j = 1, 2, \cdots, r$. That is,

$$K_1^j(p_j^0, q_j^0) \geq K_1^j(p_j, q_j^0) \text{ for all pure strategy } p_j \text{ for player I}$$

and

$$K_2^j(p_j^0, q_j^0) \geq K_2^j(p_j^0, q_j) \text{ for all pure strategy } q_j \text{ for player II}$$

We shall construct equilibrium pairs (τ_1^0, τ_2^0) for the game corresponding to T. This we shall do for the three different cases that may arise in the first move.

Case 1: The special vertex $* \in S_0 \cap A_r$ (that is, the first move is a chance move).

Let $\lambda_1, \lambda_2, \cdots, \lambda_r > 0$; $\displaystyle\sum_{j=1}^{r} \lambda_j = 1$ denote the probabilities for the alternatives $(*, a_1), (*, a_2), \cdots, (*, a_r)$ to be selected. Let x be any vertex of T. Then x is either $*$ or a_1 or $a_2 \cdots$ or a_r or any vertex of T_j for some j. Suppose that x is a vertex of T_j. Then $x \in$ of some information set U, which itself consists of a single point. Further, $U \subseteq S_0 \cap A_k$ or $U \subseteq S_1 \cap A_k$ or $U \subseteq S_2 \cap A_k$ for some $k \leq n - 1$. Suppose that $U \subseteq S_1 \cap A_k$ for some $k \leq n - 1$. Already U is an information set for T_j. Thus, define the pure strategy τ_1 in T for such a vertex x or for such an information set U by

$$\tau_1^0(x) = p_j^0(x)$$

If $U \subseteq S_2 \cap A_k$ for some $k \leq n - 1$, then define

$$\tau_2^0(x) = q_j^0(x)$$

Since $* \in S_0 \cap A_r$, we have defined the pure strategies τ_1^0, τ_2^0 for T for all information sets available to the players (remember that pure strategy is just a positive integer valued function on the available information sets for a player.)

For any pure strategy τ_1, τ_2 of the players in T, let p_j, q_j themselves be used also to denote the truncated pure strategy in T_j if no confusion is possible. Clearly

$$K_1(\tau_1, \tau_2) = \sum_j \lambda_j K_1^j(p_j, q_j)$$

$$K_2(\tau_1, \tau_2) = \sum_j \lambda_j K_2^j(p_j, q_j)$$

We have $K_1^j(p_j^0, q_j^0) \geq K_1^j(p_j, q_j^0)$ for any pure strategy p_j of player I for T_j for $j = 1, 2, \cdots, r$. Thus

$$K_1(\tau_1^0, \tau_2^0) = \sum_j \lambda_j K_1^j(p_j^0, q_j^0) \geq \sum_j \lambda_j K_1^j(p_j, q_j^0)$$

That is,

$$K_1(\tau_1^0, \tau_2^0) \geq K_1(\tau_1, \tau_2^0)$$

Here τ_1 is the pure strategy whose truncation to T_j is p_j. Since any pure strategy of player I in T can be equated to such truncations, we have

$$K_1(\tau_1^0, \tau_2^0) \geq K_1(\tau_1, \tau_2^0) \text{ for all pure strategy } \tau_1 \text{ of player I}$$

Similarly,

$$K_2(\tau_1^0, \tau_2^0) \geq K_2(\tau_1^0, \tau_2) \text{ for all pure strategy } \tau_2 \text{ of player II}$$

Thus (τ_1^0, τ_2^0) is an equilibrium pair for T.

Case 2: The vertex $* \in S_1 \cap A_r$. We define the pure strategy τ_1^0 for player I as follows: $\tau_1^0(*) = \alpha$, where α is the integer for which

$$\max_{1 \leq j \leq r} K^j(p_j^0, q_j^0) = K^\alpha(p_\alpha^0, q_\alpha^0)$$

For any other vertex $x \neq *$, τ_1^0 and τ_2^0 are defined as in case 1. We have

$$K_1(\tau_1, \tau_2) = K_1^j(p_j, q_j) \text{ if } \tau_1(*) = j$$

and

$$K_1(\tau_1^0, \tau_2) = K_1^\alpha(p_\alpha^0, q_\alpha)$$

Further

$$K_1(\tau_1^0, \tau_2^0) = K_1^\alpha(p_\alpha^0, q_\alpha^0) \geq K_1^j(p_j^0, q_j^0) \text{ for any } j = 1, 2, \cdots, r$$

But $K_1^j(p_j^0, q_j^0) \geq K_1^j(p_j, q_j^0)$ for any pure strategy p_j for player I in T_j. Further, we can think of a pure strategy τ_1 in T for player I whose truncation is p_j in T_j. Thus for any pure strategy τ_1 for player I in T,

$$K_1(\tau_1, \tau_2^0) = K_1^j(p_j, q_j^0) \text{ for some } j \text{ (here } \tau_1(*) = j)$$

and

$$K_1^j(p_j, q_j^0) \leq K_1^j(p_j^0, q_j^0) \leq K_1^\alpha(p_\alpha^0, q_\alpha^0) = K_1(\tau_1^0, \tau_2^0)$$

Thus

$$K_1(\tau_1, \tau_2^0) \leq K_1(\tau_1^0, \tau_2^0)$$

and

$$K_2(\tau_1^0, \tau_2) = K_2^\alpha(p_\alpha^0, q_\alpha) \leq K_2^\alpha(p_\alpha^0, q_\alpha^0) = K_2(\tau_1^0, \tau_2^0)$$

Hence (τ_1^0, τ_2^0) is an equilibrium pair for T.

Case 3: $* \in S_2 \cap A_r$. This is similar to case 2. The proof of the theorem is complete.

2.6 MIXED STRATEGIES

The question of finding saddle points is itself of interest, but many payoff matrices do not possess saddle points. Consider the following simple game, called "matching pennies":

Example 2.6.1: Players I and II choose heads or tails on a coin, each without revealing his choice to his opponent. Then they compare their choices, and player I gets 1 unit from player II if both had the same choice. Otherwise player II gets 1 unit from player I.

Here the payoff to player I is

II

Heads Tails

		Heads	Tails
	Heads	1	−1
I			
	Tails	−1	1

Evidently $\max_i \min_j a_{ij} < \min_j \max_i a_{ij}$ for this matrix. So one cannot claim that a particular choice, say heads, is always good for player I. Here the word "always" has a deeper significance. Suppose, for example, that this game is repeated several times. Could we possibly stick to the same pure strategy? For example, this game is played one hundred times, and player I chooses heads always. For a while player II may choose heads and tails in his own fashion. But the moment he detects the uniform choice of player I, he would choose tails continuously and get 1 unit every time, until player I is compelled to change his choices alternately. Perhaps the players could choose their pure strategies according to some chance mechanism so that detection is not possible completely. Suppose that player I chooses heads and tails by actually taking the outcome when it is tossed; then the outcome is not in his control. But still it ceases to give any hint to player II about what is going to happen next. Even if player II sticks to the same choice, say to tails every time, the expected income for player I is $\frac{1}{2}(-1) + \frac{1}{2}(1) = 0$. But perhaps player I may deviate from this way of playing for a while and teach a lesson to player II to change his strategy, and then he could come back to this sort of choice. Lastly, suppose that both try to choose by random experiment. Say that player II chooses heads and tails as follows. He throws a die (unbiased) and chooses heads if it gives 1; otherwise he chooses tails. For some time player I sticks to his strategy of tossing the coin and choosing the actual outcome of the coin. He can also keep a record of player II's choices and try to detect the probabilities of his choices (he guesses first that player II chooses according to a random experiment). Then suddenly he chooses a random experiment that would yield a more average income for this choice.

Perhaps his deviation like this for a long time may land him in trouble if player II comes back to his senses. Thus it is safer for them to choose heads and tails with probability half each. Further, this gives an assurance about their minimum average income or their maximum average loss.

Any probability distribution on the set of pure strategies is called a *mixed strategy* for the player.

2.7 MINIMAX THEOREM FOR FINITE GAMES

In this section we shall prove the fundamental theorem of von Neumann on matrix games.

Definition 2.7.1: Let $A = (a_{ij})$ be the payoff matrix of a game G in its normal form with $1 \leq i \leq m$, $1 \leq j \leq n$. The *mixed extension* of the game G is the new game $H = (X, Y, K)$, where

$$X = \{x: x = (\xi_1, \xi_2, \cdots, \xi_m): \xi_i \geq 0, \quad i = 1, 2, \cdots, m, \quad \sum_{i=1}^{m} \xi_i = 1\}$$

$$Y = \{y: y = (\eta_1, \eta_2, \cdots, \eta_n): \eta_j \geq 0, \quad j = 1, 2, \cdots, n, \quad \sum_{j=1}^{n} \eta_j = 1\}$$

$$K(x, y) = \sum_i \sum_j a_{ij} \xi_i \eta_j, \quad \text{where } x = (\xi_1, \xi_2, \cdots, \xi_m), \quad y = (\eta_1, \eta_2, \cdots, \eta_n)$$

Remark 2.7.1: The mixed extension is the new game where player I chooses a probability distribution $x = (\xi_1, \xi_2, \cdots, \xi_m)$ on his pure strategy space $P = \{1, 2, \cdots, m\}$ and player II chooses a probability distribution $y = (\eta_1, \eta_2, \cdots, \eta_n)$ on his pure strategy space $Q = \{1, 2, \cdots, n\}$ with the payoff $K(x, y) = \Sigma \Sigma a_{ij} \xi_i \eta_j$ to player I. The fundamental theorem of von Neumann asserts that the mixed extension of any finite game always has a saddle point, although the original payoff may not possess one. We are now ready to prove this fundamental minimax theorem.

Theorem 2.7.1 (von Neumann)

Let $A = (a_{ij})$ be any payoff matrix of a game G with $1 \leq i \leq m$, $1 \leq j \leq n$. Let $H = (X, Y, K)$ be its mixed extension. Then the payoff K has a saddle point; that is:

$$\min_{y \in Y} \max_{x \in X} K(x, y) = \max_{x \in X} \min_{y \in Y} K(x, y)$$

PROOF: Let $C_1, C_2, \cdots, C_n$ be the column vectors of the matrix A. For any $y = (\eta_1, \eta_2, \cdots, \eta_n) \in Y$, $\eta_1 C_1 + \eta_2 C_2 + \cdots + \eta_n C_n$ is a vector that lies in the smallest convex set S containing $C_1, C_2, \cdots, C_n$. Thus for any

$y \in Y$, we have $c(y) = \sum\limits_{j} \eta_j C_j$, a point in the convex set S. Conversely for any $s \in S$, $s = \eta_1' C_1 + \eta_2' C_2 + \cdots + \eta_n' C_n$ and $y' = (\eta_1', \eta_2', \cdots, \eta_n') \in Y$. Thus for any $y \in Y$ there is a unique $c(y) \in S$, and for any $s \in S$ we have at least one $y' \in Y$. Here

$$C_j = \begin{bmatrix} a_{1j} \\ a_{2j} \\ \vdots \\ a_{mj} \end{bmatrix} \qquad j = 1, 2, \cdots, n$$

Since

$$K(x, y) = \sum \sum a_{ij} \xi_i \eta_j = \sum_i \xi_i \left(\sum a_{ij} \eta_j \right)$$

if $s = \sum\limits_{j} \eta_j C_j$, then $s_i = \sum \eta_j a_{ij}$, and

$$K(x, y) = \sum_{j=1}^{m} \xi_i s_i$$

We have reduced the problem to the following new game:

$$R = (X, S, M)$$

where $M(x, s) = \sum\limits_{i} \xi_i s_i$, $x = (\xi_1, \xi_2, \cdots, \xi_m)$, and $s = (s_1, \cdots, s_m)$. Further, X and S are closed bounded convex sets, and M is in fact a bilinear functional on $X \times S$. Now

$$\max_{x \in X} M(x, s) = \max (s_1, s_2, \cdots, s_m) \text{ for any } s$$

Since for any $s = (s_1, s_2, \cdots, s_m) \in S$, the function $h(s) = \max (s_1, \cdots, s_m)$ is continuous, and since S is closed bounded in R^m (real m-space), $\min\limits_{s \in S} h(s) = h(s^0)$ for some $s^0 \in S$. Thus,

$$\min_{s \in S} \max_{x \in X} M(x, s) = \max (s_1^0, s_2^0, \cdots, s_m^0) \text{ for some}$$

$$s^0 = (s_1^0, s_2^0, \cdots, s_m^0) \in S$$

Let T be the convex set in R^m with

$$T = \{(t_1, t_2, \cdots, t_m): t_1 < \alpha, t_2 < \alpha, \cdots, t_m < \alpha\}$$

where $\alpha = \max (s_1^0, s_2^0, \cdots, s_m^0)$.

Here T is evidently open, and, since s^0 has at least one coordinate equal to α, $s^0 \in \overline{T}$, the closure of T. Trivially, the vector $(\alpha, \alpha, \cdots, \alpha) \in \overline{T}$. Further, $T \cap S = \emptyset$, for otherwise there would be a vector in S with all components strictly less than α, which would contradict the fact that $h(s^0) = \min h(s)$. By theorem 1.6.2 we have a nontrivial linear functional f on R^m with

$$f(x) \geq c, \qquad x \in S$$

$$f(x) \leq c, \qquad x \in T$$

for some c.

The continuity of f together with $s^0 \in \overline{T}$ gives $f(s^0) \leq c$. But $f(s^0) \geq c$ as $s^0 \in S$. Thus $f(s^0) = c$. Since any linear functional $f(x)$ on R^m is of the form $\sum_i f_i x_i$ for $x = (x_1, x_2, \cdots, x_m)$, we have

$$\sum_i f_i s_i^0 = c$$

For any $\varepsilon > 0$, $f_1 s_1^0 + \cdots + f_j(s_j^0 - \varepsilon) + f_{j+1} s_{j+1}^0 + \cdots + f_m s_m^0 = c - f_j \varepsilon$, and the vector $s^* = (s_1^0, s_2^0, s_{j-1}^0, s_j^0 - \varepsilon, s_{j+1}^0, \cdots, s_m^0) \in \overline{T}$, and $\Sigma f_i s_i^* = c - f_j \varepsilon \leq c$. Thus $f_j \geq 0$ for all j. (Here s_i^* is the ith coordinate of s^*.)

Since the linear functional f is nontrivial, some $f_j > 0$, and without loss of generality $\sum_j f_j = 1$ (because division by Σf_j would not affect the nature of the inequalities). Further, $\sum_j f_j \alpha \leq c$ implies $\alpha \leq c$.

Thus we have $f \in X$, with $\Sigma f_i s_i \geq \alpha$ for all $s \in S$. Further, $c = \Sigma f_i s_i^0 \leq \Sigma f_i \alpha$ (note that the maximum coordinate of s^0 is α), and hence $c = \alpha$. Equivalently,

$$\sum_i \sum_j f_i a_{ij} \eta_j \geq \alpha = c \text{ for all } y = (\eta_1, \cdots, \eta_n) \in Y$$

and therefore

$$K(f, y) \geq c \text{ for all } y \in Y \text{ with } K(f, y^0) = c$$

where

$$s_i^0 = \sum a_{ij} \eta_j^0, \qquad i = 1, 2, \cdots, m, \qquad y^0 = (\eta_1^0, \eta_2^0, \cdots, \eta_n^0)$$

But

$$K(x, y^0) = \sum_i \sum_j a_{ij} \xi_i \eta_j^0 = \sum_i \xi_i s_i^0 \leq \max(s_1^0, s_2^0, \cdots, s_m^0) = c$$

Therefore we have

$$K(x, y^0) \leq c \text{ for all } x \in X$$

$$K(f, y) \geq c \text{ for all } y \in Y$$

Thus

$$\max_{x \in X} K(x, y^0) \leq c \leq K(f, y) \text{ for each } y \in Y$$

and

$$\min_{y} \max_{x} K(x, y) \le \max_{x \in X} K(x, y^0) \le c \le \min_{y \in Y} K(f, y)$$
$$\le \max_{x \in X} \min_{y \in Y} K(x, y)$$

That is,

$$\min_{y} \max_{x} K(x, y) \le \max_{x} \min_{y} K(x, y)$$

The reverse inequality is always true, and hence

$$\max_{x} \min_{y} K(x, y) = \min_{y} \max_{x} K(x, y)$$

Remark 2.7.2: The constant c is called the *value* of the game, and the mixed strategies f and y^0 are called *optimal mixed strategies* for the players. The inequality

$$K(f, y) \ge \alpha \text{ for all } y \in Y$$

indicates the fact that, by choosing the rows according to the probability law $f = (f_1, f_2, \cdots, f_m)$, player I, without bothering about the choice of player II, can be assured of an average income which is at least α. Similarly, the inequality

$$K(x, y^0) \le \alpha \text{ for all } x \in X$$

indicates that, by choosing the columns according to the probability law $y^0 = (\eta_1^0, \eta_2^0, \cdots, \eta_n^0)$, player II, without bothering about the choice of player I, can be guaranteed a loss that is at most α. Thus f and y^0 are optimal in the sense that f maximizes his minimum gain and y^0 minimizes the maximum loss.

Before closing this section we wish to make some comments on this theorem.

Minimax theorem gives a sort of optimal guidance to a player's action only when one can view the game being played several times. Intuitively, one gets a level of satisfaction in maximizing his average income in a repeated play, and the optimal strategies should be adopted in such cases. The choice of the optimal strategy by player II in the sense of the theorem makes player II assume that player I is a cut-throat chap and his only intention is to maximize player II's average loss. Thus the theorem does not properly guide us in a case where player II is justified in assuming that the intention of player I is not necessarily to maximize his opponent's loss. Even in a repeated play for such cases, the optimal strategy of player II given by the theorem makes him a pessimist. Statistical problems come under this category.

There are cases where one may not know the individual entries of the payoff matrix but might know only when the players have chosen the corresponding pure strategies. For example, the entry a_{ij} in the payoff may be known only when the pure strategy i is chosen by player I and j is chosen by player II. Thus, even for repeatedly playing this game, one cannot know the optimal mixed strategy in the beginning. In these games one may have to use the players' past actions for their future choice. Such games lead to what are called adaptive competitive games. Sometimes the payoff itself may change from play to play, depending on the choices of the players in the previous plays. Such games lead to what are called stochastic games (we shall discuss stochastic games elsewhere).

In the next section we shall explicitly normalize a simplified poker game and another card game and compute the optimal mixed strategies for the players.

2.8 EXAMPLES OF FINITE GAMES

Example 2.8.1: From cards numbered 1, 2, and 3, two cards are randomly selected, and a card is given to each of two players. Now they can say low or high, but they must shout it simultaneously. Each player knows the card he has but not the card his opponent has. If both shout low, the cards are compared and the one with the higher card number gets 1 unit from his opponent. If both shout high, the cards are compared and the one with the higher card gets 2 units from his opponent. If one says low and the other says high, the one who shouted low has the option of passing and paying 1 unit outright to his opponent. Otherwise he can say high, in which case the payment is made as though they both had called high simultaneously.

Let us first form the payoff matrix. This means that we should enumerate the course of action a player has to take for any situation that could occur to him. This means deciding beforehand what actions he could take when he gets card 1 or 2 or 3. Theoretically, he can take any of the actions—namely, say high or say low and subsequently pass, or say low and subsequently bet high if an occasion arises. Let us write them symbolically as high = H, low with subsequent pass = L, and low with subsequent bet of shouting high = B. Thus a pure strategy consists of a triplet like (H, B, B), which means bet high when the card number is 1 and say low first on 2 or 3 but say high if the opponent says high. Thus there are 27 possible pure strategies for each of the players, as each coordinate can take H or L or B. Since any card is equally likely, we appropriately take the average of the possible deals that they can get.

As an example, suppose that (B, H, H) is the pure strategy chosen by player I, and that (L, B, H) is chosen by player II. Then if the cards were (1, 2) (1 for player I and 2 for player II), player II would get 1 unit; if they were (2, 1), player I would get 1 unit from player II; if they were (1, 3), player II would have 2 units from player I; etc. Then, since all the six deals are equally likely, we get the new payoff by averaging over all possible such incomes and losses for player I.

Further, the players may not choose a particular strategy in any of their play. In that case we can assume that the row or column corresponding to that pure strategy is chosen with zero probability. Effectively, we have a smaller matrix to handle. We shall eliminate such pure strategies by finding them out.

Suppose that 1 is the card a player has. He is sure that his opponent has a better card. Thus it is meaningless to say low first and refute his opponent's high by a high; that is, the pure strategies in which the first component is B would never be used. This means that we have thrown out nine pure strategies from our consideration. Suppose that a player gets the card numbered 2. He can say low first and then say high when the opponent says high, or he can say high in the beginning. Lastly he can say low and never take a second chance. We shall show that he can as well discard the action of saying high and instead use the action of saying low first and refute with a high when his opponent calls high. Let player I get card 2. Then player II has either card 1 or card 3. Suppose that he has card 1. He can choose L or H, as he knows that B is not good for him for card 1. Then let us compare $(B, L), (H, L)$ and $(B, H), (H, H)$. Pair (B, L) yields 1 unit to player I, and so does (H, L). Pair (B, H) yields 2 units to player I, and so does (H, H). Thus, whatever be the action of player II, the actions B and H yield the same income when player I has card 2 and player II has card 1. Lastly, when player I has card 2 and player II has card 3, we shall once again try to compare the pairs. First let us note that player II would never say low without rebetting when the opponent says high, for he knows that his card is the best. So the strategies with the last component taking L is never chosen by any of the players. Thus we need compare only $(B, B), (H, B)$ and $(B, H), (H, H)$. In the first case (H, B) is inferior to (B, B), but (B, H) and (H, H) are the same. Thus it is preferable to choose B rather than H, for both yield the same amount to a player, and if there is a difference then B fares better. Thus a player can restrict his pure strategies to those whose second component never takes H, for when he wants to make a choice of such a strategy he can gain in a better way by choosing B in all such cases where it fares better.

Finally, if player I gets 3 it is not advantageous to choose H rather than B, for the opponent can have only 1 or 2. Suppose that the opponent's card is 1. He would use either L or H; in this case (B, L) and (H, L) are the same. Similarly, (B, H) and (H, H) are the same. Suppose that he has card 2. Then he would use either L or B, for he knows that B is as good as H and in fact sometimes better. In these cases (B, L) and (H, L) are the same, but (B, B) is less advantageous compared to (H, B). Thus, H is always preferable to B for the third component.

Thus, when one plays he can optimally decide to play as follows. Never have the pure strategies with B in the first component. Since B is as good as

H when 2 is the card for a player and sometimes B is better, never choose pure strategies whose second component is H. Lastly never choose a pure strategy whose third component is L or B.

Thus, the players have the following strategies with L or H in the first component, L or B in the second, and always H in the third. They are

$$L\ L\ H$$
$$L\ B\ H$$
$$H\ L\ H$$
$$H\ B\ H$$

Thus, we can as well assume that they choose only one of these strategies. Let us form the expected payoff for player I for this case.

	$L\,L\,H$	$L\,B\,H$	$H\,L\,H$	$H\,B\,H$
$L\,L\,H$	0	$\frac{1}{6}$	$-\frac{1}{6}$	0
$L\,B\,H$	$-\frac{1}{6}$	0	$\frac{1}{6}$	$\frac{1}{3}$
$H\,L\,H$	$\frac{1}{6}$	$-\frac{1}{6}$	0	$-\frac{1}{3}$
$H\,B\,H$	0	$-\frac{1}{3}$	$\frac{1}{3}$	0

Observe that no player has any special preference given over the other, and that this symmetry ensures that the value $v \not> 0$ and $v \not< 0$, and that $v = 0$. See theorem 3.1.11.

If player I restricts himself to choosing LLH, LBH, and HLH with equal probability, this expected income is 0, independent of player II's choice. The same strategy is good also for player II. On card 1 a player says high $\frac{1}{3}$ times and says low otherwise. On card 2 a player rebets $\frac{1}{3}$ times when the opponent says high, and says low otherwise; players always say high on card 3.

Example 2.8.2: A number is chosen randomly from 1, 2, 3, 4, 5, and two players are asked to guess an upperbound for the chosen number. If only one of them is successful in his guess, he gets 1 unit from his opponent. If both are successful, then the one with the strictly smaller upperbound gets 1 unit from his opponent. In all other cases no one gets anything. (The players know that the number is only one of 1, 2, 3, 4, and 5 and also that it is selected randomly.)

Suppose that player I guesses i and player II guesses j. Let $K(i,j)$ be the payoff. Let the number randomly selected be x. Here $x \in \{1, 2, \cdots, 5\}$, $i \in \{1, 2, \cdots, 5\}$, $j \in \{1, 2, \cdots, 5\}$, and $P(x = k) = \frac{1}{5}$, $k = 1, 2, 3, 4,$ and 5. Let us compute the expected payoff for player I for $i < j$. Either $k \le i < j$, or $i < k \le j$, or $i < j < k$ is true.

If the first case happens, player I wins; that is, he gets 1 unit with $P(k \leq i)$. In the second case he loses 1 unit, for which the probability is $P(i < k \leq j)$. In the third case nothing happens.

Thus, for $i < j$, his expected income is $K(i,j) = P(k \leq i) - P(i < k \leq j) = 2P(k \leq i) - P(k \leq j)$. Similarly, for $i > j$ his expected income is $K(i,j) = -P(k \leq j) + P(j < k \leq i) = -2P(k \leq j) + P(k \leq i)$. Lastly $K(i,j) = 0$ for $i = j$. Since i,j's can take only one of 1, 2, 3, 4, and 5, we have the expected payoff matrix $k(i,j)$ as

$$
\begin{array}{c}
\\
\\
\\
i = 3 \\
\\
\\
\end{array}
\begin{array}{c}
j = 1 \\
1 \\
2 \\
3 \\
4 \\
5
\end{array}
\begin{bmatrix}
0 & 0 & -\frac{1}{5} & -\frac{2}{5} & -\frac{3}{5} \\
0 & 0 & \frac{1}{5} & 0 & -\frac{1}{5} \\
\frac{1}{5} & -\frac{1}{5} & 0 & \frac{2}{5} & \frac{1}{5} \\
\frac{2}{5} & 0 & -\frac{2}{5} & 0 & \frac{3}{5} \\
\frac{3}{5} & \frac{1}{5} & -\frac{1}{5} & -\frac{3}{5} & 0
\end{bmatrix}
$$

The symmetry between the players of the game demands that the value $v = 0$. For example, the mixed strategy $(0, \frac{2}{3}, 0, \frac{1}{3}, 0)$ is optimal for both. Thus if the players give 2 as the upper limit $\frac{2}{3}$ times and 4 as the upper limit $\frac{1}{3}$ times, it is optimal in a repeated play.

We leave it to the reader to solve example 2.1.1 given in the beginning.

REFERENCES

[1] Berge, C., and Ghouila-Houri, A. (1965). "Programming, Games and Transportation Networks," Methuen and Co. Ltd., London.
[2] Blackwell, D., and Girshick, M. A. (1954). "Theory of Games and Statistical Decisions," John Wiley & Sons, New York.
[3] Burger, E. (1963). "Introduction to the Theory of Games," Prentice-Hall, Englewood Cliffs, New Jersey.
[4] Dalkey, N. (1953). Equivalence of information patterns and essentially determinate games. In "Contributions to the Theory of Games," Vol. 2, pp. 217–244 (*Ann. Math. Studies No. 28*), edited by H. W. Kuhn and A. W. Tucker, Princeton University Press, Princeton, New Jersey.
[5] Dresher, M. (1961). "Games of Strategy," Prentice-Hall, Englewood Cliffs, New Jersey.
[6] Gale, D. (1960). "The Theory of Linear Economic Models," McGraw-Hill Book Company, New York.
[7] Kakutani, S. (1941). A generalization of Brouwer's fixed point theorem. *Duke Math. J.* **8**, 457–458.
[8] Karlin, S. (1959). "Matrix Games, Programming and Mathematical Economics," Addison-Wesley Publishing Co., Reading, Massachusetts.

[9] Kuhn, H. W. (1953). Extensive games and the problem of information. In "Contributions to the Theory of Games," Vol. 2, pp. 193–216 (*Ann. Math. Studies No. 28*), edited by H. W. Kuhn and A. W. Tucker, Princeton University Press, Princeton, New Jersey.

[10] Kuhn, H. W. (1950). A simplified two-person poker. In "Contributions to the Theory of Games," Vol. 1, pp. 97–103 (*Ann Math. Studies No. 24*), edited by H. W. Kuhn and A. W. Tucker, Princeton University Press, Princeton, New Jersey.

[11] Loomis, L. H. (1946). On a theorem of von Neumann. *Proc. Natl. Acad. Sci. U.S.*, **32**, 213–215.

[12] McKinsey, J. C. C. (1952). "Introduction to the Theory of Games," McGraw-Hill Book Company, New York.

[13] Luce, R. D., and Raiffa, H. (1957). "Games and Decisions," John Wiley & Sons, New York.

[14] Nash, J. F., and Shapley, L. S. (1950). A simple three-person poker game. In "Contributions to the Theory of Games," Vol. 1, pp. 105–116 (*Ann. Math. Studies No. 24*), edited by H. W. Kuhn and A. W. Tucker, Princeton University Press, Princeton, New Jersey.

[15] von Neumann, J., and Morgenstern, O. (1944). "Theory of Games and Economic Behaviour," Princeton University Press, Princeton, New Jersey.

[16] Weyl, H. (1950). Elementary proof of a minimax theorem due to von Neumann. In "Contributions to the Theory of Games," Vol. 1, pp. 19–25 (*Ann. Math. Studies No. 24*), edited by H. W. Kuhn and A. W. Tucker, Princeton University Press, Princeton, New Jersey.

Chapter Three

OPTIMAL STRATEGIES OF FINITE GAMES

In this chapter we shall try to determine some of the properties of optimal mixed strategies and their relationship to the form of the payoff matrix and the value of the game. Some theorems proved here do not use the full linear structure of the real numbers that constitute the entries in a payoff matrix. Some of the theorems concern matrices with positive entries and the relationship between their spectral properties and their values. Section 3.2 contains results on saddle-point theorems.

3.1 OPTIMAL STRATEGY SETS

Theorem 3.1.1

Let $A = (a_{ij})$, $1 \leq i \leq m$, $1 \leq j \leq n$, be the payoff matrix of a game G. Then the set of optimal mixed strategies for player I is a closed bounded convex set. The same is true also for the set of optimal mixed strategies of player II.

PROOF: Let $x^0 = (\xi_1^0, \xi_2^0, \cdots, \xi_m^0)$ and $x^* = (\xi_1^*, \xi_2^*, \cdots, \xi_m^*)$ be two optimal mixed strategies with v as the value of the game. Then

$$\sum_i a_{ij}\xi_i^0 \geq v, \qquad \sum_i a_{ij}\xi_i^* \geq v, \qquad j = 1, 2, \cdots, n$$

Clearly for any λ in $0 \leq \lambda \leq 1$, and for any j,

$$\sum_i a_{ij}(\lambda\xi_i^0 + (1 - \lambda)\xi_i^*) \geq \lambda v + (1 - \lambda)v = v$$

and

$$\sum_i (\lambda_i\xi_i^0 + (1 - \lambda)\xi_i^*) = 1$$

Thus $\lambda x^0 + (1 - \lambda)x^*$ is also an optimal mixed strategy for player I. Since the function $K(x, y) = \sum_i \sum_j a_{ij}\xi_i\eta_j$ with $x = (\xi_1, \xi_2, \cdots, \xi_m)$ and $y = (\eta_1, \eta_2, \cdots, \eta_n)$ is continuous in x, y, the set of optimal mixed strategies is a closed set. A similar proof applies to player II.

Remark 3.1.1: The set of optimal strategies (from now on optimal mixed strategies will be called optimal strategies) for player I is a compact convex subset of R^m.

By the Krein–Milman theorem (theorem 1.8.1) we have extreme points for this set. Further, since the set is in R^m and since the set lies on the plane $\Sigma \xi_i = 1$, its dimension is reduced. So it is of interest to know the dimension of the set of optimal strategies that, coupled with theorem 1.9.3, helps us to characterize all optimal strategies. Lastly, if we know that there are only a finite number of extreme points for this set, then we could as well use them all to characterize all optimal strategies. In fact, we shall show elsewhere that the set of extreme points for our set is finite.

Definition 3.1.1: A column j is *relevant* for player II in a payoff matrix (a_{ij}), $1 \le i \le m$, $1 \le j \le n$, if there exists at least one optimal strategy $y^* = (\eta_1{}^*, \eta_2{}^*, \cdots, \eta_j{}^*, \cdots, \eta_n{}^*)$ for player II with $\eta_j{}^* > 0$. A column j is *irrelevant* if it is not relevant.

Theorem 3.1.2

If a column j is relevant for player II in a payoff matrix (a_{ij}), $1 \le i \le m$, $1 \le j \le n$, then with value v for the game,

$$\sum_i a_{ij} \xi_i^0 = v$$

for all optimal $x^0 = (\xi_1^0, \xi_2^0, \cdots, \xi_m^0)$ for player I.

PROOF: Since x^0 is optimal for player I, $\sum_i a_{ik} \xi_i^0 \ge v$ for all k and in particular for $k = j$. Suppose that $\sum_i a_{ij} \xi_i^0 = v + \varepsilon > v$; then for some optimal $y^* = (\eta_1{}^*, \eta_2{}^*, \cdots, \eta_j{}^*, \cdots, \eta_n{}^*)$ with $\eta_j{}^* > 0$, we have

$$\sum_i \sum_k a_{ik} \xi_i^0 \eta_k{}^* = \sum_{k \ne j} \sum_i a_{ik} \xi_i^0 \eta_k{}^* + \sum_i a_{ij} \xi_i^0 \eta_j{}^*$$

$$\ge v \sum_{k \ne j} \eta_k{}^* + (v + \varepsilon) \eta_j{}^*$$

$$\ge v + \varepsilon \eta_j{}^* > v \text{ (since } \eta_j{}^* > 0 \text{ and } \varepsilon > 0)$$

This contradicts the optimality of y^*. Thus $\sum_i a_{ij} \xi_i^0 = v$. We can prove a similar theorem for player I.

Definition 3.1.2: A vector $a = (\alpha_1, \alpha_2, \cdots, \alpha_n)$ is said to *dominate* a vector $b = (\beta, \beta_2, \cdots, \beta_n)$ if $\alpha_i - \beta_i > 0$, $i = 1, 2, \cdots, n$.

Theorem 3.1.3

A column k in a payoff matrix (a_{ij}), $1 \le i \le m$, $1 \le j \le n$, is irrelevant for player II if the kth column vector dominates some other sth column vector of the matrix. A row r is irrelevant if the rth row vector is dominated by some other tth row vector of the matrix.

PROOF: For any optimal strategy $x^0 = (\xi_1^0, \xi_2^0, \cdots, \xi_m^0)$ of player I and $y^0 = (\eta_1^0, \eta_2^0, \cdots, \eta_n^0)$ of player II, the dominance of the kth column over the sth column implies

$$\sum_i a_{ik}\xi_i^0 > \sum_i a_{is}\xi_i^0 \geq v$$

Hence by the previous theorem we see that column k is irrelevant. Similarly, since

$$\sum_j a_{rj}\eta_j^0 < \sum_j a_{tj}\eta_j^0 \leq v$$

the rth row is irrelevant.

Remark 3.1.2: Since dominance leads to irrelevancy of a column or row, we can discard the presence of that column or row in our search for optimal strategies.

Theorem 3.1.4

If the kth column dominates some convex combination of other columns, then the kth column is irrelevant. If the rth row is dominated by some convex combination of other rows, then the rth row is irrelevant.

PROOF: By assumption we have

$$a_{ik} > \lambda_1 a_{i\alpha} + \lambda_2 a_{i\beta} + \cdots + \lambda_s a_{i\tau}, \qquad i = 1, 2, \cdots, m$$

for some columns α, β, $\cdots$, τ, with $\Sigma \lambda_t = 1$, $\lambda_t > 0$, $t = 1, 2, \cdots, s$. Clearly

$$\sum_i a_{ik}\xi_i^0 > \lambda_1 \sum_i a_{i\alpha}\xi_i^0 + \lambda_2 \sum_i a_{i\beta}\xi_i^0 + \cdots +$$

$$+ \lambda_s \sum_i a_{i\tau}\xi_i^0 \geq v \sum_t \lambda_t = v$$

for some optimal strategy $x^0 = (\xi_1^0, \xi_2^0, \cdots, \xi_m^0)$ of player I. Thus

$$\sum_i a_{ik}\xi_i^0 > v$$

This shows that the kth column is irrelevant for player II. Similarly, we can prove that the rth row is irrelevant for player I.

Definition 3.1.3: A mixed strategy $x^0 = (\xi_1^0, \xi_2^0, \cdots, \xi_m^0)$ is *completely mixed* for player I if $\xi_i^0 > 0$, $i = 1, 2, \cdots, m$. A similar definition can be given for player II.

Evidently when player I has a completely mixed optimal strategy, then every row is relevant. The following theorems concern games with completely mixed optimal strategies for a player.

Theorem 3.1.5

Let $A = (a_{ij})$, $1 \leq i \leq m$, $1 \leq j \leq n$, be the payoff matrix of a game G with value 0. If every optimal strategy for player I is completely mixed, then the rank $r(A)$ of the matrix satisfies $(m - 1) \leq r(A) \leq (n - 1)$. If the rank is exactly $(m - 1)$, then we have exactly one optimal strategy $x^0 = (\xi_1^0, \xi_2^0, \cdots, \xi_m^0)$ with $\sum_i a_{ij}\xi_i^0 = 0$, $j = 1, 2, \cdots, n$, for player I.

PROOF: If $y^0 = (\eta_1^0, \eta_2^0, \cdots, \eta_n^0)$ is an optimal strategy for player II, by theorem 3.1.2, y^0 satisfies $\sum_j a_{ij}\eta_j^0 = 0$, $i = 1, 2, \cdots, m$, and $r(A) \leq (n - 1)$.

To show that $(m - 1) \leq r(A)$, let us assume the contrary, with $u = (u_1, u_2, \cdots, u_m)$ and $v = (v_1, v_2, \cdots, v_m)$ as two linearly independent solutions to

$$\sum_i a_{ij}x_i = 0, \qquad j = 1, 2, \cdots, n$$

Let $\pi = (\pi_1, \pi_2, \cdots, \pi_m)$ be an optimal strategy for player I; then

$$\sum_i a_{ij}\pi_i \geq 0, \quad j = 1, 2, \cdots n, \quad \pi_i > 0, \quad i = 1, 2, \cdots, m, \quad \sum_1^m \pi_i = 1$$

Since u and v are linearly independent, either u and π or v and π are linearly independent. Suppose that u and π are linearly independent. Now $\sum_i u_i = 0$ or $\sum_i u_i \neq 0$ is true; let us first assume that $\sum_i u_i = c \neq 0$. Since u is a solution of the homogeneous equations considered above, $(1/c)\,u$ is also a solution and is also linearly independent of π. So we shall assume without loss of generality that $\sum_i u_i = 1$ in this case.

Let $\theta = \max_i (u_i/\pi_i)$. Since $\pi_i > 0$ for each i, and since $\sum_i u_i = 1$, for some u_{i_0}/π_{i_0} the maximum is attained, and $\theta > 0$. In fact, $\theta > 1$; for if $\theta < 1$, then $u_i/\pi_i \leq 1$, $i = 1, 2, \cdots, m$, which would imply that $u_i \leq \pi_i$, $i = 1, 2, \cdots, m$. Since $\Sigma u_i = \Sigma \pi_i = 1$, we would have $u = \pi$, which is not true by assumption. Let $\lambda = 1/(\theta - 1) > 0$. If $z_i = (1 + \lambda)\pi_i - \lambda u_i$, $i = 1, 2, \cdots, m$, then $z_i = \lambda\theta\pi_i - \lambda u_i \geq 0$, and $\sum_i z_i = (1 + \lambda)\Sigma\pi_i - \lambda\Sigma u_i = 1$.

Further, $z_{i_0} = \lambda \pi_{i_0} (\theta - u_{i_0}/\pi_{i_0}) = 0$. Now for the mixed strategy $z = (z_1, z_2, \cdots, z_m)$,

$$\sum_i a_{ij}z_i = (1 + \lambda) \sum_i a_{ij}\pi_i - \lambda \sum_i a_{ij}u_i \geq 0, \qquad j = 1, 2, \cdots, n$$

That is, z is an optimal strategy for player I with $z_{i_0} = 0$. This contradicts the assumption that every optimal strategy of player I is completely mixed. Let us consider the second case, where $\Sigma u_i = 0$. We can assume without loss of generality that some u_k is positive.

Clearly $\theta = \max_i (u_i/\pi_i)$ is positive. Let $\lambda = 1/\theta$; then $z_i = \pi_i - \lambda u_i \geq 0$, $i = 1, 2, \cdots, m$, and $\Sigma z_i = \Sigma \pi_i = 1$. Further, $\theta = u_{i_0}/\pi_{i_0}$ for some i_0, and $z_{i_0} = 0$. Now for the mixed strategy $z = (z_1, z_2, \cdots, z_m)$,

$$\sum_i a_{ij}z_i = \sum_i a_{ij}\pi_i - \lambda \sum_i a_{ij}u_i \geq 0, \qquad j = 1, 2, \cdots, n$$

That is, z is optimal for player I with $z_{i_0} = 0$, which once again contradicts our assumptions in the theorem. Thus $r(A) \not< (m - 1)$; that is, $r(A) \geq (m - 1)$. To prove the last assertion, suppose that $r(A) = (m - 1)$; then the equations

$$\sum_i a_{ij}x_i = 0, \qquad j = 1, 2, \cdots, n$$

have a unique solution up to a scalar multiple. If an optimal strategy π of player I is linearly independent of a solution u for the above equations, then as we proved earlier we can construct a strategy z that is optimal for player I but is not completely mixed. This would contradict the assumption in the theorem. Thus the condition that an optimal $\pi = (\pi_1, \pi_2, \cdots, \pi_m)$ satifies $\Sigma \pi_i = 1$ implies that π is the only optimal strategy for player I.

Theorem 3.1.6

If $A = (a_{ij})$, $1 \leq i \leq m$, $1 \leq j \leq n$, is a payoff matrix with $m > n$, then there exists an optimal strategy for player I that is not completely mixed.

PROOF: By adding a suitable constant k to each entry, we can always make the value of the new payoff matrix $B = (b_{ij})$ with $b_{ij} = a_{ij} + k$ to be zero. Evidently such an addition does not affect the optimal strategies for the two games G and G' with payoff $A = (a_{ij})$ and $B = (b_{ij})$, respectively. So without loss of generality let us assume the value $v = 0$ for the payoff $A = (a_{ij})$. If every optimal strategy of player I is completely mixed, then by the previous theorem $(m - 1) \leq r(A) \leq (n - 1)$, which would contradict $m > n$. Thus there exists at least one optimal strategy for player I that is not completely mixed.

In fact, in the proof of the minimax theorem any optimal strategy for player II determines the point s^0 in the convex span of columns. Further, s^0 is a boundary point of this set. By the remark of theorem 1.9.3, we can with a convex combination of m columns get the point s^0. That is, player II has an optimal strategy with at most min (m, n) nonzero coordinates.

Theorem 3.1.7

If $A = (a_{ij})$ is a square matrix with $1 \le i, j \le n$ and if there exists an optimal strategy for player II that is not completely mixed, then so it is for player I.

PROOF: Let $y = (\eta_1, \eta_2, \cdots, \eta_n)$ be an optimal strategy for player II that is not completely mixed, say $\eta_n = 0$. Suppose that every optimal strategy for player I is completely mixed. Without loss of generality let us assume the value of the game to be 0. By theorem 3.1.5, $(n - 1) \le r \le (n - 1)$, where r is the rank of the matrix. Thus $r = (n - 1)$. By the last assertion of theorem 3.1.5, the optimal strategy for player I is unique.

Let A_{ij} be the cofactor of a_{ij}, $1 \le i, j \le n$. Since the rank is $(n - 1)$, the determinant $|A|$ vanishes, and

$$\sum_j a_{ij} A_{kj} = 0, \qquad i, k = 1, 2, \cdots, n$$

$$\sum_{j=1}^{n} a_{ij} \eta_j = 0, \qquad i = 1, 2, \cdots, n$$

Since we have only a unique solution to within a scalar multiple for the above equations (this is because $r = n - 1$), $(A_{k1}, A_{k2}, \cdots, A_{kn}) = \alpha(\eta_1, \eta_2, \cdots, \eta_n)$ for some scalar α and for all $k = 1, 2, \cdots, n$. But $\eta_n = 0$, and $A_{1n} = A_{2n} = \cdots = A_{nn} = 0$. Thus every minor of order $(n - 1)$ of the matrix

$$B = \begin{bmatrix} a_{11} & a_{12} \cdots & a_{1,n-1} \\ a_{21} & a_{22} & a_{2,n-1} \\ a_{n1} & a_{n2} & a_{n,n-1} \end{bmatrix}$$

vanishes. This shows that the rank of the matrix B is at most $(n - 2)$, and that the system of equations

$$\sum_i a_{ij} x_i = 0, \qquad j = 1, 2, \cdots, n - 1$$

has at least two linearly independent solutions, $u = (u_1, u_2, \cdots, u_n)$ and $v = (v_1, v_2, \cdots, v_n)$. At least one of u, v is linearly independent of the unique optimal strategy $x^0 = (\xi_1^0, \xi_2^0, \cdots, \xi_m^0)$, satisfying

$$\sum_i a_{ij}\xi_i^0 = 0, \qquad j = 1, 2, \cdots, n$$

Let u be linearly independent of x^0. Now

$$\sum_i a_{in}u_i \neq 0$$

For otherwise the linearly independent u, ξ^0 would both satisfy the homogeneous equations

$$\sum_i a_{ij}x_i = 0, \qquad j = 1, 2, \cdots, n$$

which would imply that the rank $r \leq (n - 2)$.

Now let us assume without loss of generality that $\sum_i u_i = 1$ or $\sum_i u_i = 0$.

Suppose that $\Sigma u_i = 1$. Then either $\sum_i a_{in}u_i < 0$ or $\Sigma a_{in}u_i > 0$. If

$\sum_i a_{in}u_i < 0$, choose $z_i = (1 + \lambda)\,\xi_i^0 - \lambda u_i$, $\lambda > 0$, $z_i \geq 0$, $i = 1, 2, \cdots, n$,

and $z_{i_0} = 0$ for some i_0. We can use the same sort of technique as was used in theorem 3.1.5 to get z_i's.

If $\sum_i a_{in}u_i > 0$, we can choose $\lambda < 0$ with $z_i = (1 + \lambda)\,\xi_i^0 - \lambda u_i > 0$,

$i = 1, 2, \cdots, n$, and $z_{i_0} = 0$ for some i_0. Similarly, if $\Sigma u_i = 0$, either

$\sum_i a_{in}u_i > 0$ or $\sum_i a_{in}u_i < 0$. But in this case we can always assume without

loss of generality that $\sum_i a_{in}u_i < 0$.

Thus we can find $z_i = \xi_i^0 - \lambda u_i$, $\lambda > 0$, with $z_i = \xi_i^0 - \lambda u_i \geq 0$, $i = 1, 2, \cdots, n$, and $z_{i_0} = 0$ for some i_0. Lastly,

$$\sum_i a_{ij}z_i = 0, \qquad j = 1, 2, \cdots, n - 1 \text{ in all these cases}$$

$$> 0, \qquad j = n$$

This shows that $z = (z_1, z_2, \cdots, z_n)$ is optimal for player I, which is not completely mixed, a contradiction to our assumption. Hence the theorem.

Definition 3.1.4: A game G with payoff $A = (a_{ij})$, $1 \leq i \leq m, 1 \leq j \leq n$, is completely mixed if and only if every optimal strategy for both players is completely mixed.

Theorem 3.1.8

Let a game G with payoff $A = (a_{ij})$, $1 \le i \le m$, $1 \le j \le n$, have value 0. Then a set of necessary and sufficient conditions for G to be completely mixed is (1) $m = n$; (2) rank of $A = (n - 1)$; and (3) all cofactors A_{ij} of a_{ij}'s are of the same sign and are different from zero.

PROOF: Let G be completely mixed with value 0. Then by theorems 3.1.5 and 3.1.6 we have $m = n$ with rank of $A = (n - 1)$. Since the rank is $(n - 1)$ and since the optimal strategies are completely mixed, we have

$$\sum_i a_{ij}\xi_i^0 = 0, \qquad j = 1, 2, \cdots, n$$

$$\sum_j a_{ij}A_{kj} = 0, \qquad i, k = 1, 2, \cdots, n$$

$$\sum_j a_{ij}\eta_j^0 = 0, \qquad i = 1, 2, \cdots, n$$

$$\sum_i a_{ij}A_{il} = 0, \qquad j, l = 1, 2, \cdots, n$$

for the unique optimal strategies $x^0 = (\xi_1^0, \xi_2^0, \cdots, \xi_n^0)$ of player I and $y^0 = (\eta_1^0, \eta_2^0, \cdots, \eta_n^0)$ of player II. Moreover, they are unique solutions to within scalar multiples of the respective equations and

$$(A_{1j}, A_{2j}, \cdots, A_{nj}) = \alpha_j(\xi_1^0, \xi_2^0, \cdots, \xi_n^0) \text{ for some } \alpha_j$$
$$\text{for each } j = 1, 2, \cdots, n$$

and

$$(A_{i1}, A_{i2}, \cdots, A_{in}) = \beta_i(\eta_1^0, \eta_2^0, \cdots, \eta_n^0) \text{ for some } \beta_i$$
$$\text{for each } i = 1, 2, \cdots, n$$

Since ξ_i^0's and η_j^0's are all positive, we have the property that A_{ij}'s are of the same sign. Even if one of them is zero, all are zero, which would contradict the fact that the rank of A is $(n - 1)$.

To prove the sufficiency part, let us consider a matrix $A = (a_{ij})$ satisfying the three conditions. Then

$$\sum_j a_{ij}A_{1j} = 0, \qquad i = 1, 2, \cdots, n$$

Let

$$\eta_j^0 = \frac{A_{1j}}{A_{11} + A_{12} + \cdots + A_{1n}}$$

We have $\Sigma\, \eta_j^0 = 1$. Then $\displaystyle\sum_j a_{ij}\eta_j^0 = 0$, $i = 1, 2, \cdots, n$, and $\eta_j^0 > 0$,

$j = 1, 2, \cdots, n$. Similarly we can find

$$\xi_i^0 = \frac{A_{i1}}{A_{11} + A_{21} + \cdots + A_{n1}} > 0, \qquad i = 1, 2, \cdots, n$$

such that

$$\sum_i a_{ij}\xi_i^0 = 0, \qquad j = 1, 2, \cdots, n$$

Thus the value of the game is 0, and these optimal strategies are completely mixed. Any other optimal strategy $x' = (\xi_1', \xi_2', \cdots, \xi_n')$ for player I also satisfies

$$\sum_i a_{ij}\xi_i' = 0, \qquad j = 1, 2, \cdots, n \text{ (because } \eta^0 \text{ is completely mixed)}$$

This, coupled with the fact that the rank of A is $(n - 1)$, implies that $\xi^0 = \xi'$. Hence the optimal strategy for player I is unique and is completely mixed. A similar argument applies for player II.

Theorem 3.1.9

Let $A = (a_{ij})$, $1 \le i, j \le n$, be the payoff of a completely mixed game G. Let v be the value of the game. Then

$$v = \frac{|A|}{\displaystyle\sum_i \sum_j A_{ij}}$$

(Here $|A|$ is the determinant of A.)

PROOF: *Case 1:* $|A| \ne 0$. Since the game is completely mixed, for any optimal $y^0 = (\eta_1^0, \eta_2^0, \cdots, \eta_n^0)$ of player II,

$$\sum_j a_{ij}\eta_j^0 = v, \qquad i = 1, 2, \cdots, n$$

If $a^{(ij)}$ is the ijth entry of the inverse matrix A^{-1} (which exists in our case), then

$$a^{ij} = \frac{A_{ij}}{|A|}$$

where A_{ij} is the cofactor of a_{ij}. Further

$$\eta_j^0 = v\sum_i a^{(ij)} = \frac{v}{|A|}\sum_i A_{ij}$$

and

$$1 = \sum_j \eta_j^0 = \frac{v}{|A|} \sum_j \sum_i A_{ij}$$

Thus

$$v = \frac{|A|}{\sum_i \sum_j A_{ij}}$$

Case 2: $|A| = 0$. We shall show that $v = 0$. Suppose that $v < 0$. Then, since the game is completely mixed and $|A| = 0$, we have, for any optimal $y^0 = (\eta_1^0, \eta_2^0, \cdots, \eta_n^0)$,

$$\sum_j a_{ij}\eta_j^0 = v < 0, \qquad i = 1, 2, \cdots, n$$

$$\sum_j a_{ij}\pi_j = 0, \qquad i = 1, 2, \cdots, n \text{ for some } \pi = (\pi_1, \pi_2, \cdots, \pi_n)$$

If p and y^0 are linearly dependent, then $v < 0$ would be contradicted. If they are linearly independent, then we could construct an optimal strategy for player II (as in theorem 3.1.5) that is not completely mixed. This would contradict our assumption in the theorem. Thus $v < 0$ is not true. Similarly, we can prove that $v > 0$ is not true. Thus $v = 0$. This, coupled with the previous theorem, implies that A_{ij}'s are of the same sign and are different from zero and that

$$0 = v = \frac{|A|}{\sum_i \sum_j A_{ij}}$$

Theorem 3.1.10

If $m > n$ in a game G with payoff $A = (a_{ij})$, $1 \le i \le m$, $1 \le j \le n$, then the value $v(G)$ of the game is given by $v(G) = \max_i v(G_i)$, where $\{G_i\}$ runs over the $n \times n$ square submatrices of G. If $m < n$, then $v(G) = \min_j v(G_j)$, where $\{G_j\}$ runs over the $m \times m$ square submatrices of G.

PROOF: The last assertions in theorem 3.1.6 tell us that each player has an optimal strategy that concentrates on at most $\min(m, n)$ rows or columns. If $m > n$, then $\min(m, n) = n$, and player I omits $(m - n)$ rows in some optimal strategy. If $v(G_i^*)$ is the value of this square matrix, clearly $v(G) = v(G_i^*) = \max_i v(G_i)$. A similar argument yields $v(G) = v(G_j^*) = \min_j v(G_j)$.

Theorem 3.1.11

A payoff $A = (a_{ij})$, $1 \le i, j \le n$, with $a_{ij} = -a_{ji}$ for all i, j, has the value 0. Further, any optimal strategy for one player is optimal also for the other player.

PROOF: Let v be the value of the game. We shall show that $v \leq 0$. Otherwise let $v > 0$ and $x^0 = (\xi_1^0, \xi_2^0, \cdots, \xi_n^0)$ be optimal for player I. Then

$$\sum_i a_{ij}\xi_i^0 \geq v > 0, \qquad j = 1, 2, \cdots, n$$

That is,

$$-\sum_i a_{ji}\xi_i^0 > 0 \quad \text{and} \quad \sum_i a_{ji}\xi_i^0 < 0, \qquad j = 1, 2, \cdots, n$$

Since i is the second suffix now, x^0 is a strategy that assures player II an expected loss less than 0 independent of what player I does. This contradicts $v > 0$. Thus $v \leq 0$. By considering an optimal strategy y^0 of player II, we can show that $v \geq 0$; that is, $v = 0$. Thus $\sum_i a_{ij}\xi_1^0 \geq 0$ implies that

$$\sum_i a_{ji}\xi_i^0 \leq 0.$$ Hence x^0 is optimal for player II if it is optimal for player I.

Theorem 3.1.12

If a payoff $A = (a_{ij})$, $1 \leq i, j \leq n$, is skew-symmetric (that is, $a_{ij} = -a_{ji}$ as above for all $1 \leq i, j \leq n$) and if n is even, then there exists an optimal strategy for the players that is not completely mixed.

PROOF: Since $v = 0$ and n is even, if the game were to be completely mixed, then by theorem 3.1.8 the rank of A would be $(n - 1)$, which is an odd number. But for skew-symmetric matrices the rank is always even. This shows that the game is not completely mixed. This with theorem 3.1.7 proves our assertion.

Theorem 3.1.13

Let $A = (a_{ij})$, $1 \leq i, j \leq n$, be the payoff for a game G that is not completely mixed. Let G_{ij} correspond to the game with the payoff matrix formed out of A by omitting the ith row and the jth column. Then the value matrix $(v(G_{ij}))$, $1 \leq i, j \leq n$, of the values of games G_{ij} has a saddle point.

PROOF: Since the game is not completely mixed and since the payoff is a square matrix by theorem 3.1.7, each player has an optimal strategy that is not completely mixed. By theorem 3.1.10 we know that, if player I omits the ith row in an optimal strategy, then the number of columns is greater and the value v is

$$v = \min_j v(G_{ij})$$

Since player I can omit one of the rows, he would omit the ith row only when the value of the corresponding rectangular matrix is the maximum among all such matrices obtained from the original matrix with one row deleted. That is,

$$v = \max_k v(G_k) \quad \text{and} \quad v(G_k) = \min_j v(G_{kj})$$

Thus

$$v = \max_{k} \min_{j} v(G_{kj})$$

A similar argument shows that $v = \min_{j} \max_{k} v(G_{kj})$; that is, the matrix $(v(G_{ij}))$ has a saddle point.

Definition 3.1.5: If x^0, y^0 are the optimal mixed strategies for a payoff $A = (a_{ij})$, $1 \le i \le m$, $1 \le j \le n$, with v as the value and

$$\sum_{i} a_{ij}\xi_i^0 = v, \qquad j = 1, 2, \cdots, n, \text{ where } x^0 = (\xi_1^0, \xi_2^0, \cdots, \xi_m^0)$$

$$\sum_{j} a_{ij}\eta_j^0 = v, \qquad i = 1, 2, \cdots, m, \text{ where } y^0 = (\eta_1^0, \eta_2^0, \cdots, \eta_n^0)$$

then the pair (x^0, y^0) is called a *simple solution* to A.

Theorem 3.1.14 (Shapley-Snow)

Let $A = (a_{ij})$, $1 \le i \le m$, $1 \le j \le n$, be the payoff of a game G with value $v \ne 0$. The set of extreme optimal strategies (extreme points of the set of optimals) for the players is finite. A strategy $x^0 = (\xi_1^0, \xi_2^0, \cdots, \xi_m^0)$ is an extreme optimal strategy for player I, and a strategy $y^0 = (\eta_1^0, \eta_2^0, \cdots, \eta_n^0)$ is an extreme optimal strategy for player II, if and only if A has a nonsingular square submatrix B for which (x_B^0, y_B^0) constitutes a simple solution for B, where x_B^0 is the vector obtained from x^0 by deleting those components corresponding to the rows deleted to obtain B from A, and y_B^0 is the vector obtained from y^0 in a similar fashion for the columns.

PROOF (NECESSITY): By changing the rows, one does not affect the nature of the game, and one just changes the labels of the pure strategies. A similar permutation in columns also changes just the labels of the pure strategies for player II. Thus by a permutation of rows and columns one can assume without loss of generality that

$$x^0 = (\xi_1^0, \xi_2^0, \cdots, \xi_p^0, 0, 0, \cdots, 0), \qquad p \le m$$

$$y^0 = (\eta_1^0, \eta_2^0, \cdots, \eta_q^0, 0, 0, \cdots, 0), \qquad q \le n$$

Thus we assume without loss of generality that the first p rows are chosen by player I with positive probability and the rest of the rows are optimally omitted, and that the first q columns are chosen with positive probability by player II and the rest of the columns are optimally omitted. Let

$$\sum_{i} a_{ij}\xi_i^0 = v, \qquad j = 1, 2, \cdots, \bar{q}, \qquad q \le \bar{q}$$

and

$$\sum_j a_{ij}\eta_j^0 = v, \qquad i = 1, 2, \cdots, \bar{p}, \qquad p \le \bar{p}$$

($q \le \bar{q}$ and $p \le \bar{p}$ follows from the fact that columns $j = 1, 2, \cdots, q$ are relevant, and rows $i = 1, 2, \cdots, p$ are relevant.)

Let C be the submatrix obtained from A by deleting the last $m - \bar{p}$ rows and $n - \bar{q}$ columns; that is,

$$C = \begin{bmatrix} a_{11} & a_{12} & \cdots & a_{1q} & a_{1\bar{q}} \\ a_{p1} & a_{p2} & & a_{pq} & a_{p\bar{q}} \\ a_{\bar{p}1} & a_{\bar{p}2} & & a_{\bar{p}q} & a_{\bar{p}\bar{q}} \end{bmatrix}$$

We claim that the first p rows are linearly independent in C. For otherwise we have $t = (\tau_1, \tau_2, \ldots, \tau_p, \tau_{p+1}, \tau_{\bar{p}}, \ldots, \tau_m)$ with $(\tau_1, \tau_2, \ldots, \tau_p) \ne 0$, $(\tau_{p+1}, \ldots, \tau_{\bar{p}}, \ldots, \tau_m) = 0$, and

$$\sum_{i=1}^{\bar{p}} a_{ij}\tau_i = 0, \qquad j = 1, 2, \cdots, \bar{q}$$

Since

$$\sum_{j=1}^{\bar{q}} a_{ij}\eta_j^0 = v, \qquad i = 1, 2, \cdots, \bar{p}$$

by assumption we have

$$\sum_{i=1}^{\bar{p}} \tau_i \sum_{j=1}^{\bar{q}} a_{ij}\eta_j^0 = v \sum_{i=1}^{\bar{p}} \tau_i = \sum_{j=1}^{\bar{q}} \eta_j^0 \sum_{i=1}^{\bar{p}} a_{ij}\tau_i = 0 \cdot \sum_{j=1}^{\bar{q}} \eta_j^0 = 0$$

But $v \ne 0$, and

$$\sum_{i=1}^{\bar{p}} \tau_i = 0; \text{ that is, } \sum_{i=1}^{m} \tau_i = 0$$

Thus for $|\varepsilon|$ sufficiently small, the mixed strategy $x^{(\varepsilon)} = x^0 + \varepsilon t$ satisfies

$$\sum_{i=1}^{m} a_{ij}\xi_i^0 \pm \varepsilon \sum_{i=1}^{m} a_{ij}\tau_i = v \text{ for } j = 1, 2, \cdots, \bar{q}$$

$$> v \text{ for } j > \bar{q}$$

(since $\sum_{i=1}^{m} a_{ij}\xi_i^0 > v, j > \bar{q}$, and $|\varepsilon|$ is very small). Thus $x^{(\varepsilon)}$ and $x^{(-\varepsilon)}$ are two optimal strategies defined by $x^{(\varepsilon)} = x^0 + \varepsilon t$ and $x^{(-\varepsilon)} = x^0 - \varepsilon t$, with

$$x^0 = \frac{x^{(\varepsilon)} + x^{(-\varepsilon)}}{2}$$

That is, x^0 is not an extreme optimal strategy, which is a contradiction to our assumption. Thus the first p rows and similarly the first q columns in C are linearly independent. Let r be the rank of C. Then $r \geq \max(p, q)$. Thus there exists a nonsingular square submatrix B of C of order $r \times r$, with an optimal strategy x_B^0 for player I where B consists of components including $1, 2, \cdots, p$, and an optimal strategy y_B^0 for player II where B consists of components including $1, 2, \cdots, q$. We also have

$$\sum_i a_{ij} x_i^0 = v, \quad j = 1, 2, \cdots, q, j \in J$$

$$\sum_j a_{ij} \eta_j^0 = v, \quad i = 1, 2, \cdots, p, i \in I$$

where J consists of $r - q$ other indices j in $q < j \leq \bar{q}$, and I consists of $r - p$ other indices i in $p < i \leq \bar{p}$.

SUFFICIENCY: Let B be an $r \times r$ nonsingular submatrix of A obtained by deleting some $m - r$ rows and $n - r$ columns such that (x_B^0, y_B^0) is a simple solution of B. We claim that x^0 is an extreme optimal strategy for player I and y^0 is an extreme optimal strategy for player II. Suppose that x^0 is not an extreme optimal strategy. Then there exist optimal strategies $x' \neq x^*$ to player I, with $x^0 = \lambda x' + (1 - \lambda) x^*$ for some $0 < \lambda < 1$. Without loss of generality we can assume that the submatrix B is the principal $r \times r$ submatrix—that is, the matrix (a_{ij}), $1 \leq i \leq r, 1 \leq j \leq r$. Now

$$\sum_{i=1}^m a_{ij} \xi_i^* = v, \quad j = 1, 2, \cdots, r, \text{ where } x^* = (\xi_1^*, \xi_2^*, \cdots, \xi_m^*)$$

$$\sum_{i=1}^m a_{ij} \xi_i' = v, \quad j = 1, 2, \cdots, r, \text{ where } x' = (\xi_1', \xi_2', \cdots, \xi_m')$$

This equality holds because (x^0, y^0) is a simple solution. Further, since $x^0 = \lambda x^* + (1 - \lambda) x'$, $0 < \lambda < 1$, and $\xi_i^0 = 0$, $i > r$, we have $\xi_i^* = 0$ for $i > r$ and $\xi_i' = 0$ for $i > r$.

The above equations imply by these considerations that

$$\sum_{i=1}^r a_{ij}(\xi_i^* - \xi_i') = 0, \quad j = 1, 2, \cdots, r$$

Since the matrix $B = (a_{ij})$, $1 \leq i, j \leq r$, is nonsingular, $\xi_i^* - \xi_i' = 0$, $i = 1, 2, \cdots, r$, and $x^* = x'$. This is a contradiction. Hence the sufficiency part. A similar argument applies for y^0.

Remark 3.1.3: The assumption that $v \neq 0$ is not serious, as we can always make it so by adding a large positive constant to each entry in the payoff without affecting the optimal strategy sets for the players.

Theorem 3.1.15

If $A = (a_{ij})$, $1 \leq i \leq m$, $1 \leq j \leq n$, is the payoff for a game G with value $v_A > 0$, then the skew-symmetric matrix B of order $m + n + 1$ by $m + n + 1$ given by

$$
\begin{array}{|ccccccc|cccc|c|}
\hline
0 & 0 & 0\cdots 0 \cdots 0 & & | & a_{11} & a_{12} \cdots a_{1n} & | & -1 \\
\cdots\cdots\cdots\cdots\cdots\cdots & & | & \cdots\cdots\cdots\cdots\cdots & | & -1 \\
& & | & & | & \cdots \\
0 & 0 & 0\cdots 0 \cdots 0 & & | & a_{m1} & a_{m2} \cdots a_{mn} & | & -1 \\
\hline
-a_{11} & -a_{21} \cdots -a_{m1} & | & 0 & 0 & 0 & | & 1 \\
\cdots\cdots\cdots\cdots\cdots & | & & | \\
-a_{1n} & -a_{2n} & -a_{mn} & | & 0 & 0 & 0 & | & 1 \\
\hline
1 & 1 & 1 & | & -1 & -1 & -1 & | & 0 \\
\hline
\end{array}
$$

and written compactly in the matrix form

$$
B = \begin{vmatrix} 0 & A & -1' \\ -A' & 0 & \underset{\sim}{1} \\ \underset{\sim}{1} & -1 & 0 \end{vmatrix}
$$

has optimal strategies $w^0 = (u_1, u_2, \cdots, u_m, v_1, v_2, \cdots, v_m, \theta)$, where $\Sigma u_i = a > 0$, $\Sigma v_i = a > 0$, with $\left(\dfrac{u_1}{a}, \dfrac{u_2}{a}, \cdots, \dfrac{u_m}{a}\right)$, $\left(\dfrac{v_1}{a}, \dfrac{v_2}{a}, \cdots, \dfrac{v_m}{a}\right)$ optimal for players I and II for the payoff A. Also, $v_A = \theta/a$.

PROOF: Since B is skew-symmetric, the value of the payoff B is zero, and the optimal strategy w^0 for players I and II for this game will satisfy

$$\sum_j a_{ij}v_j - \theta \leq 0, \qquad i = 1, 2, \cdots, m \tag{1}$$

$$-\sum_i a_{ij}u_i + \theta \leq 0, \qquad j = 1, 2, \cdots, n \tag{2}$$

and

$$\sum_i u_i - \sum_j v_j \leq 0 \tag{3}$$

Clearly, $\Sigma v_j > 0$, for otherwise inequality (3) with u_i's ≥ 0 implies that $u_i = 0$ for all i, and that $u = 0$, $v = 0$, where $u = (u_1, u_2, \cdots, u_m)$, $v = (v_1, v_2, \cdots, v_n)$. Now $u = 0$, $v = 0$, implies that $\theta = 0$, from (1) and (2). This contradicts the fact that w^0 is a mixed strategy. Thus $a = \Sigma v_j > 0$, and $(1/a) v$ is a mixed strategy for player II for the payoff matrix A, with

$$\sum_j a_{ij}\left(\frac{v_j}{a}\right) \leq \frac{\theta}{a} \qquad i = 1, 2, \cdots, m$$

Therefore the value $\leq 0/a$. Further, $0 > 0$, for otherwise (1) would yield $(1/a)$ v as a mixed strategy with value ≤ 0 for the game A. This contradicts the assumption that $v_A > 0$.

Multiplying (1) by u_i and (2) by v_j and summing them up, we have

$$\sum_i \sum_j a_{ij} u_i v_j - 0 \sum_i u_i \leq 0$$

$$-\sum_i \sum_j a_{ij} u_i v_j + 0 \sum_j v_j \leq 0$$

and

$$0 \sum_j v_j \leq \sum_i \sum_j a_{ij} u_i v_j \leq 0 \sum_i u_i$$

This shows by the positivity of 0 and by (3) that

$$\sum_j v_j = \sum_i u_i = a$$

and $(1/a)$ v, $(1/a)$ u are optimal for players I and II with value $0/a$ for the payoff A. Conversely, if $x^0 = (\xi_1^0, \xi_2^0, \cdots, \xi_m^0)$, $y^0 = (\eta_1^0, \eta_2^0, \cdots, \eta_n^0)$, are optimal for the payoff A with value $v_A > 0$, then $w^0 = \dfrac{1}{2 + v_A} (\xi_1^0, \xi_2^0, \cdots, \xi_m^0,$ $\eta_1^0, \cdots, \eta_n^0, v_A)$ is optimal for B.

Theorem 3.1.16

Let $A = (a_{ij})$, $1 \leq i \leq m$, $1 \leq j \leq n$, be the payoff of a game G with value v. For every optimal strategy $y^0 = (\eta_1^0, \eta_2^0, \cdots, \eta_n^0)$ of player II, let

$$\sum_j a_{kj} \eta_j^0 = v$$

Then there exists an optimal strategy $x^0 = (\xi_1^0, \xi_2^0, \cdots, \xi_m^0)$ for player I with $\xi_k^0 > 0$.

PROOF: Without loss of generality let $v = 0$. Suppose that we find non-negative $u_1, u_2, \cdots, u_m$, with $u_k = 1$, satisfying $\sum_i a_{ij} u_i \geq 0, j = 1, 2, \cdots, n$; then by normalizing the u_i's we get the required optimal strategy for player I.

Consider the vectors

$$C_1 = (a_{11}, a_{12}, \cdots, a_{1n}, 0), \quad C_2 = (a_{21}, a_{22}, \cdots, a_{2n}, 0), \cdots, C_{k-1}$$
$$= (a_{k-1,1}, a_{k-1,2}, \cdots, a_{k-1,n}, 0), \quad C_k = (a_{k1}, a_{k2}, \cdots, a_{kn}, 1), C_{k+1}$$
$$= (a_{k+1,1}, a_{k+1,2}, \cdots, a_{k+1,n}, 0), \cdots, C_m = (a_{m1}, a_{m2}, \cdots, a_{mn}, 0),$$
$$x_1 = (-1, 0, \cdots, 0), \quad x_2 = (0, -1, \cdots, 0), \cdots,$$
$$x_n = (0, 0, \cdots, -1, 0)$$
$$b = (0, 0, \cdots, 0, 1) \text{ in } R^{n+1}$$

If there exists $u_i \geq 0$, $i = 1, 2, \cdots, m$, $\lambda_j \geq 0$, $j = 1, 2, \cdots, n$, with

$$\sum_1^m u_r C_r + \sum_1^n \lambda_j x_j = b$$

then equating the jth coordinate on both sides, we have

$$\sum_{i=1}^m u_i a_{ij} - \lambda_j = 0, \qquad j = 1, 2, \cdots, n$$

$$u_k = 1 \quad \text{(by equating the } (n+1)\text{st coordinate)}$$

The u_i's would give us the required optimal strategy after normalization. Suppose that no such u_i's ≥ 0, λ_j's ≥ 0, exist; then $b \notin K$, where K is defined by

$$K = \{z : z = \Sigma u_r C_r + \Sigma \lambda_s x_s;\ u_r \geq 0,\ \lambda_s \geq 0,\ r = 1, 2, \cdots, m,\ s = 1, 2, \cdots, n\}$$

Evidently K is a closed convex set, and $0 \in K$. Since $b \notin K$, there exists by theorem 1.6.2 a linear functional $f = (f_1, f_2, \cdots, f_n, f_0)$ with $(f, z) \leq 0$ for $z \in K$ and $(f, b) > 0$; that is,

$$(f, C_i) \leq 0, \qquad i = 1, 2, \cdots, m$$

$$(f, x_j) \leq 0, \qquad j = 1, 2, \cdots, n$$

$$(f, b) > 0$$

This shows that

$$\sum_j a_{ij} f_j \leq 0, \qquad i = 1, 2, \cdots, m, \qquad i \neq k$$

$$\sum_j a_{kj} f_j + f_0 \leq 0$$

and

$$-f_j \leq 0, \qquad j = 1, 2, \cdots, n$$

$$f_0 > 0$$

Thus $f^* = (f_1, f_2, \cdots, f_n)$ is an n-vector satisfying

$$\sum_j a_{ij} f_j \leq 0, \qquad i \neq k, i = 1, 2, \cdots, m$$

$$\sum_j a_{kj} f_j < 0$$

and

$$f_j \geq 0, \qquad j = 1, 2, \cdots, n$$

The inequality $\sum_j a_{kj} f_j < 0$ implies $f^* \neq 0$. Thus by normalizing f^* we get an optimal strategy $y^0 = (\eta_1^0, \eta_2^0, \cdots, \eta_n^0)$ for player II with $\sum_j a_{kj}\eta_j^0 < 0$. This contradicts our assumption in the theorem. Thus $b \in K$; hence the theorem.

Remark 3.1.4: Suppose that in our theorem the equality $\sum_j a_{ij}\eta_j^0 = v$ is satisfied for $i \in D$ and for every $y^0 = (\eta_1^0, \eta_2^0, \cdots, \eta_n^0)$ optimal for player II; then the average of the optimal strategies satisfying the conditions of the theorem would be an optimal strategy $x^0 = (\xi_1^0, \xi_2^0, \cdots, \xi_n^0)$ for player I with $\xi_i^0 > 0, i \in D$.

3.2 SADDLE-POINT THEOREMS

Theorem 3.2.1

Let $A = (a_{ij})$, $1 \le i \le m$, $1 \le j \le n$, be the payoff matrix of a game G. If every 2×2 submatrix obtained by deleting $(m - 2)$ rows and $(n - 2)$ columns has a saddle point, then A has a saddle point.

PROOF: Let v be the value of the game. Suppose that A has no saddle point. Then in each column there is some element strictly greater than v, and in each row there is some element strictly less than v. Now for each column j, let θ_j be the number of elements strictly greater than v, and let $\min_j \theta_j = \theta_k$. By assumption $\theta_k \ge 1$, and thus for some i, $a_{ik} > v$. Since the whole ith row cannot have all elements $> v$, for some r, $a_{ir} \le v$. Consider the columns k and r. They are

$$\begin{bmatrix} a_{1k} \\ \vdots \\ a_{ik} > v \\ \vdots \\ a_{mk} \end{bmatrix} \quad \begin{bmatrix} a_{1r} \\ \vdots \\ a_{ir} \le v \\ \vdots \\ a_{mr} \end{bmatrix}$$

Since θ_k is the minimum of θ_j's, there are at least θ_k entries in the rth column that are strictly greater than v. Further, $a_{ik} > v$ and $a_{ir} \le v$. Therefore there should exist some s with $a_{sk} \le v$ and $a_{sr} > v$. Now the 2×2 matrix

$$\begin{bmatrix} a_{ik} > v & a_{ir} \le v \\ a_{sk} \le v & a_{sr} > v \end{bmatrix}$$

has no saddle point. This contradicts our assumption in the theorem, and thus we have a saddle point for A.

Theorem 3.2.2

Let every 2×2 submatrix of a payoff matrix $A = (a_{ij})$, $1 \leq i \leq m$, $1 \leq j \leq n$, obtained by deleting $(m - 2)$ rows and $(n - 2)$ columns have a saddle point. Then there exists either two distinct rows or two distinct columns such that the minimum entry in one is not smaller than the maximum entry in the other.

PROOF: By the previous theorem we have saddle points for A and for the transpose A'. Let the values for A and A' be $v(A)$ and $v(A')$. If $v(A) < v(A')$, then for some row of A', or equivalently for some column of A, every entry is $\geq v(A')$. But the value of A is $v(A)$, and we have a column in A with every entry in it $\leq v(A)$. The two columns are distinct, and they satisfy the required condition. If $v(A) > v(A')$, by similar argument we can get two rows satisfying the conditions of the theorem. Suppose that $v(A) = v(A')$. Then either we have two columns or rows satisfying the conditions of the theorem, or the matrices A and A' have a common saddle point. In that case for some row r and column s, $a_{rj} = v(A)$, $j = 1, 2, \cdots, n$, and $a_{is} = v(A)$, $i = 1, 2, \cdots, m$. Let B be the matrix obtained from A by deleting the rth row and the sth column. Since the conditions of the previous theorem are fulfilled also for B, B has a saddle point. If the value $v(B)$ of this matrix is $< v(A)$, then for some column j of B every entry in the jth column $\leq v(B) < v(A)$. Thus the columns $j, s, j \neq s$ for A serve the purpose. If $v(B) \geq v(A)$ by a similar argument, we can get a row $k \neq r$ such that the kth and rth rows serve our purpose.

Theorem 3.2.3

Let $A = (a_{ij})$, $1 \leq i \leq m$, $1 \leq j \leq n$, be a payoff matrix with no two entries in a row or column equal. Let $2 \leq p \leq m$, $2 \leq q \leq n$. If every $p \times q$ submatrix of A obtained by deleting $m - p$ rows and $n - q$ columns has a saddle point, then A has a saddle point.

PROOF: For $p = q = 2$, we have the result by theorem 3.2.1. Let us prove the theorem only for the case $p = m \geq 2$ and $q = n - 1 \geq 2$. The rest will follow by induction and symmetry.

Let $A(s)$ be the submatrix obtained from A by deleting the sth column. By assumption, $A(s)$ has a saddle point. It is unique, since no two entries in a row or column are equal. Let (i_s, j_s) be the row and column index where it occurs. For each s, if the column j_s is different, then by varying s over $1, 2, \cdots, n$ we have n distinct columns j_s where the unique saddle point occurs. Now the maximum entry in the matrix A occurs in some column j_s which is evidently the unique saddle point of $A(s)$. Suppose that (i_s, j_s) is the saddle point. Then $a_{i_s j_s}$ is the minimum entry in the i_sth row of A_s. But no two entries in a row of A_s are equal. Thus it is the strict minimum. But it is also the maximum of all entries in A. Further, $A(s)$ has at least two columns. Hence we get a contradiction to the assumption that the j_s's are distinct. Thus for some $s \neq t$, $j_s = j_t$; that is, $a_{i_s j_s} = a_{i_t j_t}$, and (i_s, j_s) are the row and column indices where the saddle point for A occurs.

Remark 3.2.1: The theorem is not true in general if we do not assume that the elements in a row or column are distinct. The following is a counter example with $p = 3$ and $q = 2$.

$$A = \begin{bmatrix} 4 & 2 & 3 \\ 5 & 6 & 2 \\ 1 & 5 & 3 \end{bmatrix}$$

The matrix A has no saddle point, although every 3×2 matrix of A has a saddle point.

3.3 ORDERING OF PAYOFFS

Definition 3.3.1: Let $a = (\alpha_1, \alpha_2, \cdots, \alpha_n)$, $b = (\beta_1, \beta_2, \cdots, \beta_n)$, be two vectors. We say that $a \sim b$ if for each

$$i, j, 1 \leq i, j \leq n, \alpha_i \gtrless \alpha_j \text{ implies } \beta_i \gtrless \beta_j.$$

Two payoff matrices A and B of the same order are equivalent if each row or column vector of A is equivalent ($\sim$) to the corresponding row or column vector of B. Let a be the equivalence class of such matrices. This could be thought of as a partial ordering ($\lesssim$) on the index pairs (i, j) where we compare the collinear points of the index pairs; say that (i, j) and (i, k) are comparable. Let $I(a)$ be the collection of such equivalence classes. Namely, $I(a)$ is the collection of index pairs with the partial order $\lesssim$. We say that $(i, j) < (i, k)$ if $(i, j) \lesssim (i, k)$ and $(i, j) \nsim (i, k)$.

Definition 3.3.2: Let $K \subset I(a)$ consist of the elements $\{(i_1, j_1), (i_2, j_2), \cdots, (i_k, j_k)\}$. Here $i_1, i_2, \cdots, i_k$ are row indices (some of them may be equal), and $j_1, j_2, \cdots, j_k$ are column indices.

Let $K_1 = \{i_1, i_2, \cdots, i_k\}$ and $K_2 = \{j_1, j_2, \cdots, j_k\}$. Let $\overline{K} = K_1 \times K_2$ be the smallest rectangular matrix whose row and column indices contain K. For example, if $K = \{(1, 2), (1, 3), (2, 4)\}$, then $\overline{K} = \{(1, 2), (1, 3), (1, 4), (2, 2), (2, 3), (2, 4)\}$. K is *rectangular* if $K = \overline{K}$.

Definition 3.3.3: Let a be a rectangular order matrix. A rectangular set $\overline{K} = K_1 \times K_2 \subset I(a)$ is called a *generalized saddle point* (GSP) if for each $i \notin K_1$ there exists a $p \in K_1$ such that $(i, j) < (p, j)$ for all $j \in K_2$, and for each

$j \notin K_2$ there exists a $q \in K_2$ with $(i, q) < (i, j)$ for all $i \in K_1$. (Here the strict inequalities are to be maintained.) A minimal GSP (that is, no strict subset of it is a GSP) is called a *saddle*.

Theorem 3.3.1

Let $A = (a_{ij})$, $1 \le i \le m$, $1 \le j \le n$, be the payoff of a game G with value v. Further, let the strict submatrix $A_0 = (a_{ij})$, $1 \le i \le r$, $1 \le j \le s$, of A satisfy the following properties: (1) For each $i > r$ there exists a $p \le r$ such that $a_{pj} > a_{ij}$, $1 \le j \le s$; (2) for each $j > s$ there exists a $q \le s$ such that $a_{iq} < a_{ij}$, $1 \le i \le r$. Then any optimal strategies for player I and II are of the form

$$x^0 = (\xi_1^0, \xi_2^0, \cdots, \xi_r^0, 0, 0, \cdots, 0), \qquad y^0 = (\eta_1^0, \eta_2^0, \cdots, \eta_s^0, 0, 0, \cdots, 0)$$

PROOF: Suppose that there exists an optimal strategy x^0 for player I with $x^0 = (\xi_1^0, \xi_2^0, \ldots, \xi_r^0, \xi_{r+1}^0, \ldots, \xi_m^0)$ with $\xi_i^0 > 0$ for some $i > r$. Let R be the set of row indices i with $i > r$ and $\xi_i^0 > 0$. Now for each $i \in R$ we have by condition (1) a $p \ge r$ such that

$$a_{pj} > a_{ij}, \qquad 1 \le j \le s$$

Now let $\xi_p^* = \xi_p^0 + \xi_i^0$. Consider the ξ_i^*'s for each $i > r$, and define $x^* = (\xi_1^*, \xi_2^*, \cdots, \xi_r^*, 0, 0, \cdots, 0)$, where $\xi_p^* = \xi_p^0 + \xi_i^0$ for the domination corresponding to $i > r$. Clearly

$$\sum_{i=1}^m a_{ij}\xi_i^* > \sum_{i=1}^m a_{ij}\xi_i^0 \ge v, \qquad 1 \le j \le s, \text{ by property (1)}$$

Now for $j > s$, by property (2) there exists a $q \ge s$ with $a_{iq} < a_{ij}$, $1 \le i \le r$; that is,

$$\sum_i a_{ij}\xi_i^* > \sum_i a_{iq}\xi_i^* \ge v \text{ for } j > s$$

Thus

$$\sum_i a_{ij}\xi_i^* > v, \qquad 1 \le j \le n$$

This contradicts the fact that the value of the game is v. Hence we have no optimal strategy for player I that chooses with positive probability any of the last $m - r$ rows. A similar proof applies for player II.

Remark 3.3.1: Every GSP of a payoff matrix A consists of the row indices i on which alone the optimal strategy is concentrated, and it consists of the column indices j on which alone the optimal strategy of player II is concentrated.

Theorem 3.3.2

Every element of the equivalence class of order matrices has a unique saddle.

PROOF: Since there could be only a finite number of GSP's, if we show that the intersection of any two GSP's is again a GSP, then the intersection of all GSP's of a would yield the unique saddle. Now let $\overline{K} = K_1 \times K_2$ and $\overline{L} = L_1 \times L_2$ be two GSP's. Clearly, by the remark above, the optimal strategy is concentrated only over K_1 and L_1. Further, $K_1 \cap L_1 \neq \emptyset$, for otherwise, by theorem 3.3.1 above, every optimal strategy for player I would omit the row indices in L_1, which would contradict the GSP property of $\overline{L}$. Similarly, $K_2 \cap L_2 \neq \emptyset$. Let the rectangular $\overline{K} \cap \overline{L} = (K_1 \cap L_1) \times (K_2 \cap L_2)$ be considered. We shall show that $\overline{K} \cap \overline{L}$ is a GSP.

Let $i \notin K_1 \cap L_1$—say, $i \notin K_1$. By assumption we have a $p \in K_1$ with $(i, j) < (p, j)$ for all $j \in K_2$ and in particular for $j \in K_2 \cap L_2$. If $p \in L_1$, then we have proved half the theorem. The other half follows by applying this argument for column indices.

Suppose that $p \notin L_1$; then by the GSP property of L_1 we have a $q \in L_1$ with $(p, j) < (q, j)$ for $j \in K_2 \cap L_2$. Thus $(i, j) < (p, j) < (q, j), j \in K_2 \cap L_2$. This has to terminate in a finite number of steps, and we have a $p_0 \in K_1 \cap L_1$ such that $(i, j) < (p_0, j), j \in K_2 \cap L_2$.

Remark 3.3.2: The essential submatrix is contained in the saddle. One could think of weak GSP's, but in that case we cannot assure that the intersection of two weak GSP's is a weak GSP.

A one-element saddle is the ordinary (strict) saddle point, and it could be easily found out in a large matrix. But it is not easy to locate the saddle in a large matrix. The following theorem helps for the rapid search of the saddle.

Theorem 3.3.3

Let a be an equivalence class element of order matrices, with at least two columns. Let (p, q) be maximal with respect to $\lesssim$. Let a^q denote an equivalence class element of order matrices obtained from a by deleting the qth column. Then the saddle $S(a^q) \subset S(a)$.

PROOF: Let $S = S(a) = S_1 \times S_2$, and $S^q = S_1 \times (S_2 - \{q\})$. If $q \notin S_2$, then S^q is a GSP of a^q. If $q \in S_2$, then $p \in S_1$; for otherwise we would have an $r \in S_1$ with $(p, q) < (r, q)$, which would contradict the maximality of (p,q). Thus if $q \in S_2$, the row condition is always satisfied in S^q, and we have difficulty only in the case of the column condition.

By the maximality of (p, q), no column outside S_2 could dominate the qth column, and therefore in the saddle $S = S_1 \times S_2$ the qth column is never used for minorization. Hence S^q is a GSP; that is, $S(a^q) \subset S^q \subset S$.

One could easily prove a similar theorem for the deletion of rows, containing minimal elements. One could easily prove that $S(\mathscr{B}) \subset S(a)$, where $\mathscr{B}$ is obtained from a by omitting a strictly majorized row or a strictly minorized column.

For any given A, an equivalence element of a, we can apply the above majorization and minorization technique and theorem 3.3.3 above and end up with a single element in $m + n - 2$ steps. The element we obtain by such a process is called a *residual* of a. Let $R(a)$ be the collection of residuals of a. Since the saddle is rectangular, $R(a) \subset \overline{R}(a) \subset S(a)$.

Theorem 3.3.4

Let $A = (a_{ij}) \in a$ be an equivalence class element from order matrices. If $\min_j \max_i a_{ij} = a_{i^0 j^0}$, then (i^0, j^0) is a residual of a. If $\max_i \min_j a_{ij} = a_{i^* j^*}$, then (i^*, j^*) is a residual of a.

PROOF: By assumption, if there are at least two columns of A, a column q other than j^0 would contain a maximal element (p, q) with respect to $\lesssim$. Even if we omit the qth column and start with A^q, (i^0, j^0) would still remain the min max element. Proceeding inductively we can omit all but the jth column, and by row deletions in the column j^0 we can reach (i^0, j^0). A similar proof applies for (i^*, j^*).

The saddle could be obtained by the following algorithm. Let K^0 be a submatrix known to be contained in the saddle. Having obtained $K^\alpha \subset S(a)$, $\alpha = 0, 1, 2, \cdots$, the new $K^{\alpha+1}$ is obtained by adjoining either (1) a new row that is not strictly majorized in $I_1(a) \times K_2^\alpha$ or (2) a new column that does not strictly majorize any column in $K_1^\alpha \times I_2(a)$. Then $K^{\alpha+1} \subset S(a)$. If neither (1) nor (2) holds, then $K^\alpha = S(a)$.

Definition 3.3.4: Let $C_1(A)$ denote the set of row indices i in $A = (a_{ij})$ which are chosen with positive probability by player I in some optimal strategy of A. Let $C_2(A)$ be defined similarly for the column indices. Let $C(A) = C_1(A) \times C_2(A)$. Then the set

$$C(a) = \bigcup_{A \in a} C(A)$$

will be called the center of a. The center is always contained in the saddle, and it is not necessarily rectangular. One can hope that $R(a) \subset C(a)$, but this may not always be true (see [9]).

Suppose that we modify our concept of the order matrix for the sequence $\{a^{(k)}\}$ of equivalence class elements that determine the residual, retaining noncollinear comparisons from a whenever applicable. Using such comparisons in a, we can arrive at a new collection $R^*(a) \subset R(a)$, called the restricted residual set. We shall in fact prove that $R^*(a) \subset C(a)$ when a represents the equivalence class element in order matrices with the property that no two elements in a row or column in any matrix of a are equal.

Theorem 3.3.5

If $(r, s) \in C(A)$ for a payoff $A = (a_{ij})$, $1 \le i \le m$, $1 \le j \le n$, then every neighborhood of A contains a matrix B with a unique solution, and $(r, s) \in C(B)$.

PROOF: Without loss of generality, let $v \ne 0$, and let (x^0, y^0) be an extreme optimal pair with $\xi_r^0 > 0$, $\eta_s^0 > 0$. By theorem 3.1.14, and by relabeling the row and column indices, we can assume the existence of a nonsingular submatrix of the form (a_{ij}), $1 \le i, j \le k$, for some k, and

$$x^0 = (\xi_1^0, \xi_2^0, \cdots, \xi_k^0, 0, \cdots, 0)$$

$$y^0 = (\eta_1^0, \eta_2^0, \cdots, \eta_k^0, 0, \cdots, 0)$$

Here $\xi_i^0 > 0$,

$$i = 1, 2, \cdots, \bar{r}, \qquad \eta_j^0 > 0$$

$$j = 1, 2, \cdots, \bar{s},$$

(that is, in the new label $r \le \bar{r}$, $s \le \bar{s}$). Clearly

$$\sum_i a_{ij}\xi_i^0 = v, \quad j = 1, 2, \cdots, k, \quad \text{and} \quad \sum_j a_{ij}\eta_j^0 = v, \quad i = 1, 2, \cdots, k$$

Let Q be the $k \times k$ matrix with entries

$$(q_{ij}) = a_{ij} - \sum_{i=1}^{k} a_{ij}\xi_i^* - \sum_{j=1}^{k} a_{ij}\eta_j^*$$

$$+ 2 \sum_{i=1}^{k} \sum_{j=1}^{k} a_{ij}\xi_i^*\eta_j^*$$

If $x^* = (\xi_1^*, \xi_2^*, \cdots, \xi_{\bar{r}}^*, 0, \cdots, 0)$ and $y^* = (\eta_1^*, \eta_2^*, \cdots, \eta_{\bar{s}}^*, 0, \cdots, 0)$ are mixed strategies chosen in the neighborhood of x^0, y^0, then $\xi_i^* > 0$, $\eta_j^* > 0$, for $i \le \bar{r}$, $j \le \bar{s}$. Further, we can also assume Q to be nonsingular, in which case it has x^*, y^* as the unique optimal strategies. By suitably perturbing the entries a_{ij} of A for $i > k$, $j > k$, we can construct a new matrix $B = (b_{ij})$, $1 \le i \le m$, $1 \le j \le n$, whose principal $k \times k$ minor is

Q, and every optimal strategy for player I or II omits all rows $i > k$ and columns $j > K$. Since the value of Q is in the neighborhood of the value of A, such a perturbation is possible. This shows that x^*, y^* are unique optimal strategies for B, as $(\xi_1^*, \xi_2^*, \cdots, \xi_k^*)$, $(\eta_1^*, \eta_2^*, \cdots, \eta_k^*)$ are unique for Q. Clearly $(r, s) \in C(B)$.

Theorem 3.3.6

Let $A = (a_{ij})$, $1 \le i \le m$, $1 \le j \le n$, $n > 2$, be a payoff with value v. Let A^q be the matrix obtained from A by deleting the qth column.

Let $A(w)$ be the matrix obtained from A by adding w to the entry a_{pq}. Then $\lim_{w \to \infty} \text{value}\,[A(w)] = \text{value}\,(A^q) = v$.

PROOF: Let $x^0 = (\xi_1^0, \xi_2^0, \cdots, \xi_m^0)$ be optimal for player I for A^q. Let $\alpha = \max_{ij} |a_{ij}|$. Let $x = (\xi_1, \xi_2, \cdots, \xi_m)$, $\xi_p > 0$, be a mixed strategy with $\|x - x^0\| < \varepsilon$. If $a_{ij}(w)$ is the ijth component of $A(w)$, then

$$\sum_i a_{ij}(w)\xi_i = \sum_i a_{ij}(w)\xi_i^0 + \sum_i a_{ij}(w)\,(\xi_i - \xi_i^0)$$

$$\ge v - \varepsilon\alpha, \qquad j \ne q$$

$$\ge -\alpha + w\xi_p, \qquad j = q$$

Thus for large w, player I can get at least $v - \varepsilon\alpha$ in the game $A(w)$. Hence $v - \varepsilon\alpha \le \text{value}\,A(w) \le v$, and the proof is complete.

The following theorems involve the modified order matrix concept in which $\underset{a}{\lesssim}$ and $\underset{\mathscr{B}}{\lesssim}$ may include also noncollinear comparisons.

Theorem 3.3.7

Let a be an equivalence class element of order matrices with no two elements in any row or column of $A \in \mathrm{a}$ being equal. Let (p, q) be maximal with respect to $\underset{a}{\lesssim}$. Let the elements of $\mathscr{B}$ be the elements of a with column q deleted. Then $C(\mathscr{B}) \subset C(\mathrm{a})$.

PROOF: Let $(r, s) \in C(\mathscr{B})$. Using theorem 3.3.5, coupled with the fact that the elements in each row and column of any $A \in \mathrm{a}$ are unequal, we can find $A^q \in \mathscr{B}$, which has unique solution (x^*, y^*) with $x_r^* > 0$, $y_s^* > 0$. Let

$$\varphi(x) = \min_{j \ne q} \sum_i a_{ij}x_i; \qquad \psi(y) = \max_i \sum_{j \ne q} a_{ij}y_j$$

Let X be the set of all mixed strategies for player I, and let Y be the set of all mixed strategies for player II with the last component zero.

We have $\max_X \varphi(x) = \varphi(x^*)$, $\min_Y \psi(y) = \psi(y^*)$, at the unique points x^*, y^*. Thus there exists an $\varepsilon > 0$ such that

$$\varphi(x) \leq \text{value } [A^q] - \varepsilon \, \|x^* - x\| \text{ for all } x \in X \cdots \qquad \text{**}$$

Similarly there exists a $\delta > 0$ such that

$$\psi(y) \geq \text{value } [A^q] + \delta \, \|y^* - y\| \text{ for all } y \in Y \cdots \qquad \text{***}$$

Now let $A(w)$ be the matrix obtained from A by adding w to a_{pq}, and let $(x(w), y(w))$ be optimal for players I and II. Clearly

$$\text{Value } [A(w)] \leq \varphi(x(w))$$

and from inequality ** above

$$\varepsilon \, \|x^* - x(w)\| \leq \text{value } [A^q] - \text{value } [A(w)]$$

By the previous theorem, $x(w) \to x^*$ as $w \to \infty$. Hence $\xi_r(w)$, the rth component of $x(w)$, is positive for large w. To prove that $\eta_s(w)$, the sth component of $y(w)$ is positive for large w, first note that

$$\sum_j a_{pj}\eta_j(w) + w\eta_q(w) \leq \text{value } [A(w)] \leq \max_{ij} [a_{ij}] = \alpha$$

Hence

$$\eta_q(w) \leq \frac{2\alpha}{w} \cdots \qquad *$$

Then we have

$$\text{Value } [A(w)] = \max_i \sum_j a_{ij}(w) \, \eta_j(w)$$

$$\geq \max_i \left\{ \sum_{j \neq q} a_{ij}\eta_j(w) + a_{is}\eta_q(w) + (a_{iq} - a_{is}) \, \eta_q(w) \right\}$$

$$\geq \max_i \left\{ \sum_{j \neq q} a_{ij}\eta_j(w) + a_{is}\eta_q(w) \right\} - 2\alpha\eta_q(w)$$

$$\geq \psi(y^1(w)) - \frac{4\alpha^2}{w} \text{ by } *$$

where $y^1(w)$ is like $y(w)$ except for $\eta_q^1(w) = 0$ (that is, the qth coordinate of $y^1(w)$), and $\eta_s^1(w) = \eta_s(w) + \eta_q(w)$. Hence by the inequality *** above

$$\text{Value } [A(w)] \geq \text{value } [A^q] + \delta \, \|y^* - y^1(w)\| - 4\frac{\alpha^2}{w}$$

and $y^1(w) \to y^*$ by the previous theorem. Since $\eta_q(w) \to 0$, we have $\eta_s(w) \to \eta_s^* > 0$. Thus both r and s are active in $A(w)$ for large w. But $A(w) \in \mathfrak{a}$ for all positive w. Hence $(r, s) \in C(\mathfrak{a})$.

Remark 3.3.3: If $\mathscr{B}$ is obtained from a by omitting a strictly majorized row or a strictly minorized column, the $C(\mathscr{B}) \subseteq C(a)$ for such a row or column is never chosen in any optimal strategy.

Theorem 3.3.8

If a is an equivalent class element of order matrices and if no two elements in any row or column of $A \in a$ are the same, then $R^*(a) \subset C(a)$.

PROOF: The proof follows from the definition of $R^*(a)$ and the previous remark and the theorem above.

3.4 POSITIVE GAME MATRICES

Since the strategy sets are unaffected by adding a constant to each entry, and since in any payoff matrix $A = (a_{ij})$, $1 \leq i \leq m$, $1 \leq j \leq n$, each player has an optimal strategy using at most $\min(m, n)$ rows or columns as the case may be, we shall assume in this section that

$$A = (a_{ij}), a_{ij} > 0, 1 \leq i, j \leq n$$

Trivially the value of the payoff A is positive. We write $A > 0$ for such matrices. The first theorem does not assume that $A > 0$.

Theorem 3.4.1

Let $A = (a_{ij})$ be an $n \times n$ symmetric payoff matrix. If every optimal strategy of player II omits a column j_0, then there exists a vector $f = (\varphi_1, \varphi_2, \cdots, \varphi_n)$, $f > 0$, and a constant $c > 0$ such that

$(f, x^0) \geq c$ for all optimal x^0 of player I with strict inequality for some x^*

$(f, y^0) \leq c$ for all optimal y^0 of player II

That is, a positive linear fuctional separates the convex sets of optimal strategies of the player.

PROOF: By adding a large constant $N > 0$ to each entry in A, we get a new matrix which is also symmetric and in which each entry is positive. Without loss of generality, let $A > 0$. It is well known by the theorem of Perron–Frobenius [5] that A has a positive eigenvalue λ with an eigenvector $f > 0$. We can choose f to be a mixed strategy. Since j_0 is omitted in every optimal strategy of player II by theorem 3.1.2 and theorem 3.1.16, we have an optimal $x^* = (\xi_1^*, \xi_2^*, \cdots, \xi_n^*)$ for player I with

$$\sum_i a_{ij}\xi_i^* \geq v, \quad j \neq j_0$$

$$> v, \quad j = j_0$$

(Here v is the value of A.) Since $f > 0$, $\Sigma \varphi_j = 1$,

$$\sum_i \sum_j a_{ij} \xi_i^* \varphi_j > v$$

That is, $(x^*, Af) = \lambda(x^*, f) > v$. Here the ith component of Af is given by $(Af)_i = \sum_j a_{ij} \varphi_j$. Thus $(x^*, f) > v/\lambda$ (since $\lambda > 0$). Further, $(x^0, f) \geq v/\lambda$ for all optimal x^0 of player I. Now

$$(y^0, f) = \left(y^0 \frac{1}{\lambda} Af \right) = \frac{1}{\lambda} (y^0, Af) = \frac{1}{\lambda} (Ay^0, f)$$

(by the symmetry of A)

Since $\sum_j a_{ij} \eta_j^0 \leq v$ for all optimal $y^0 = (\eta_1^0, \eta_2^0, \cdots, \eta_n^0)$ of player II, we have

$$v \geq \sum_i \sum_j a_{ij} \eta_j^0 \varphi_i = \lambda \sum_j \varphi_j \eta_j^0 = \lambda(f, y^0)$$

That is, $(f, y^0) \leq v/\lambda$. Hence the theorem.

Theorem 3.4.2

Let $\{A_t, t \in T\}$ be a commuting family of positive matrices $\{A_t > 0, t \in T\}$ with the following properties: (1) Any $(n + 1)$ of them have a common optimal strategy for player I. (2) Player II has a completely mixed optimal strategy for each A_t, $t \in T$. Then the ratio of the values v_t, v_s of A_t and A_s is the same as the ratio of their Perron–Frobenius eigenvalues λ_t and λ_s.

PROOF: Let R_t be the set of optimal strategies for player I for A_t. The R_t's are closed bounded convex sets in R^n. By (1), any $(n + 1)$ of them intersect in a nonempty set, and therefore by Helly's theorem (theorem 1.9.6) we have a common optimal strategy for any finite subfamily of $\{R_t, t \in T\}$. By the compactness of the set of mixed strategies, we have a common optimal strategy for player I for all the A_t's. Let $x^0 = (\xi_1^0, \xi_2^0, \cdots, \xi_n^0)$ be a common optimal strategy. Now

$$\sum_i a_{ij}(t) \xi_i^0 = v_t, \qquad j = 1, 2, \cdots, n$$

The Perron–Frobenius root is also geometrically simple; that is, any two eigenvectors of the eigenvalue λ_t for A_t differ only by a scalar multiple [5]. Let z be a positive eigenvector of A_t for the eigenvalue λ_t. Further, without loss of generality let z be a mixed strategy. Now for any A_s,

$$A_t(A_s z) = A_s(A_t z) = \lambda_t(A_s z)$$

That is, $A_s z$ is an eigenvector for A_t. By the geometric simplicity of λ_t, $A_s z = \alpha_s z$ for some $\alpha_s > 0$. In fact, $\alpha_s = \lambda_s$. For, suppose that $\alpha_s \neq \lambda_s$; then if $u > 0$ is an eigenvector for the transpose A_s' for the eigenvalue λ_s, we have

$$\lambda_s(u, z) = (A_s'u, z) = (u, A_s z) = \alpha_s(u, z)$$

Further, $\alpha_s \neq \lambda_s$ implies that $(u, z) = 0$, which contradicts $u > 0$, $z > 0$. Thus we have $A_t z = \lambda_t z$, $t \in T$. Now

$$(x^0, z) = \frac{1}{\lambda_t}(x^0, A_t z) = \frac{1}{\lambda_t}(A_t'x_0, z)$$

By assumption (2), $(A_t'x_0, z) = v_t \Sigma z_i = v_t$. Thus $v_t/\lambda_t = (x^0, z)$ for all $t \in T$. Hence the theorem.

Recently Thompson and Weil [11, 12] have studied the interrelationship between game theory and eigenvalues for general matrices.

REFERENCES

[1] Arrow, K. J., Barankin, E. W., and Blackwell, D. (1953). Admissible points of convex sets. In "Contributions to the Theory of Games," Vol. 2, pp. 87–91 (*Ann. Math. Studies No. 28*), edited by H. W. Kuhn and A. W. Tucker, Princeton University Press, Princeton, New Jersey.

[2] Blackwell, D., and Girschick, G. A. (1954). "Theory of Games and Statistical Decisions," John Wiley & Sons, New York.

[3] Bohnenblust, H., Karlin, S., and Shapley, L. S. (1950). Solutions of discrete two-person games. In "Contributions to the Theory of Games," Vol. 1, pp. 51–72 (*Ann. Math. Studies No. 24*), edited by H. W. Kuhn and A. W. Tucker, Princeton University Press, Princeton, New Jersey.

[4] Dresher, M. (1961). "Games of Strategy," Prentice-Hall, Englewood Cliffs, New Jersey.

[5] Gantmacher, F. (1959). "Application of the Theory of Matrices," Vol. 2, Interscience Publishers, New York.

[6] Gale, D. (1960). "The Theory of Linear Economic Models," McGraw-Hill Book Company, New York.

[7] Kaplansky, I. (1945). A contribution to von Neumann's theory of games. *Ann. Math.* **46**, pp. 474–479.

[8] Raghavan, T. E. S. (1965). On positive game matrices and their extensions. *J. London Math. Soc.* pp. 467–477.

[9] Shapley, L. S. (1964). Some topics in two-person games. "Advances in Game Theory," pp. 1–28, Princeton University Press, Princeton, New Jersey.

[10] Shapley, L. S., and Snow, R. N. (1950). Basic solutions of discrete games. In "Contributions to the Theory of Games," Vol. 1, pp. 27–35 (*Ann. Math. Studies No. 24*), edited by H. W. Kuhn and A. W. Tucker, Princeton University Press, Princeton, New Jersey.

[11] Thompson, G. S. L., and Weil, R. L. (1968). Further relations between game theory and eigensystems. Management Science report No. 136, Carnegie Mellon University.

[12] Weil, R. L., (1968). Game Theory and eigensystems. SIAM Review, **10**, pp. 360–367.

Chapter Four

ON COMPUTING OPTIMAL STRATEGIES

In this chapter we shall describe in detail the simplex algorithm for finding a pair of optimal strategies for a payoff matrix. Since this is closely linked to the problem of linear programming, we shall discuss some aspects of this more general problem. We shall also digress a bit and discuss the theorem of Kuhn and Tucker on nonlinear programming.

4.1 LINEAR INEQUALITIES FOR THE GAME

For any payoff $A = (a_{ij})$, $1 \leq i \leq m$, $1 \leq j \leq n$, we can add a positive constant, c, to all its entries, so that the new matrix has all its entries positive. By doing so we do not affect the sets of optimal strategies. If v is the value for A, $v + c$ is the value for the new matrix. Thus, we can always assume without loss of generality that $v > 0$ and that $A = (a_{ij})$, $a_{ij} > 0$, $1 \leq i \leq m$, $1 \leq j \leq n$. In this chapter we shall always assume the payoff A to be a matrix with positive entries when we refer to the case of games.

Of all probability vectors $y = (\eta_1, \eta_2, \cdots, \eta_n)$ satisfying

$$\sum_j a_{ij} \eta_j \leq \alpha, \qquad i = 1, 2, \cdots, m \tag{A}$$

any optimal strategy $y^0 = (\eta_1^0, \eta_2^0, \cdots, \eta_n^0)$ is the one for which

$$\sum_j a_{ij} \eta_j^0 \leq v, \qquad i = 1, 2, \cdots, m$$

and the value v is the smallest such α satisfying condition A.

Thus, to find the value $v > 0$ and optimal strategy $y^0 = (\eta_1^0, \eta_2^0, \cdots \eta_n^0)$ for player II, it is sufficient to find the minimum α satisfying condition A or equivalently max $1/\alpha$, where α satisfies $\sum_j a_{ij} \eta_j \leq \alpha, i = 1, 2, \cdots, m$, for some vector

$$y = (\eta_1, \eta_2, \cdots, \eta_n), \eta_1, \eta_2, \cdots, \eta_n \geq 0, \sum_i^n \eta_j = 1.$$

If $y_j = \eta_j/\alpha$, $j = 1, 2, \cdots, n$, then we want to maximize $1/\alpha$ where

$$\max \frac{1}{\alpha} = \max \sum_{j=1}^n y_j$$

subject to $\sum_{j} a_{ij} y_j \leq 1$, $i = 1, 2, \cdots, m$, and $y_1, y_2, \cdots, y_n \geq 0$. Similarly,

we can consider the problem of $\min \sum_{i}^{m} x_i$ subject to $\sum_{i=1}^{m} a_{ij} x_i \geq 1$,

$j = 1, 2, \cdots, n$, $x_1, x_2, \cdots, x_m \geq 0$.

Since the optimal strategies exist, we know that both problems have solutions $x^0 = (x_1^0, x_2^0, \cdots, x_m^0)$ and $y^0 = (y_1^0, y_2^0, \cdots, y_n^0)$, which are the optimal solutions in the sense of minimizing Σx_i and maximizing Σy_j subject to their respective constraints. When we define $x_i = \xi_i / \alpha$ for any $i = 1, 2, \cdots, m$, where $(\xi_1, \xi_2, \cdots, \xi_m)$ is a probability vector satisfying $\sum_{i} a_{ij} \xi_i \geq \alpha$, $j = 1, 2, \cdots, n$, we find the maximum of α, or equivalently the minimum of $1/\alpha = \Sigma x_i$ is attained at the point $x^0 = (x_1^0, x_2^0, \cdots, x_m^0)$, where $x_i^0 = \xi_i^0 / \alpha$ and $(\xi_1^0, \xi_2^0, \ldots \xi_m^0)$ is optimal for player I. Further, we know that this $\min \sum_{i} x_i$ is $1/v = \max \sum_{i} y_j$. The two problems are known as dual problems. In the next section we shall consider the more general dual problems.

4.2 LINEAR PROGRAMMING

Consider the following two problems.

PROBLEM I:

$$\max_{y \in T} \sum_{j=1}^{n} q_j y_j$$

where

$$T = \left\{ y : y = (y_1, y_2, \cdots, y_n); y_1, y_2, \cdots, y_n \geq 0, \sum_{j=1}^{n} a_{ij} y_j = p_i, i = 1, 2, \cdots, m \right\}.$$

PROBLEM II:

$$\min_{x \in S} \sum_{i=1}^{m} p_i x_i$$

where $S = \left\{ x : x = (x_1, x_2, \cdots, x_m); \sum_{i} a_{ij} x_i \geq q_j, j = 1, 2, \cdots, n \right\}$. (Note that the x_i's need not be nonnegative.)

The functions $\Sigma q_j y_j$ and $\Sigma p_i x_i$ are called the *objective functions* for the respective problems, and the points in S and T are called *feasible solutions* to the problems. A point $x^0 \in T$ which maximizes the objective function is called an *optimal solution* to problem I. A similar definition holds for problem II.

Theorem 4.2.1

If $x^0 \in S$, $y^0 \in T$, then $\Sigma q_j y_j^0 \leq \Sigma p_i x_i^0$. If the x^0, y^0 further satisfy $\Sigma q_j y_j^0 = \Sigma p_i x_i^0$, they are in fact the optimal solutions to the problems.

PROOF: Since $y^0 \in T$, we have $\sum_j a_{ij} y_j^0 = p_i$, $i = 1, 2, \cdots, m$, and

$$\sum_i \left(\sum_j a_{ij} y_j^0 \right) x_i^0 = \sum_i p_i x_i^0. \text{ Since } \sum_i a_{ij} x_i^0 \geq q_j, j = 1, 2, \cdots, n, \text{ and}$$

since $y_j^0 \geq 0$, $j = 1, 2, \cdots, n$, we have

$$\sum_i p_i x_i^0 = \sum_i \left(\sum_j a_{ij} y_j^0 \right) x_i^0 = \sum_j y_j^0 \left(\sum_i a_{ij} x_i^0 \right) \geq \sum_j q_j y_j^0$$

Hence the first part of the theorem. For the second part,

$$\sum_i p_i x_i^0 = \sum_j q_j y_j^0 \leq \max_{y \in T} \sum_j q_j y_j \leq \min_{x \in S} \sum_i p_i x_i \leq \sum_i p_i x_i^0$$

(by the first part of the theorem)

Thus x^0 is an optimal solution to problem II, and y^0 is an optimal solution to problem I.

Definition 4.2.1: Let C_1, C_2, $\cdots$, C_n, p be $n + 1$ vectors in R^m. Let $y_1 C_1 + y_2 C_2 + \cdots + y_n C_n = p$ for some y_1, y_2, $\cdots$, $y_n \geq 0$. Let y_{α_1}, y_{α_2}, $\cdots$, $y_{\alpha_k} > 0$, $1 \leq \alpha_1, \alpha_2, \cdots, \alpha_k \leq n$, and the rest of the y's be zero. Then $y = (y_1, y_2, \cdots, y_n)$ is called a *basic solution* if $C_{\alpha_1}, C_{\alpha_2}, \cdots, C_{\alpha_k}$ are linearly independent.

Theorem 4.2.2

Let y^0 be an optimal solution to problem I. Then we always have a basic optimal solution, where the C_j's are the columns of the matrix

$$\begin{bmatrix} a_{11} & a_{12} & \cdots & a_{1n} \\ \cdots & \cdots & \cdots & \cdots \\ \cdots & \cdots & \cdots & \cdots \\ a_{m1} & a_{m2} & \cdots & a_{mn} \end{bmatrix}$$

and the vector p is $p = \begin{bmatrix} p_1 \\ \vdots \\ p_m \end{bmatrix}$

PROOF: Let $y^0 = (y_1^0, y_2^0, \cdots, y_n^0)$ have the first s components positive and the rest zero. Then we have

$$y_1^0 C_1 + y_2^0 C_2 + \cdots + y_s^0 C_s = p$$

If $C_1, C_2, \cdots, C_s$ are linearly independent, we have nothing to prove. Otherwise we would find a new optimal solution involving $C_1, C_2, \cdots, C_s$ with at most $(s-1)$ of its coordinates positive. We can check whether the corresponding C_j's are linearly independent, or else we can once again continue with our procedure and construct a new optimal solution with at most $(s-2)$ coordinates positive, and so on until we arrive at a basic optimal solution. Since we have only a finite number of C_j's we shall eventually terminate in a finite number of steps.

Let $C_1, C_2, \cdots, C_s$ be linearly dependent. Then there exist $\lambda_1, \lambda_2, \cdots, \lambda_s$, not all zero with $\sum_{j=1}^{s} \lambda_j C_j = 0$. Without loss of generality we can have some $\lambda_j > 0$. If we define $\alpha = \max_{1 \le j \le s} \dfrac{\lambda_j}{y_j^0}$, and $y_j' = y_j^0 - \dfrac{1}{\alpha} \cdot \lambda_j, j = 1, 2, \cdots, s$, then $\alpha > 0$, $y_j' \ge 0$, $j = 1, 2, \cdots, s$, with $y_k' = 0$ for some k in $1 \le k \le s$. Further,

$$\sum_{j=1}^{s} y_j' C_j = \sum_{j=1}^{s} y_j^0 C_j - \frac{1}{\alpha} \sum_{j=1}^{s} \lambda_j C_j = p$$

Thus the point $y' = (y_1', y_2', \cdots, y_s', 0, 0, \cdots, 0) \in T$ has at most $(s-1)$ coordinates positive. Further $\sum_{j=1}^{s} y_j' q_j = \sum_{j=1}^{s} y_j^0 q_j - \left(\dfrac{1}{\alpha}\right) \sum_{j=1}^{s} \lambda_j q_j$. If $\sum_{j=1}^{s} \lambda_j q_j = 0$, then y' is also an optimal solution to the problem. Since y^0 is optimal, $\Sigma \lambda_j q_j$ cannot be <0, in which case y' would strictly improve the objective function. The only possibility is $\sum_j \lambda_j q_j > 0$. We shall show that this cannot happen.

Suppose that $\Sigma \lambda_j q_j > 0$. Since some $\lambda_j > 0$, by assumption we can have either all the λ_j's ≥ 0 or some $\lambda_t < 0$. The first case cannot happen, for otherwise $\Sigma \lambda_j q_j > 0$, $\lambda_1, \lambda_2, \cdots, \lambda_s \ge 0$, $\Sigma \lambda_j C_j = 0$, would imply

$$y^* = (y_1^0 + \lambda_1, y_2^0 + \lambda_2, \cdots, y_s^0 + \lambda_s, 0, 0, \cdots, 0) \in T$$

with $\sum_i q_j(y_j^0 + \lambda_j) > \sum_j q_j y_j^0$, a contradiction to the optimality of y^0.

In the second case we have some $\lambda_j > 0$ and some $\lambda_t < 0$ with $\Sigma \lambda_j q_j > 0$. If we define $\mu_1 = -\lambda_1, \cdots, \mu_t = -\lambda_t, \cdots, \mu_s = -\lambda_s$, then $\Sigma \mu_j C_j = 0$, and $\Sigma \mu_j q_j < 0$. As we constructed the y', we can construct a y'' using the μ's which would again contradict the optimality of y^0 when $\Sigma \mu_j q_j < 0$.

Thus we always have $\sum_{j=1}^{s} \lambda_j q_j = 0$, and y' is an optimal solution with at most $(s-1)$ of the coordinates positive. Hence the theorem.

Theorem 4.2.3

If $e_1, e_2, \cdots, e_m$ is a basis in R^m and f is a vector in R^m with $f = \sum_1^m \alpha_j e_j$ and $\alpha_r \neq 0$, then $e_1, e_2, \cdots, e_{r-1}, f, e_{r+1}, \cdots, e_m$ is a basis in R^m.

PROOF: Let $\tau_1 e_1 + \cdots + \tau_{r-1} e_{r-1} + \tau f + \tau_{r+1} e_{r+1} + \cdots + \tau_m e_m = 0$. That is,

$$\sum_{\substack{j=1 \\ j \neq r}}^m \tau_j e_j + \tau \sum_{j=1}^m \alpha_j e_j = 0$$

Since the e_j's form a basis, the coefficients $\tau_j + \tau\alpha_j = 0, j \neq r$, and $\tau\alpha_r = 0$. By assumption, $\alpha_r \neq 0$, and hence $\tau = 0$; that is, $\tau_j = 0, j \neq r$, and $\tau = 0$. Hence the theorem.

Suppose that $C_1, C_2, \cdots, C_n$ are n vectors in R^m, and $e_1, e_2, \cdots, e_m$ is a basis in R^m. If

$$C_k = \sum_{i=1}^m \alpha_{ik} e_i, \qquad k = 1, 2, \cdots, n$$

then this could be represented by the following matrix table.

$$
\begin{array}{c}
 \\
e_1 \\
e_2 \\
e_r \\
e_m
\end{array}
\begin{array}{c}
C_1 \quad C_2 \cdots \; C_k \ldots C_n \\
\begin{bmatrix}
\alpha_{11} & \alpha_{12} & \cdots & \alpha_{1k} & \cdots & \alpha_{1n} \\
\alpha_{21} & \alpha_{22} & \cdots & \alpha_{2k} & \cdots & \alpha_{2n} \\
\alpha_{r1} & \alpha_{r2} & \cdots & \alpha_{rk} & \cdots & \alpha_{rn} \\
\alpha_{m1} & \alpha_{m2} & \cdots & \alpha_{mk} & \cdots & \alpha_{mn}
\end{bmatrix}
\end{array}
$$

Here the jth column gives the coordinates of the vector C_j with respect to the basis $e_1, e_2, \cdots, e_m$. Suppose that $\alpha_{rk} \neq 0$; then $e_1, e_2, e_{r-1}, C_k, e_{r+1}, \cdots, e_m$ is a new basis (this follows from the previous theorem), and the new table with respect to the new basis $e_1, e_2, \cdots, e_{r-1}, C_k, e_{r+1}, \cdots, e_m$, given by

$$
\begin{array}{c}
 \\
e_1 \\
e_2 \\
e_{r-1} \\
C_k \\
e_{r+1} \\
e_m
\end{array}
\begin{array}{c}
C_1 \quad C_2 \; \cdots \; C_s \quad C_n \\
\begin{bmatrix}
\alpha_{11}' & \alpha_{12}' & \cdots & \alpha_{1s}' & \alpha_{1n}' \\
\cdot & \cdot & \cdot & \cdot & \cdot \\
\cdot & \cdot & \cdot & \cdot & \cdot \\
\alpha_{r1}' & \alpha_{r2}' & \cdots & \alpha_{rs}' & \alpha_{rn}' \\
\cdot & \cdot & \cdot & \cdot & \cdot \\
\alpha_{m1}' & \alpha_{m2}' & \cdots & \alpha_{ms}' & \alpha_{mn}'
\end{bmatrix}
\end{array}
$$

can be written from the old table by the following computations. For any s,

$$C_s = \sum_{\substack{i=1 \\ i \neq r}}^{m} \alpha_{is}' e_i + \alpha_{rs}' C_k$$

$$= \sum_{\substack{i=1 \\ i \neq r}}^{m} \alpha_{is} e_i + \alpha_{rs}' \sum_{i=1}^{m} \alpha_{ik} e_i$$

$$= \sum_{\substack{i=1 \\ i \neq r}}^{m} (\alpha_{is}' + \alpha_{rs}' \alpha_{ik}) e_i + \alpha_{rs}' \alpha_{rk} e_r$$

But

$$C_s = \sum_{i=1}^{m} \alpha_{is} e_i$$

Thus

$$\alpha_{is}' + \alpha_{rs}' \alpha_{ik} = \alpha_{is}, \qquad i \neq r$$

$$\alpha_{rs}' \alpha_{rk} = \alpha_{rs}$$

The new coefficients are given by

$$\alpha_{rs}' = \frac{\alpha_{rs}}{\alpha_{rk}}, \qquad s = 1, 2, \cdots, n$$

$$\alpha_{is}' = \alpha_{is} - \frac{\alpha_{ik}}{\alpha_{rk}} \cdot \alpha_{rs}, \qquad i \neq r, \qquad s = 1, 2, \cdots, n$$

The new table can be easily constructed from the old table by the following rule of thumb.

Subtract suitable multiples of the rth row from any of the other rows so that the kth component of them becomes zero. Finally, divide the rth row by α_{rk}. The matrix of α_{ij}''s would look like the matrix below.

$$
\begin{array}{c}
\\ e_1 \\ \\ e_{r-1} \\ \\ C_k \\ \\ e_{r+1} \\ \\ e_n
\end{array}
\begin{array}{cccc}
C_1 & C_2 & \cdots & C_k \quad C_n \\
\left[\begin{array}{cccc}
\alpha_{11}' & \alpha_{12}' & \cdots 0 & \alpha_{1n}' \\
& & \cdots \ 0 & \cdots \\
\alpha_{r1}' & & \cdots \ 1 & \cdots \alpha_{rn}' \\
& & \cdots \ 0 & \cdots \\
\alpha_{m1}' & & \cdots \ 0 & \cdots \alpha_{mn}'
\end{array}\right]
\end{array}
$$

One can easily check the rule of thumb with the formulas derived above.

4.3 SIMPLEX ALGORITHM

Suppose that for the problem $\max_{T} \Sigma q_j y_j$, where

$$T = \{(y_1, y_2, \cdots, y_n) = y_1, y_2, \cdots, y_n \geq 0,$$
$$\Sigma a_{ij} y_j = p_i, \qquad i = 1, 2, \cdots, m\}$$

we have an optimal solution. Then we should like to evolve a computational procedure that would yield the optimal solution in a finite number of steps. The so-called simplex algorithm will be illustrated for our problem with the following assumptions throughout.

(*Nondegeneracy assumption*): Any point y of T has at least m components positive, and the rank of the matrix (a_{ij}), $1 \leq i \leq m$, $1 \leq j \leq n$, is m.

Our computational procedure would in fact start with a feasible solution which is also basic. By theorem 4.2.2, when we have an optimal solution we can even get a basic optimal solution. Since the rank is m, any $m + 1$ columns are linearly dependent. Thus, we would have at most m components of any basic feasible solution, y, positive. Our nondegeneracy assumption demands that at least m components of any feasible solution be positive. Thus every basic feasible solution has exactly m components positive. Since there can be only a finite number of distinct bases that could be chosen from the columns of the matrix A, and since any basic optimal solution would correspond to one such basis, we could terminate in a finite number of steps, by enumerating all possible bases and choosing the right one among them. This is an extremely tedious process to determine the basic optimal solution in a systematic way.

Our procedure would be as follows. First we would start with a basic feasible solution (in all our game theoretic problems this is very easy). Our aim would be to check whether we could, by throwing out some element from the basis and bringing some other column of the matrix into the basis, get a new basic feasible solution that strictly improves the value of the objective function. If it fails, then we would stop at that stage and say that this is the optimal solution. Surprisingly our algorithm would tell us the following: (1) whether our nondegeneracy assumption is true; (2) whether the problem has an optimal solution. Consider the following problem.

PROBLEM I: $\max_{y \in T} \Sigma q_j y_j$

$$T = \left\{ (y_1, y_2, \cdots, y_n): y_1, y_2, \cdots, y_n \geq 0, \right.$$

$$\left. \sum_{j=1}^{n} a_{ij} y_j = p_i, \qquad i = 1, 2, \cdots, m \right\}$$

Let $C_1, C_2, \cdots, C_n$ denote the n columns of the matrix

$$\begin{bmatrix} a_{11} & a_{12} & \cdots & a_{1n} \\ \cdots\cdots\cdots\cdots \\ \cdots\cdots\cdots\cdots \\ a_{m1} & a_{m2} & \cdots & a_{mn} \end{bmatrix}$$

and

$$p = \begin{bmatrix} p_1 \\ p_2 \\ \vdots \\ p_m \end{bmatrix}$$

Further, let us assume that the first m columns of the matrix, namely $C_1, C_2, \cdots, C_m$, are linearly independent, and we have a feasible point $y = (y_1, y_2, \cdots, y_m, 0, 0, \cdots, 0) \in T$, with $y_1, y_2, \cdots, y_m > 0$. We can start with the following table:

	C_1	C_2	$\cdots$	C_m	$\cdots C_k \cdots$	C_n	y
C_1	1	0	$\cdots$	0	α_{1k}	α_{1n}	y_1
C_2	0	1	$\cdots$	0	α_{2k}	α_{2n}	y_2
C_r	0	0	$\cdots$	0	α_{rk}	α_{rn}	y_r
C_m	0	0	$\cdots$	1	α_{mk}	α_{mn}	y_m

Here, as we have explained earlier, $C_k = \sum_{i=1}^{m} \alpha_{ik} C_i$, $k = 1, 2, \cdots, n$. The last column outside the matrix gives $p = \sum_{i=1}^{m} y_i C_i$. If some $\alpha_{rk} \neq 0$, we know by theorem 4.2.3 that C_r can always be thrown out of the basis, and C_k can be brought into the basis. Thus by this process we can get new bases from the old.

For any $1 \leq r \leq m$ and $k > m$ with $\alpha_{rk} \neq 0$, our procedure is worth the trouble only when the following are satisfied: (1) p is a nonnegative linear combination of $C_1, C_2, \cdots, C_{r-1}, C_k, C_{r+1}, \cdots, C_m$; (2) the new solution y' strictly improves the value of the objective function.

Let us suppose that the new table is constructed. We have

$$
\begin{array}{c}
\begin{array}{cccccc} C_1 & C_2 & \cdots & C_k & C_n & \qquad p \end{array}\\
\begin{array}{c} C_1 \\ C_2 \\ C_{r-1} \\ C_k \\ C_{r+1} \\ \vdots \\ C_m \end{array}
\begin{bmatrix}
\alpha_{11}' & \alpha_{12}' & \cdots & 0 & \alpha_{1n}' \\
 & & 0 & & \\
 & & & & \\
 & & 1 & & \\
 & & 0 & & \\
 & & & & \\
\alpha_{m1}' & \alpha_{m2}' & \cdots & 0 & \alpha_{mn}'
\end{bmatrix}
\begin{bmatrix}
y_1' \\ y_2' \\ y_{r-1}' \\ \cdot \\ \cdot \\ \cdot \\ y_m'
\end{bmatrix}
\end{array}
$$

The rule of thumb used to get the α_{ij}''s can be stretched also for the column corresponding to p, and one can easily check that

$$
y_i' = y_i - y_r \frac{\alpha_{ik}}{\alpha_{rk}}, \qquad i \neq r
$$

$$
y_r' = \frac{y_r}{\alpha_{rk}}
$$

with

$$
\sum_{\substack{i=1 \\ i \neq r}}^{m} y_i' C_i + y_r' C_k = p
$$

If the new solution $y' = (y_1', y_2', \cdots, y_r', \cdots, y_n')$ were to be feasible, we should have

$$
y_i' \geq 0, \qquad i = 1, 2, \cdots, m
$$

That is,

$$
y_i \geq y_r \frac{\alpha_{ik}}{\alpha_{rk}}, \qquad i \neq r, \qquad i = 1, 2, \cdots, m
$$

and

$$
\frac{y_r}{\alpha_{rk}} \geq 0
$$

Thus we need $\alpha_{rk} > 0$ for $y_r' > 0$. Since $y_i > 0$, $i = 1, 2, \cdots, m$, if, for any i, $\alpha_{ik} < 0$, $i \neq r$, $i = 1, 2, \cdots, m$, then the condition

$$
y_i \geq \frac{y_r}{\alpha_{rk}} \alpha_{ik}
$$

is satisfied (when $\alpha_{rk} > 0$) and $y_i' \geq 0$ in that case for $i \neq r$. The only case that can create trouble is the case when $\alpha_{ik} > 0$. To make $y_i' \geq 0$ in such a case, we can proceed as follows. For any fixed k, let

$$\min_{\substack{i \\ \alpha_{ik}>0}} \frac{y_i}{\alpha_{ik}} = \frac{y_r}{\alpha_{rk}}$$

Then all our required conditions are satisfied. Thus we envisage the following situation.

Having decided to bring C_k into the basis, choose to throw out that C_r for which

$$\min_{\substack{i \\ \alpha_{ik}>0}} \frac{y_i}{\alpha_{ik}} = \frac{y_r}{\alpha_{rk}}$$

The next question is, is it worth all the trouble? Yes! It is if it improves the objective function—that is, if

$$q_1 y_1' + q_2 y_2' + \cdots + q_{r-1} y_{r-1}' + q_k y_r' + q_{r+1} y_{r+1}' +$$
$$+ \cdots + q_m y_m' > \sum_{j=1}^{m} q_j y_j$$

Substituting for y_j''s the values in y_j's, we want

$$\sum_{j \neq r} q_j \left(y_j - y_r \frac{\alpha_{jk}}{\alpha_{rk}} \right) + q_k \frac{y_r}{\alpha_{rk}} > \sum_{j \neq r} q_j y_j + q_r y_r$$

Canceling $\displaystyle\sum_{j \neq r} q_j y_j$ on both sides would amount to the requirement that

$$\frac{y_r}{\alpha_{rk}} \sum_{j \neq r} q_j \alpha_{jk} + q_r y_r - q_k \frac{y_r}{\alpha_{rk}} < 0$$

Thus C_k can be thought of as a possible candidate for entering a new basis in the place of C_r for the improvement of the objective function, provided that

$$\frac{y_r}{\alpha_{rk}} \left(\sum_{j=1}^{m} q_j \alpha_{jk} - q_k \right) < 0$$

Earlier we demanded that $\alpha_{rk} > 0$ for C_k to enter the basis and for C_r to be thrown out. Since $y > 0$ (by the nondegeneracy assumption), we can for any r, k, bring C_k into and throw C_r out of the basis when

(1) $\alpha_{rk} > 0$

(2) $\displaystyle\sum_{i=1}^{m} q_j \alpha_{jk} - q_k < 0$

(3) r is that index for which $\alpha_{rk} > 0$ and

$$\min_{\substack{i \\ \alpha_{ik}>0}} \frac{y_i}{\alpha_{ik}} = \frac{y_r}{\alpha_{rk}}.$$

Since each time we have a new basis that strictly improves the objective function, we would, by our earlier discussion on the number of bases, eventually terminate in a finite number of steps.

Suppose that condition (2) fails to hold. Then it is well and good, and the present basic feasible solution is actually optimal. This is contained in the following theorem.

Theorem 4.3.1

Let $\displaystyle\sum_{j=1}^{m} q_j \alpha_{jk} - q_k \geq 0$ for all $k = 1, 2, \cdots, n$. Then the solution y^0 that we have at present is an optimal solution.

PROOF: By assumption,

$$y_1^0 C_1 + y_2^0 C_2 + \cdots + y_m^0 C_m = p, \qquad y_1^0, y_2^0, \ldots, y_m^0 > 0$$

and $C_1, C_2, \cdots, C_m$ are linearly independent.
Thus we always have a solution x^0 to the system of equations

$$(x^0, C_j) = q_j, \qquad j = 1, 2, \cdots, m$$

That is,

$$\sum_{j=1}^{m} a_{ij} x_i^0 = q_j, \qquad j = 1, 2, \cdots, m$$

Moreover,

$$(x^0, C_k) = \left(x^0, \sum_{j=1}^{m} \alpha_{jk} C_j\right) \qquad \text{for any } k > m$$

and

$$(x^0, C_k) = \sum_{j=1}^{m} \alpha_{jk}(x^0, C_j) = \sum_{j=1}^{m} \alpha_{jk} q_j \geq q_k \qquad \text{(by our assumption}$$

$$\text{in the theorem)}$$

Thus

$$\sum_i a_{ij} x_i^0 = q_j, \qquad j = 1, 2, \cdots, m$$

$$\sum_i a_{ij} x_i^0 \geq q_j, \qquad j = m + 1, m + 2, \cdots, n$$

This shows that x^0 is a feasible solution to the dual problem II that we have been considering. Now with the solution $y^0 = (y_1^0, y_2^0, \cdots, y_m^0, 0, 0, \cdots, 0)$

for our maximum problem and $x^0 = (x_1^0, x_2^0, \cdots, x_m^0)$ for the minimum problem (problem II), we have

$$\sum_{i=1}^{m}\sum_{j=1}^{n} a_{ij}x_i^0 y_j^0 = \sum_{i=1}^{m} x_i^0 \left(\sum_{j=1}^{n} a_{ij}y_j^0\right) = \sum_{i=1}^{m} x_i^0 p_i$$

Further,

$$\sum_{i=1}^{m}\sum_{j=1}^{n} a_{ij}x_i^0 y_j^0 = \sum_{j=1}^{n} y_j^0 \left(\sum_{i=1}^{m} a_{ij}x_i^0\right) = \sum_{j=1}^{m} y_j^0 \left(\sum_{i=1}^{m} a_{ij}x_i^0\right)$$

(This is because $y_{m+1}^0 = y_{m+2}^0 = \cdots = y_n^0 = 0$.)

But for $j = 1, 2, \cdots, m$ we have $\sum_{i=1}^{m} a_{ij}x_i^0 = q_j$, and therefore

$$\sum_{i=1}^{m}\sum_{j=1}^{n} a_{ij}x_i^0 y_j^0 = \sum_{j=1}^{n} q_j y_j^0 = \sum_{i=1}^{m} p_i x_i^0$$

By theorem 4.2.1, y^0 is optimal for the problem.

Although condition (2) may be satisfied for some k, there may not be any $\alpha_{ik} > 0$. When we face this situation we can claim that the objective function is unbounded. The following theorem settles the issue.

Theorem 4.3.2

Let $\sum_{j=1}^{m} \alpha_{jk}q_j - q_k < 0$ for some k with $\alpha_{jk} \le 0$, $j = 1, 2, \cdots, m$. Then the objective function is unbounded over the set T of feasible solutions.

PROOF: Let y be our present solution. Then

$$\sum_{j=1}^{m} y_j C_j = p, \qquad y_1, y_2, \cdots, y_m > 0, \qquad y_{m+1} = y_{m+2} = \cdots = y_n = 0$$

and

$$\sum_{j=1}^{m} \alpha_{jk}C_j - C_k = 0$$

For any $\mu > 0$, we have

$$\sum_{j=1}^{m} (y_j - \mu\alpha_{jk}) C_j + \mu C_k = p$$

Further, by the nonpositivity of α_{jk}'s,

$$(y_1 - \mu\alpha_{1k}, y_2 - \mu\alpha_{2k}, \cdots, y_m - \mu\alpha_{mk}, 0, \cdots, 0, \mu, 0, \cdots, 0)$$

$$\downarrow$$

$$k\text{th coordinate}$$

is a feasible solution.

The value of the objective function at this point is

$$\sum_{j=1}^{m} q_j y_j - \mu \sum_{j=1}^{m} \alpha_{jk} q_j + \mu q_k = \sum_{j=1}^{m} q_j y_j - \mu \left(\sum_{j=1}^{m} \alpha_{jk} q_j - q_k \right)$$

Since $\displaystyle\sum_{j=1}^{m} \alpha_{jk} q_k - q_k < 0$, by increasing μ we can have as large a value as we like. Hence the objective function is unbounded over T.

Degeneracy: When we are faced with degeneracy, it can be due to two reasons. The rank of the matrix (a_{ij}) for the problem of maximization may be less than m (this never occurs in our game theoretic problems). There may be points of T with less than m positive components, in which case we may change the basis without actually improving the objective function. In practice this rarely occurs, and after a few trials we generally approach the optimal solution. By an elaborate selection rule we can eliminate this (see [2]).

We shall solve the following problem by the simplex method after a suitable reduction.

Example 4.3.1: Two players, one (player I) with a silent gun and the other (player II) with a noisy gun, approach a target with equal accuracy in five steps. The accuracy at the sth step is $s/5$, $s = 0, 1, 2, \cdots, 5$. Each player has only one bullet in his gun, and player I alone can hear whether player II has used his bullet or not. At the end of reaching the target, the one who successfully hits the target gets 1 unit from his opponent. Solve the game, and find a pair of optimal strategies for the two players.

Let player I decide to hit at the ith step and player II at the jth step. Then the expected income to player I is

$$K(i,j) = \frac{i}{5} - \left(1 - \frac{i}{5}\right)\frac{j}{5}, \qquad i < j$$

$$= 0, \qquad i = j$$

$$= \left(1 - \frac{j}{5}\right) - \frac{j}{5} = 1 - \frac{2j}{5}, \qquad i > j$$

The payoff matrix is

$i =$ \\ $j =$	0	1	2	3	4	5
0	0	-5	-10	-15	-20	-25
1	25	0	-3	-7	-11	-15
$\frac{1}{25} \times$ 2	25	15	0	1	-2	-5
3	25	15	5	0	7	5
4	25	15	5	-5	0	15
5	25	15	5	-5	-15	0

Before using the simplex technique, we can easily eliminate a few rows and columns from consideration, by simple domination arguments (theorem 3.1.3).

Since the first two rows are weakly dominated by the third row ($i = 2$), the first two rows can be omitted optimally by player I. Since the first two columns dominate the third column, they can be omitted optimally. Further, the last row is weakly dominated by the fourth row ($i = 3$). Hence, the last row could also be omitted optimally by player I. Thus, we are left with the submatrix (we neglect the constant factor of $^1/_{25}$ for the time being).

	$j = 2$	$j = 3$	$j = 4$	$j = 5$
$i = 2$	0	1	-2	-5
$i = 3$	5	0	7	5
$i = 4$	5	-5	0	15

We can add 5 to all entries so that the matrix has all entries nonnegative. We get the new matrix:

$$B = \begin{bmatrix} 5 & 6 & 3 & 0 \\ 10 & 5 & 12 & 10 \\ 10 & 0 & 5 & 20 \end{bmatrix}$$

Since the second row has all entries positive, the value $v_0 > 0$ for this matrix. Let us find an optimal strategy for player II. From our earlier discussion we know that this is equivalent to the problem

$$\max (x + y + z + w)$$

subject to

$$5x + 6y + 3z + 0w \leq 1$$
$$10x + 5y + 12z + 10w \leq 1$$
$$10x + 0y + 5z + 20w \leq 1$$
$$x, \quad y, \quad z, \quad w \geq 0$$

or, by introducing slack variables, this is equivalent to

$$\max (0 \cdot t + 0 \cdot u + 0 \cdot v + 1 \cdot x + 1 \cdot y + 1 \cdot z + 1 \cdot w)$$

subject to

$$1 \cdot t + 0 \cdot u + 0 \cdot v + 5x + 6y + 3z + 0w = 1$$
$$0 \cdot t + 1 \cdot u + 0 \cdot v + 10x + 5y + 12z + 10w = 1$$
$$0 \cdot t + 0 \cdot u + 1 \cdot v + 10x + 0y + 5z + 20w = 1$$
$$t, \quad u, \quad v, \quad x, \quad y, \quad z, \quad w \geq 0$$

This is like problem I and we can start with the basic feasible solution

$$(t = 1, u = 1, v = 1, x = 0, y = 0, z = 0, w = 0)$$

If $C_1, C_2, C_3, \ldots, C_7$ are the columns of the above equations in its matrix form, then the initial table is given by

	C_1	C_2	C_3	C_4	C_5	C_6	C_7	p
C_1	1	0	0	5	6	3	0	1
C_2	0	1	0	10	5	12	10	1
C_3	0	0	1	10	0	5	20	1
	0	0	0	−1	−1	−1	−1	0

The last row between the ruled lines gives $\sum_{j=1}^{m} \alpha_{js} q_j - q_s$, where C_1, C_2, C_3 are our basis vectors, (α_{js}) is our given matrix $q = (0, 0, 0, 1, 1, 1, 1)$, and $p = (1, 1, 1)$. The last component zero of the last row gives the value of the objective function at the present solution.

We can bring C_7 into the basis. Since $\min (^1/_{10}, ^1/_{20}) = {}^1/_{20}$, C_3 leaves the basis. The new table is given by

	C_1	C_2	C_3	C_4	C_5	C_6	C_7	p
C_1	1	0	0	5	6	3	0	1
C_2	0	1	$-\frac{1}{2}$	5	5	$\frac{19}{2}$	0	$\frac{1}{2}$
C_7	0	0	$\frac{1}{20}$	$\frac{1}{2}$	0	$\frac{1}{4}$	1	$\frac{1}{20}$
	0	0	$\frac{1}{20}$	$-\frac{1}{2}$	−1	$-\frac{3}{4}$	0	$\frac{1}{20}$

The computation of $\Sigma \alpha_j s q_j - q_s$ with j running over the indices of the basis vectors (in this case 1, 2, 7) is easily computed and given in the last row between the ruled lines as follows.

Add suitable multiples of the third row to the last row (between the ruled lines) so that the entry corresponding to the column C_7 in this row is zero. In fact, we do the same sort of operation for this row as we did for the other rows of the matrix.

We can now bring C_5 into the basis, since the entry in the last row corresponding to C_5 is negative (here it is -1). We have $\min\left({}^1/_6,\ {}^1/_2 \cdot {}^1/_5\right) = {}^1/_2 \cdot {}^1/_5$ which occurs corresponding to C_2. Thus C_2 leaves the basis. The new table is given by

	C_1	C_2	C_3	C_4	C_5	C_6	C_7	
C_1	1	$-\frac{6}{5}$	$\frac{6}{10}$	-1	0	$-\frac{42}{5}$	0	$\frac{4}{10}$
C_5	0	$\frac{1}{5}$	$-\frac{1}{10}$	1	1	$\frac{19}{10}$	0	$\frac{1}{10}$
C_7	0	0	$\frac{1}{20}$	$\frac{1}{2}$	0	$\frac{1}{4}$	1	$\frac{1}{20}$
	0	$\frac{1}{5}$	$-\frac{1}{10}$	$\frac{1}{2}$	0	$\frac{46}{40}$	0	$\frac{3}{20}$

We throw C_1 out of the basis, and C_3 enters the basis. The table is given by

C_3	$\frac{10}{6^2}$	-2	1	$-\frac{10}{6^2}$	0	-14	0	$\frac{2}{3}$
C_5	$\frac{1}{6}$	0	0	$\frac{5}{6}$	1	$\frac{5}{10}$	0	$\frac{1}{6}$
C_7	$-\frac{1}{12}$	$\frac{1}{10}$	0	0	$\frac{7}{12}$	$\frac{19}{20}$	1	$\frac{1}{60}$
	$\frac{1}{12}$	$\frac{1}{10}$	0	$\frac{5}{12}$	0	$\frac{9}{20}$	0	$\frac{11}{60}$

Since all the entries corresponding to $C_1, C_2, \cdots, C_7$ in the last row (between ruled lines) are nonnegative, no more improvement is possible, and by theorem 4.3.1 we have reached the optimal solution. Thus we get

$$tC_1 + uC_2 + vC_3 + xC_4 + yC_5 + zC_6 + wC_7 = p$$

where C_3, C_5, and C_7 are used with positive combinations ${}^2/_3$, ${}^1/_6$, and ${}^1/_{60}$, as seen in the last column. That is, $v = {}^2/_3$, $y = {}^1/_6$, and $w = {}^1/_{60}$, and the value of the objective function is ${}^{11}/_{60}$ at this point.

By our earlier discussions, we find $\dfrac{1}{v(B)} = \dfrac{11}{60}$, where $v(B)$ is the value of the payoff matrix B.

If (ξ, η, ζ, W) is an optimal strategy for player II, then $\dfrac{\xi}{v(B)} = x$,

$\dfrac{\eta}{v(B)} = y$, $\dfrac{\zeta}{v(B)} = z$, $\dfrac{W}{v(B)} = w$ is true for one such optimal strategy. In our case the solution is

$$(t = 0, u = 0, v = \tfrac{2}{3}, x = 0, y = \tfrac{1}{6}, z = 0, w = \tfrac{1}{60})$$

for the linear programming problem. Therefore we get

$$v(B) = \tfrac{60}{11} \qquad \xi = x \cdot v(B) = 0 \qquad \eta = y \cdot v(B) = \tfrac{1}{6} \cdot \tfrac{60}{11} = \tfrac{10}{11}$$

$$\zeta = z \cdot v(B) = 0 \cdot \tfrac{60}{11} = 0 \qquad W = w \cdot v(B) = \tfrac{1}{60} \cdot \tfrac{60}{11} = \tfrac{1}{11}$$

Thus player II chooses columns 1, 2, 3, and 4 in the matrix B with probabilities 0, $^{10}/_{11}$, 0, $^{1}/_{11}$.

Correspondingly, for our main problem, player II with his noisy gun never shoots in the beginning, in the first step ($j = 1$), or in the second step ($j = 2$). He shoots in the third step with probability $^{10}/_{11}$, never shoots in the fourth step, and shoots in the fifth step with probability $^{1}/_{11}$. The value of the game is $v(B) - 5 = {}^{60}/_{11} - 5 = {}^{5}/_{11}$. Thus, $v(A) = {}^{1}/_{25} \cdot {}^{5}/_{11} = {}^{1}/_{55}$. Surprisingly, he wants to omit the fourth step optimally and decides to wait until the end with a positive probability (this is not intuitive).

We can determine the optimal strategies of the first player from the same table. The entries between ruled lines corresponding to C_1, C_2, C_3 are $^{1}/_{12}$, $^{1}/_{10}$, 0. One can check, for that first linear programming problem, their optimality for its dual. The dual is min $(a + b + c)$, subject to

$$5a + 10b + 10c \geq 1$$

$$6a + 5b + 0c \geq 1$$

$$3a + 12b + 5c \geq 1$$

$$0a + 10b + 20c \geq 1$$

$$a, \qquad b, \qquad c \geq 0$$

Here $a = {}^{1}/_{12}$, $b = {}^{1}/_{10}$, $c = 0$ satisfy all the constraints with $a + b + c = {}^{11}/_{60}$. Hence (a, b, c) is optimal for the dual. If α, β and γ are the optimal probabilities for player I for the selection of rows 1, 2, and 3, then

$$\alpha = \tfrac{1}{12} \cdot \tfrac{60}{11} = \tfrac{5}{11}, \qquad \beta = \tfrac{1}{10} \cdot \tfrac{60}{11} = \tfrac{6}{11}, \qquad \gamma = 0 \cdot \tfrac{60}{11} = 0$$

Thus player I (the silent gunner) never shoots in the beginning or in the first step, but shoots with probability $^{5}/_{11}$ in the second step and $^{6}/_{11}$ in the third step, and never shoots in the fourth or fifth steps.

The solution which we got for the dual is no fluke. It is true in general and is seen by the following argument. Consider the problem:

$$\max_{y \in T} \Sigma q_j y_j$$

where

$$T = \left\{ y = (y_1, y_2, \cdots, y_n) : y_1, \cdots, y_n \geq 0, \sum_{j=1}^{n} a_{ij} y_j = p_i, \quad i = 1, 2, \cdots, m \right\}$$

When the nondegeneracy hypothesis is true and when the problem has an optimal solution y^0, then we get this from the simplex algorithm. Let the columns $C_{r_1}, C_{r_2}, \cdots, C_{r_m}$ be used as the basis vectors in the final table, and let $C_1, C_2, \cdots, C_m$ be the basis vectors in the initial table. The initial table and the final table are:

	C_1	$C_2 \cdots C_m$	C_s	$\cdots$	C_n		p
C_1	1	0 0	α_{1s}		α_{1n}		y_1
C_2	0	1 0	α_{2s}		α_{2n}		y_2
$\cdots$							
C_m	0	0 1	α_{ms}		α_{mn}		y_m
	0	0 0	$\sum_{j=1}^{m} \alpha_{js} q_j - q_s$	$\sum_{j=1}^{m} \alpha_{jn} q_j - q_n$			$\sum_{j=1}^{m} y_j q_j$

and

	C_1	$C_2 \cdots C_m$		$C_s \cdots C_n$		
C_{r_1}	β_{11}	β_{12} β_{1m}	β_{1s}	β_{1n}		$y^0_{r_1}$
						$\vdots$
C_{r_2}						$y^0_{r_2}$
$\vdots$						$\vdots$
C_{r_m}	β_{m1}	β_{m2} β_{mn}	β_{ms}	β_{mn}		$y^0_{r_m}$
	$\zeta_1 - q_1$	$\zeta_2 - q_2$		$\zeta_n - q_n$		ζ_0

Here $\zeta_j - q_j = q_{r_1} \beta_{1j} + q_{r_2} \beta_{2j} + \cdots + q_{r_m} \beta_{mj} - q_j, \quad j = 1, 2, \cdots, n$

$$\zeta_0 = q_{r_1} y^0_{r_1} + \cdots + q_{r_m} y^0_{r_m}$$

We also know that $\zeta_j - q_j \geq 0, j = 1, 2, \cdots, n$, in the final table. If we consider the β_{ij}'s for submatrix corresponding to $(C_{r_1}, C_{r_2}, \cdots, C_{r_m})$, this would be the identity matrix in the final table. Since coefficients of the matrices are obtained only from row operations, and since the original submatrix corresponding to $(C_{r_1}, C_{r_2}, \cdots, C_{r_2} \cdots C_{r_m})$, which we shall call L is reduced to the identity matrix I in the final table purely by row operations, the matrix in the final table is $L^{-1}A$, where A is the original matrix (a_{ij}). Let L^{-1} the inverse of L be

$$L^{-1} = \begin{pmatrix} l^{11} & l^{12} \cdots l^{1m} \\ \cdots\cdots\cdots\cdots \\ \cdots\cdots\cdots\cdots \\ l^{m1} & l^{m2} \cdots l^{mm} \end{pmatrix}$$

(Note that in our case the first m columns of the final table form the L^{-1} matrix.) Thus the β_{ij} matrix is

$$\begin{bmatrix} l^{11} & l^{12}\cdots l^{1m} \\ \cdots\cdots\cdots\cdots \\ \cdots\cdots\cdots\cdots \\ l^{m1} & l^{m2}\cdots l^{mm} \end{bmatrix} \begin{bmatrix} a_{11} & a_{12}\cdots a_{1n} \\ \cdots\cdots\cdots\cdots \\ \cdots\cdots\cdots\cdots \\ a_{m1} & a_{m2} & a_{mn} \end{bmatrix}$$

Further

$$(q_{r_1}, q_{r_2}, \cdots, q_{r_m}) \begin{bmatrix} \beta_{11} & \beta_{12}\cdots\beta_{1n} \\ \cdots\cdots\cdots\cdots \\ \cdots\cdots\cdots\cdots \\ \beta_{m1} & \beta_{m2}\cdots\beta_{mn} \end{bmatrix} = (\zeta_1, \zeta_2, \cdots, \zeta_n)$$

Equivalently

$$(q_{r_1}, q_{r_2}, \cdots, q_{r_m}) \begin{bmatrix} l^{11} & l^{12}\cdots l^{1m} \\ \cdots\cdots\cdots\cdots \\ \cdots\cdots\cdots\cdots \\ l^{m1} & l^{m2}\cdots l^{mm} \end{bmatrix} \begin{bmatrix} a_{11} & a_{12}\cdots a_{1n} \\ \cdots\cdots\cdots\cdots \\ \cdots\cdots\cdots\cdots \\ a_{m1} & a_{m2}\cdots a_{mn} \end{bmatrix}$$
$$= (\zeta_1, \zeta_2, \cdots, \zeta_n)$$

If

$$(q_{r_1}, q_{r_2}, \cdots, q_{r_m}) \begin{bmatrix} l^{11} & l^{12}\cdots l^{1m} \\ \cdots\cdots\cdots\cdots \\ \cdots\cdots\cdots\cdots \\ l^{m1} & l^{m2}\cdots l^{mm} \end{bmatrix} = (x_1^0, x_2^0, \cdots, x_m^0)$$

then

$$(x_1^0, x_2^0, \cdots, x_m^0) \begin{bmatrix} a_{11} & a_{12}\cdots a_{1n} \\ \cdots\cdots\cdots\cdots \\ \cdots\cdots\cdots\cdots \\ a_{m1} & a_{m2}\cdots a_{mn} \end{bmatrix} = (\zeta_1, \zeta_2, \cdots, \zeta_n)$$

(Note that in our case the first m columns of A form the identity matrix and that $x_j^0 = \zeta_j$, $j = 1, 2, \cdots, m$.) We also know that $\zeta_j - q_j \geq 0$ for all $j = 1, 2, \cdots, n$ in the final table, and that

$$\sum_{j=1}^{m} a_{ij}x_i^0 = \zeta_j \geq q_j, \qquad j = 1, 2, \cdots, n$$

Thus $(x_1^0, x_2^0, \cdots, x_m^0)$ is a feasible solution to the dual. If

$$L = \begin{pmatrix} l_{11} & l_{12} \cdots l_{1m} \\ \cdots\cdots\cdots\cdots \\ l_{m1} & l_{m2} \cdots l_{mm} \end{pmatrix}$$

is the matrix formed by columns $(C_{r_1}, C_{r_2}, \cdots, C_{r_m})$, then we know that

$$\sum_{i=1}^{m} a_{ij}x_i^0 = \zeta_j = q_j, j = r_1, r_2, \cdots, r_m,$$ in the final table. Thus

$$\sum_i a_{ij}x_i^0 = q_j, \qquad j = r_1, r_2, \cdots, r_m$$

and

$$\sum_i a_{ij}x_j^0 \geq q_j \qquad \text{otherwise}$$

Since $y_{r_1}^0, y_{r_2}^0, \cdots, y_{r_m}^0$ alone are positive for the original problem of maximization,

$$\sum_{i=1}^{m}\sum_{j=1}^{n} a_{ij}x_i^0 y_j^0 = \sum_{j=1}^{m} y_{r_j}^0 \sum_{i=1}^{m} a_{ij}x_i^0 = \sum_{j=1}^{m} q_{r_j} y_{r_j}^0$$

Further,

$$\sum_{j=1}^{n} a_{ij}y_j^0 = p_i, \qquad i = 1, 2, \cdots, m, \qquad \text{and} \qquad \sum_i \sum_j a_{ij}x_i^0 y_j^0 = \sum_i p_i x_i^0$$

Thus x^0 is optimal for the dual. In game theoretic cases the slack variables are introduced, and they carry 0 weight in the objective function. Thus $\zeta_1 - q_1 = \zeta_1 \geq 0 \cdots \zeta_m - q_m = \zeta_m \geq 0$; that is, $x_1^0, x_2^0, \cdots, x_m^0 \geq 0$.

4.4 NONLINEAR PROGRAMMING

In this section we deviate from our main track to suggest a minimax theorem that occurs in a natural fashion in a nonlinear programming problem. We end up with an iterative procedure for solving a class of nonlinear optimization problems with linear constraints.

Suppose that a concave function is to be maximized subject to satisfying constraints imposed by n concave functions. Then this problem can be thought of as a game problem. This is contained in the following celebrated theorem of Kuhn and Tucker.

Theorem 4.4.1

Let $g, h_1, h_2, \cdots, h_k, h_{k+1}, \cdots, h_n$ be concave functions on R^m. Let $h_{k+1}, \ldots, h_n$ alone be linear affine functions among the h_i's (that is, the h_i's are of the form $h_i(x) = (a_i, x) + c_i, i = k + 1, k + 2, \cdots, n$). If there exists an $x^* \in R^m$ with $h_i(x^*) > 0$, $i = 1, 2, \cdots, k$, and $h_i(x^*) \geq 0$, $i = k + 1, \cdots, n$, then the following three problems are equivalent.

PROBLEM I
$$\max_{x \in T} g(x)$$

where $T = \{x: h_i(x) \geq 0, i = 1, 2, \cdots, n, x \in R^m\}$. Here T is convex by theorem 1.10.1.

PROBLEM II: Find an $x \in T$ above such that for any $u \in R^m$ there exists a constant $r_0 > 0$, and when $0 \leq r \leq r_0$, $x + ru \in T$ implies $g(x + ru) \leq g(x)$.

PROBLEM III: Find a saddle point for the function

$$H(x, y) = g(x) + \sum_{i=1}^{n} y_i h_i(x)$$

where $x \in R^m$, and $y = (y_1, \cdots, y_n)$ satisfies $y_1, y_2, \cdots, y_n \geq 0$.

Remark 4.1.1: Problem III is the game problem (no doubt an infinite zero-sum game). Player I chooses a point $x \in R^m$. Player II chooses a point $y = (y_1, y_2, \cdots, y_n)$ in the positive orthant of R^n, and player I receives $H(x, y)$ from player II. The theorem asserts that the players have optimal pure strategies x^0, y^0 when problem I has a solution. Here both the pure strategy spaces are noncompact, but $H(x, y)$ is concave in x and linear in y. Now we shall prove the theorem.

PROOF: *Problem I ⇔ Problem II*

Let x^0 be a solution to problem I. Trivially it is a solution to problem II. Conversely, let x^0 be a solution to problem II. Let $x \in T$. For any $0 < r < 1$,

$$g(rx + (1 - r) x_0) \geq rg(x) + (1 - r) g(x_0) \text{ (by the concavity of } g)$$

Further $rx + (1 - r) x_0 = x_0 + r(x - x_0)$. If $u = (x - x_0)$, then ·by assumption there exists $r_0 > 0$ such that, when $0 \leq r \leq r_0$ and $x_0 + ru \in T$, we have $g(x_0 + ru) \leq g(x_0)$. Choosing $0 < r < 1$ suitably small, we have

$$g(x_0) \geq g(x_0 + ru) = g(rx + (1 - r)x_0) \geq rg(x) + (1 - r) g(x_0)$$

Thus $rg(x) \leq rg(x_0)$, and $g(x) \leq g(x_0)$ for any $x \in T$. Hence x^0 is a solution to problem I.

Problem I ⇔ Problem III

Let (x^0, y^0) be a saddle point to $H(x, y)$. Then

$$H(x, y^0) \leq H(x^0, y^0) \leq H(x^0, y) \qquad \text{for all } x \in R^m$$

$$y = (y_1, y_2, \cdots, y_n)$$

$$y_1 \geq 0, \cdots, y_n \geq 0$$

Here

$$y^0 = (y_1^0, y_2^0, \cdots, y_n^0) \qquad y_1^0, y_2^0, \cdots, y_n^0 \geq 0$$

We have $g(x) + \Sigma y_i^0 h_i(x) \leq g(x^0) + \Sigma y_i^0 h_i(x^0) \leq g(x^0) + \Sigma y_i h_i(x^0)$. Putting $y = 0$, we get

$$g(x) + \Sigma y_i^0 h_i(x) \leq g(x^0)$$

Since $h_i(x) \geq 0$, $i = 1, 2, \cdots, n$, for $x \in T$, by the nonnegativity of $y_1^0, y_2^0, \cdots, y_n^0$ we have

$$g(x) \leq g(x^0) \text{ for } x \in T$$

To show that $x^0 \in T$, we should prove that $h_i(x^0) \geq 0$, $i = 1, 2, \cdots, n$. Suppose that some $h_s(x^0) < 0$ for some $1 \leq s \leq n$. Let $y = (y_1, y_2, \cdots, y_n)$, and let y_s be sufficiently large so that

$$g(x^0) + \Sigma y_i^0 h_i(x^0) \leq g(x^0) + \Sigma y_i h_i(x^0)$$

is contradicted. Thus $x^0 \in T$ and is a solution to problem I.

Conversely, let x^0 be a solution to problem I. Since we know that T contains x^* with

$$h_i(x^*) > 0, \qquad i = 1, 2, \cdots, k$$

$$h_i(x^*) \geq 0, \qquad i = k + 1, \cdots, n$$

and

$$g(x^0) - g(x) \geq 0 \qquad \text{for all } x \in T$$

we have no solution $x \in R^m$ satisfying

$$g(x) - g(x^0) > 0$$

$$h_i(x) \geq 0, \qquad i = 1, 2, \cdots, n$$

but we have $x = x^*$, satisfying

$$h_i(x) > 0, \qquad i = 1, 2, \cdots, k$$

$$\geq 0, \qquad i = k + 1, k + 2, \cdots, n$$

By theorem 1.10.6 of Minkowski-Farkas, we have $y_1^0, y_2^0, \cdots, y_n^0 \geq 0$, such that

$$g(x) - g(x^0) + \Sigma y_i^0 h_i(x) \leq 0 \qquad \text{for all } x \in R^m$$

That is,

$$H(x, y^0) \leq g(x^0) \leq g(x^0) + \Sigma y_i h_i(x^0) = H(x^0, y)$$

for any $y = (y_1, \cdots, y_n)$, $y_1, y_2, \cdots, y_n \geq 0$. Putting $x = x^0$, $y = y^0$, we get $H(x^0, y^0) = g(x^0)$. Thus $H(x, y^0) \leq H(x^0, y^0) \leq H(x^0, y)$ for all $x \in R^m$, and $y = (y_1, y_2, \cdots, y_n)$, $y_1, y_2, \cdots, y_n \geq 0$.

Thus problem I $\Rightarrow$ problem III, and the theorem is complete.

4.5 MAXIMA OF CONCAVE FUNCTIONS

Definition 4.5.1: A function $g(x)$, $x \in R^m$, is differentiable at x_0 if, for any $x_0 \in R^m$, there exists a vector, $\nabla g(x_0)$, such that

$$\underset{\|h\| \to 0}{lt} \frac{g(x_0 + h) - g(x_0) - (\nabla g(x_0), h)}{\|h\|} = 0$$

If a function $g(x)$ has continuous partial derivatives in each of the variables, by the mean-value theorem for several variables one can easily show that the function is differentiable and

$$\nabla g(x_0) = \left(\frac{\partial g}{\partial x_1}, \frac{\partial g}{\partial x_2}, \cdots, \frac{\partial g}{\partial x_m} \right), \quad \text{at} \quad x^0 = (x_1^0, x_2^0, \cdots, x_m^0)$$

We shall give an iterative technique for solving a class of nonlinear programming problems in this section. We need the following theorem in the sequel.

Theorem 4.5.1

Let $g(x)$ be a concave function in R^m with continuous partial derivatives in each of its variables. Then for any $y, z \in R^m$,

$$(\nabla g(z), y - z) \geq g(y) - g(z)$$

PROOF: The function $\psi(\lambda) = g(\lambda y + (1 - \lambda) z)$ is concave in λ in $0 \leq \lambda \leq 1$ for fixed y, z. Let $\lambda y + (1 - \lambda) z = z + \lambda(y - z) = x_0$. For any h real,

$$\psi(\lambda + h) - \psi(\lambda) = g(z + (\lambda + h)(y - z)) - g(z + \lambda(y - z))$$
$$= g(x_0 + h(y - z)) - g(x_0)$$

Since g has continuous partial derivatives, g is differentiable, and

$$\underset{h \to 0}{lt} \frac{g(x_0 + h(y - z)) - g(x_0) - (\nabla g(x_0), h(y - z))}{\|h(y - z)\|} = 0$$

That is,

$$g(x_0 + h(y - z)) - g(x_0) = h(\nabla g(x_0), y - z) + 0(h)$$

Thus

$$\psi'(\lambda) = (\nabla g(x_0), y - z)$$

Since $\psi(\lambda)$ is concave in λ in $0 \leq \lambda \leq 1$,

$$\psi(\lambda) = \psi(\lambda \cdot 1 + (1 - \lambda) 0) \geq \psi(0) + \lambda(\psi(1) - \psi(0))$$

Thus

$$\underset{\lambda \to 0^+}{lt} \frac{\psi(\lambda) - \psi(0)}{\lambda} \geq \psi(1) - \psi(0)$$

That is,

$$\psi'(0) \geq \psi(1) - \psi(0)$$

Similarly,

$$\psi'(1) \leq \psi(1) - \psi(0)$$

Thus,

$$\psi'(1) \leq \psi(1) - \psi(0) \leq \psi'(0)$$

Substituting for $\psi'(1)$, $\psi'(0)$ from the expression we have derived earlier, we have

$$\psi'(0) = (\nabla g(z), y - z) \geq \psi'(1) = (\nabla g(y), y - z)$$

and

$$\psi'(0) = (\nabla g(z), y - z) \geq \psi(1) - \psi(0) = g(y) - g(z)$$

Hence the theorem.

The algorithm: Let $g(x)$ be a concave function in R^m to be maximized over a convex set formed by finite linear inequalities of the variables. Let g possess continuous partial derivatives at all x, and further let $(\nabla g(x), y)$ be bounded for every x, y satisfying the inequalities (that is, it is bounded on the domain under consideration). Then the following iterative solution converges to the optimal solution.

Iteration procedure: Fix a $\lambda > 0$ to start with. Let us take any x_1 satisfying the linear inequality constraints. Consider the linear function $(\nabla g(x_1), z)$ in z. This has an optimal solution when z ranges over the linear-inequality constraints (this follows from the assumption that $(\nabla g(x_1), z)$ is bounded on our restricted domain). We have an extreme point which is an optimal solution, and this can be found even by the simplex method. Let z_1 be one such solution obtained. Let x_2 be chosen on the segment $[x_1, z_1]$ (that is, $x_2 = \alpha x_1 + (1 - \alpha) z$, for some $0 \leq \alpha \leq 1$), such that

$$g(x_2) \geq (1 - \lambda) g(x_1) + \lambda \max_{[x_1, z_1]} g(x)$$

Consider $(\nabla g(x_2), z)$ and get an extreme point, z_2, that maximizes this linear function over the domain. Let $x_3 = \beta x_2 + (1 - \beta) z_2$ such that

$$g(x_3) \geq (1 - \lambda) g(x_2) + \lambda \max_{[x_2, z_2]} g(x)$$

The points $x_1, x_2, \cdots$ belong to the convex set generated by x_1 and the extreme points of the convex domain. Since the domain is formed by a finite number of linear inequalities, the number of extreme points is finite, and the $x_1, x_2, \cdots$ lie in a closed bounded convex set. Further, $g(x_n)$, $n = 1, 2, \cdots$, is nondecreasing, and bounded by the continuity of g and by the closed boundedness of this convex set. Hence $g(x_n) \uparrow M$.

We shall prove that $M = \max_T g(x)$ over the constraint set T under consideration. The sequence x_n is bounded and hence has a convergent subsequence. Let $x_{n_j} \to x_*$. Then $x_* \in T$. Clearly $g(x_{n_j}) \to g(x_*) = M$. Now $\max (\nabla g(x_{n_j}, z)$ is attained at an extreme point of T for each j. Since there are only finitely many extreme points for T, we have an extreme point $z = z_0$ at which $(\nabla g(x_{n_j}, z)$ is maximum over T for infinitely many j. Without loss of generality, let it be so for the chosen subsequence $\{x_{n_j}\}$. We have, for any $0 \leq \alpha \leq 1$,

$$g(\alpha z^0 + (1 - \alpha) x_{n_j}) - g(x_{n_j}) = g(x_{n_j} + \alpha(z^0 - x_{n_j})) - g(x_{n_j})$$

$$\leq \frac{g(x_{n_j+1}) - g(x_{n_j})}{\lambda}$$

In the limit

$$g(x_* + \alpha(z^0 - x_*)) - g(x_*) \leq 0$$

But

$$0 \geq \frac{g(x_* + \alpha(z^0 - x_*)) - g(x_*)}{\alpha} = (\nabla g(x_*), z^0 - x_*) + 0(\alpha)$$

and as $\alpha \to 0^+$, we get

$$(\nabla g(x_*), z^0 - x_*) \leq 0$$

Now for any $x \in T$

$$(\nabla g(x_{n_j}), x - x_{n_j}) \leq (\nabla g(x_{n_j}), z^0 - x_{n_j}) \text{ (by the optimality of } z^0)$$

in the limit

$$(\nabla g(x_*), x - x_*) \leq (\nabla g(x_*), z^0 - x_*) \leq 0$$

But by theorem 4.5.1 above

$$g(x) - g(x_*) \leq (\nabla g(x_*) x - x_*) \leq 0$$

Thus $g(x) \leq g(x_*)$ for any $x \in T$. The proof is complete.

REFERENCES

[1] Berge, C. (1963). "Topological Spaces—Vector Spaces and Convexity," Oliver & Boyd Co., Edinburgh and London.
[2] Dantzig, G. B. (1963). "Linear Programming and Extensions," Princeton University Press, Princeton, New Jersey.
[3] Gale, D. (1960). "Theory of Linear Economic Models," McGraw-Hill Book Company, New York.
[4] Luce, R. D. and Raiffe, H. (1957). "Games and Decisions," John Wiley & Sons, New York.

Chapter Five

GENERAL MINIMAX THEOREMS

It was John von Neumann [14] who proved the well-known minimax theorem for finite games. His theorem can be stated mathematically as follows. If X and Y are compact convex subsets of R^m and R^n, respectively, and if K is a bilinear function on $X \times Y$, then K has a saddle point. That is, $\max_X \min_Y K(x, y) = \min_Y \max_X K(x, y)$. Ville and Wald have generalized von Neumann's result when X and Y are allowed to be certain infinite dimensional linear spaces and K is assumed to be bilinear. It was Kneser [5], Ky Fan [6], and Berge [1] who proved minimax theorems for concave–convex functions when one of the spaces is compact or conditionally compact in a suitable topology. Sion [10] has proved minimax theorems for quasi concave–convex functions that are appropriately semi-continuous in the variables. Very recently Teh-Tjoe Tie [12] has proved a general minimax theorem for functions that are concave–convex-like.

In this chapter we shall prove minimax theorems one of which will be a generalization of Wald's theorem while the rest will be general minimax theorems that are extensions of Teh-Tjoe Tie and Ky Fan. The extensions are apparently known, but so far they have not been explicitly stated and proved. In all the theorems, compactness or conditional compactness of one of the spaces has played a crucial role. More precisely, the assumption of compactness or conditional compactness enables one to reduce the problem to a finite dimensional case under appropriate continuity assumptions on the kernel $K(x, y)$ which in general will be unbounded. Finally, we shall prove a minimax theorem on the unit square. In short, we shall prove the following theorems in this chapter. We will also give a characterization of games with finite optimals and characterize extreme optimals of certain continuous games (5.5, 5.6).

Theorem 5.2.1

Let $K(x, y)$ be a real valued function (not necessarily bounded) defined on $X \times Y$ such that X is conditionally compact in the semi-intrinsic topology. Further, suppose that $K(x, y)$ is bounded in y for every fixed x, and let

$$K(\mu, \lambda) = \iint K(x, y) \, d\mu(x) \, d\lambda(y) = \iint K(x, y) \, d\lambda(y) \, d\mu(x)$$

112

for all $\mu \in m_X^S$ and $\lambda \in m_Y^S$. Then the game is strictly determined. That is,

$$\inf_{m_Y^S} \sup_{m_X^S} K(\mu,\lambda) = \sup_{m_X^S} \inf_{m_Y^S} K(\mu, \lambda)$$

Theorem 5.3.2

If X is conditionally compact in the (S) topology and $K(x, y)$ is concave–convex-like, then

$$\inf_Y \sup_X K(x, y) = \sup_X \inf_Y K(x, y)$$

Theorem 5.3.3

Let $K(x, y)$ be defined over $X \times Y$. Suppose that X is (S)-conditionally compact. Then the following conditions are equivalent:

(1) Given any $\varepsilon > 0$, and any finite subsets A and B of X and Y, respectively, there exists $x_0 \in X$, $y_0 \in Y$, such that $K(x_i, y_0) \leq K(x_0, y_j) + \varepsilon$ for all $x_i \in A$ and $y_j \in B$.

(2) $\sup_X \inf_Y K(x, y) = \inf_Y \sup_X (x, y)$.

(3) $\inf_Y \max_A K(x, y) \leq \sup_X \inf_Y K(x, y)$ for every finite set $A \subset X$.

(4) $\sup_X \min_B K(x, y) \geq \inf_Y \sup_X K(x, y)$ for every finite set $B \subset Y$.

Theorem 5.3.5

Let X be a compact space, and let $K(x, y)$ be upper semicontinuous in x for every fixed y. Then the following statements are equivalent:

(1) $\inf_Y \sup_X K(x, y) = \sup_X \inf_Y K(x, y)$.

(2a) Given any $\varepsilon > 0$, any finite subsets A and B of X and Y, respectively, there exists x_0, y_0 such that $K(x_i, y_0) \leq K(x_0, y_j) + \varepsilon$ for all $x_i \in A$, $y_j \in B$.

(2b) For any $C < \inf_Y \sup_X K(x, y)$ there exists a finite set A of X such that for every y there exists an $x \in A$ with $K(x, y) \geq C$.

Theorem 5.4.1

Let $K(x, y)$ be bounded on the unit square $0 \leq x, y \leq 1$, and let all points of discontinuity lie on a finite number of curves of the form $y = \varphi_K(x)$, $K = 1, 2, \cdots, n$, where $\varphi_K(x)$ are continuous functions. Further, suppose that

$$K(\mu, \lambda) = \int_0^1 \int_0^1 K(x, y) \, d\mu((x) \, d\lambda(y) = \int_0^1 \int_0^1 K(x, y) \, d\lambda(y) \, d\mu(x)$$

Then

$$\max_{m_X} \inf_{A_Y} K(\mu, \lambda) = \inf_{A_Y} \max_{m_X} K(\mu, \lambda)$$

For the sake of completeness we shall give an elegant proof of von Neumann's theorem. This proof is due to Kneser.

PROOF OF VON NEUMANN'S THEOREM

Let $v = \min_{B} \max_{A} K(x, y)$. Without loss of generality assume $v = 0$. If possible, let

$$\max_{A} \min_{B} K(x, y) < 0$$

Since A is compact and $K(x, y)$ is continuous, it follows that there exists a finite $[y_1, y_2, \cdots, y_s]$ such that for any x one y_i can be chosen with $K(x, y_i) < 0$. In other words, $\min_{1 \leq i \leq s} K(x, y_i) < 0$ for all $x \in A$.

From theorem 1.10.5 we have a probability vector $\lambda = (\lambda_1, \lambda_2, \cdots, \lambda_s)$, $\Sigma \lambda_i K(x, y_i) < 0$ for all x. That is, $K(x, y^1) < 0$ for all x where $y^1 = \sum_i \lambda_i y_i$.

Therefore $\min_{B} \max_{A} K(x, y) < 0$, which contradicts our assumption. This completes the proof of the theorem.

5.1 GAMES OF STRATEGY

If X and Y are nonempty sets and if $K(x, y)$ is a finite valued function on the Cartesian product $X \times Y$, then the triple (X, Y, K) is called a game. In a game (X, Y, K) the elements of X are called pure strategies of player 1, and the elements of Y are called pure strategies of player 2. The pairs (x, y), where $x \in X$, $y \in Y$, will be called plays. For $x \in X$, $y \in Y$, the number $K(x, y)$ is the payoff in the play (x, y).

Let $K(x, y)$ be bounded. We shall now define the neighborhood systems for every point x belonging to X.

$$I_{x_0, \varepsilon} = \left\{ x \colon \sup_{Y} \left| K(x, y) - K(x_0, y) \right| < \varepsilon \right\}$$

where $\varepsilon > 0$ and $x_0 \in X$. This class of neighborhood will form a base that will induce a topology which we shall call the intrinsic topology or simply I topology for X. Actually this topology is induced by the following pseudo metric.

$$d(x_1, x_2) = \sup_{Y} \left| K(x_1, y) - K(x_2, y) \right|$$

But one can actually convert this into a metric by defining X^1 as the class of all sets of the form

$$\bigcap_{y \in Y} \{x: K(x, y) = K(x_0, y)\}, \ x_0 \in X$$

and K^1 (x^1, y) as the function on $X^1 \times Y$ satisfying $K^1(x^1, y) = K(x, y)$ for $x \in x^1$.

Let B_X^I and m_X^I denote the smallest σ-field containing I open sets of X and the class of all probability distributions on X, respectively. Similarly B_Y^I and m_Y^I can be defined for Y. We shall assume without loss of generality that B_X^I and B_Y^I include pure strategies. Elements of $m_X^I(m_Y^I)$ are called mixed strategies.

Definition 5.1.1: X is said to be (I)-conditionally compact if, for every $\varepsilon > 0$, there exists a finite $(x_1, x_2, \cdots, x_n) \subset X$ such that $\overset{n}{\underset{i=1}{\cup}} I_{x_i, \varepsilon} = X$.

Remark 5.1.1: X is (I)-conditionally compact if and only if Y is (I)-conditionally compact. For a proof, refer to Wald [15, 17]. We shall now state the theorem of Wald without proof.

Theorem 5.1.1

Let X be (I)-conditionally compact and let $K(\mu, \lambda) = \iint K(x, y) \, d\mu(x) \, d\lambda(y)$, where $\mu \in m_X^I$ and $\lambda \in m_Y^I$. Then

$$\sup_{m_X^I} \inf_{m_Y^I} K(\mu, \lambda) = \inf_{m_Y^I} \sup_{m_X^I} K(\mu, \lambda)$$

In the next section we shall prove a generalization of theorem 5.1.1 by weakening the topological assumptions.

5.2 GENERALIZATION OF THEOREM 5.1.1

We are now going to introduce topologies for X and Y that will lead to more general theorems than theorem 5.1.1. Let $K(x, y)$ be a not necessarily bounded real valued function defined on $X \times Y$. Then the class S of the subsets

$$S(x_0, \varepsilon) = \left\{ x: \sup_Y [K(x, y) - K(x_0, y)] < \varepsilon \right\}, \ \varepsilon > 0, x_0 \in X$$

is a base for a topology for X. Similarly,

$$S(y_0, \varepsilon) = \left\{ y: \sup_X [K(x, y_0) - K(x, y)] < \varepsilon \right\}, \ \varepsilon > 0, y_0 \in Y$$

is a base for a topology for Y. We shall refer to these topologies as the semi-intrinsic topologies, or simply the S topology. This was first considered by Teh-Tjoe Tie [12].

Definition 5.2.1: X is said to be (S)-conditionally compact if and only if, given any $\varepsilon > 0$, there exists a finite set $[x_1, x_2, \cdots, x_n]$ such that
$$\bigcup_{i=1}^{n} S(x_i, \varepsilon) = X.$$

Remark 5.2.1: Every S-open set of X is I-open, and consequently every (I)-conditionally compact set is (S)-conditionally compact. However, (S)-conditional compactness need not imply (I)-conditional compactness. This is evident from the following example. Let X be the set of positive integers and Y the class of all subsets of X. Define $K(x, y) = -1$ if $x \in y$, and $K(x, y) = 1$ if $x \notin y$. It is readily seen that the Y-space is (S)-conditionally compact but is not (I)-conditionally compact. Also, it is evident from this example that (S)-conditional compactness of one space need not imply the (S)-conditional compactness of the other space, although this is true in the I topology. Let $\mathscr{B}_X^S$ denote the smallest σ-field containing S-open sets of X and the pure strategies. Let m_X^S stand for the class of all probability distributions on X; $\mathscr{B}_Y^S$ and m_Y^S are defined similarly. Let $\mathscr{B}$ be the smallest σ-algebra containing the rectangles $C \times D$, $C \in \mathscr{B}_X^S$, $D \in \mathscr{B}_Y^S$. If X is (I)-separable, then $\mathscr{B}_X^S = \mathscr{B}_Y^S$. We shall now prove the following known lemma [12], which is needed in the sequel.

LEMMA 5.2.1: If X is (I)-separable, then $K(x, y)$ is (S)-measurable.

PROOF: We define:

$$Z = \{(x, y): K(x, y) > a\}, a \text{ real}$$

$$C(x_1, r) = \{x: d(x_1, x) < r\}, r \text{ positive rational}$$

$$X_0 = \{x: K(x, y) > a \text{ for at least one } y \in Y\}$$

Since $K(x, y)$ is (I)-continuous for each fixed y, the set X_0 is I-open. Let X_1 be any fixed countable dense subset in X_0, and let (x_0, y_1) be a point of Z. Then there exists a point $x_1 \in X_1$ and a positive rational r such that

$$d(x_0, x_1) < r \quad \text{and} \quad r < K(x_1, y_1) - a$$

Let ε be a positive number such that $0 < r + \varepsilon < K(x_1, y_1) - a$. Then if (x, y) is a point of the rectangle $C(x_1, r) \times S(y_1, \varepsilon)$, we have

$$K(x_1, y_1) - K(x, y) \le | K(x_1, y_1) - K(x, y_1) | +$$
$$+ [K(x, y_1) - K(x, y)] < r + \varepsilon < K(x_1, y_1) - a$$

It follows that $K(x, y) > a$, or $C(x_1, r) \times S(y_1, \varepsilon)$ is a subset of Z. If $Y(x_1, r)$ is defined to be the set

$$Y(x_1, r) = \cup \{S(y, \varepsilon): y \in Y, \varepsilon > 0, \quad \text{and} \quad 0 < r + \varepsilon < K(x_1, y) - a\}$$

then it has been proved that for any point $(x, y) \in Z$ there exists a point $x_1 \in X_1$ and a rational $r > 0$ such that

$$(x, y) \in C(x_1, r) \times Y(x_1, r) \subset Z$$

Since $C(x_1, r) \in \mathscr{B}_X^I$ and the class of rectangles $C(x_1, r) \times Y(x_1 r)$ is countable, the set Z is S-measurable. We shall now state and prove the following theorem [8].

Theorem 5.2.1

Let $K(x, y)$ be a real valued function (not necessarily bounded) defined on $X \times Y$ such that X is (S)-conditionally compact. Further, suppose that $K(x, y)$ is bounded in y for every fixed x, and

$$K(\mu, \lambda) = \iint K(x, y) \, d\mu(x) \, d\lambda(y) = \iint K(x, y) \, d\lambda(y) \, d\mu(x)$$

for all $\mu \in m_X^S$, $\lambda \in m_Y^S$; then

$$\sup_{m_X^S} \inf_{m_Y^S} K(\mu, \lambda) = \inf_{m_Y^S} \sup_{m_X^S} K(\mu, \lambda)$$

PROOF: Given any $\varepsilon > 0$, since X is (S)-conditionally compact, there is a finite set A_ε such that for every x there exists an $x_i \in A_\varepsilon$ with

$$K(x, y) < K(x_i, y) + \varepsilon \text{ for all } y$$

Hence it follows that

$$\inf_{m_Y^S} \sup_{m_Y^S} K(\mu, \lambda) \le \inf_{m_Y^S} \max_{A_\varepsilon} K(x_i, \lambda) + \varepsilon$$

Write

$$v = \inf_{m_Y^S} \max_{A_\varepsilon} K(x_i, \lambda)$$

Now we shall establish that there exists a probability vector

$$p = (\xi_1, \xi_2, \cdots, \xi_n)$$

such that

$$\sum_i^n \xi_i K(x_i, \lambda) \ge v \quad \text{for all} \quad \lambda$$

Suppose that it is not true; it means that for every p there exists a λ such that

$$\Sigma \xi_i K(x_i, \lambda) < v$$

Since the set of probability vectors is compact, we can find a finite number of $\lambda_1, \lambda_2, \cdots, \lambda_m$ such that for every p there exists one λ_j with $\sum_i \xi_i K(x_i, \lambda_j) < v$.

In other words, for every p there exists some probability vector

$$q = (\eta_1, \eta_2, \cdots, \eta_m) \quad \text{with} \quad \sum_i \sum_j \xi_i \eta_j K(x_i, \lambda_j) < v$$

Let $A(p, q) = \sum_i \sum_j \xi_i \eta_j K(x_i, \lambda_j)$. It follows that $\max_p \min_q A(p, q) < v$. By von Neumann's theorem we have

$$\min_q \max_p A(p, q) < v$$

That is, $\sum_i \sum_j \eta_j^0 \xi_i K(x_i, \lambda_j) < v$ for all p and for some η^0. This means that $\Sigma \xi_i K(x_i, \lambda^0) < v$ for all p where $\lambda^0 = \Sigma \eta_j^0 \lambda_j$. Therefore, $K(x_i, \lambda^0) < v$ for $i = 1, 2, \cdots, n$. It follows that

$$\inf_{{}_m S_Y} \max_{A_\varepsilon} K(x_i, \lambda) < v$$

which is impossible. Hence, for some p,

$$\sum_i \xi_i K(x_i, \lambda) \geq v \text{ for all } \lambda$$

From this it follows that

$$\sup_{{}_m S_X} \inf_{{}_m S_Y} K(\mu, \lambda) \geq v$$

which in turn implies the conclusion of theorem 5.2.1. We shall now quote the result due to Teh-Tjoe Tie.

Theorem 5.2.2

Let $K(x, y)$ be a real valued function on $X \times Y$ such that X and Y are both (S)-conditionally compact. Further, suppose that

$$K(\mu, \lambda) = \iint K(x, y) \, d\mu(x) \, d\lambda(y) = \iint K(x, y) \, d\lambda(y) \, d\mu(x)$$

Then

$$\sup_{{}_m S_X} \inf_{{}_m S_Y} K(\mu, \lambda) = \inf_{{}_m S_Y} \sup_{{}_m S_X} K(\mu, \lambda)$$

We shall now give an example to show that theorem 5.2.1 is not included in theorem 5.2.2. In other words, theorem 5.2.1 is logically independent of theorem 5.2.2.

Example 5.2.1: Let X be the space of positive integers and Y the class of all subsets of X. Define $K(x, y) = -1$ if $x \in y$ and $K(x, y) = 1$ if $x \notin y$. It is readily seen that Y is (S)-conditionally compact while X is not. As X is

I-separable, from lemma 5.2.1 it follows that $K(x, y)$ is (S)-measurable. Further, $K(x, y)$ is bounded, and hence

$$\iint K(x, y)\, d\mu(x)\, d\lambda(y) = \iint K(x, y)\, d\lambda(y)\, d\mu(x)$$

All the conditions of theorem 5.2.1 are satisfied, and it is quickly seen that sup inf $K(\mu, \lambda) = $ inf sup $K(\mu, \lambda) = -1$.
Theorem 5.2.2 cannot be applied here, as X is not (S)-conditionally compact.

Remark 5.2.2: It is not known whether the conclusion of theorem 5.2.1 remains true if one omits the assumption—namely, that $K(x, y)$ is bounded in y for every fixed x.

5.3 GENERAL MINIMAX THEOREMS

Definition 5.3.1: $K(x, y)$ is upper semicontinuous in x for every fixed y if $\{x: K(x, y) < r\}$ is open for every real number r. $K(x, y)$ is lower semicontinuous in y for every fixed x if, for every real number r, the set $\{y: K(x, y) > r\}$ is open.

Definition 5.3.2: $K(x, y)$ is said to be concave in x if, for any $t \in [0, 1]$ and any $x_1, x_2 \in X$, there exists an element $tx_1 + (1 - t) x_2 \in X$ such that

$$K(tx_1 + (1 - t) x_2, y) \geq tK(x_1, y) + (1 - t) K(x_2, y) \text{ for every } y$$

$K(x, y)$ is said to be convex in y if, for any $t \in [0, 1]$ and any $y_1, y_2 \in Y$, there exists an element $ty_1 + (1 - t) y_2 \in Y$ such that

$$K(x, ty_1 + (1 - t) y_2) \leq tK(x, y_1) + (1 - t) K(x, y_2) \text{ for every } x$$

This definition is meaningful only if the spaces X and Y have linear structures. We shall now generalize this concept so that it may be valid for spaces that do not possess any linear structure. This was first done by Ky Fan.

Definition 5.3.3: A real valued function $K(x, y)$ is said to be concave-like in x (convex-like in y) if, for any $t \in [0, 1]$ and any two $x_1, x_2 \in X (y_1, y_2 \in Y)$, there exists an $x_0 \in X (y_0 \in Y)$ such that

$$tK(x_1, y) + (1 - t) K(x_2, y) \leq K(x_0, y) \text{ for all } y$$

$$(tK(x, y_1) + (1 - t) K(x, y_2) \geq K(x, y_0) \text{ for all } x)$$

$K(x, y)$ is concave–convex-like if it is concave-like in x and convex-like in y.
A concept that is equivalent to I-conditional compactness is that of almost periodic functions defined as follows [6]. A real valued bounded function $K(x, y)$ defined on the product of $X \times Y$ is left almost periodic if,

given any $\varepsilon > 0$, there exists a finite set $(x_1, x_2, \cdots, x_n)$ of X such that for any $x \in X$ there is some x_i for which $|K(x, y) - K(x_i, y)| < \varepsilon$ for all y. It is not hard to check that left almost periodicity of K implies and is implied by right almost periodicity (which can be defined in an obvious way). Ky Fan has proved the following theorem [6].

Theorem 5.3.1

If $K(x, y)$ is left almost periodic and concave–convex-like, then $\sup_X \inf_Y K(x, y) = \inf_Y \sup_X K(x, y)$. We shall now prove the following theorem [7].

Theorem 5.3.2

Let $K(x, y)$ be a real valued (not necessarily bounded) function defined over $X \times Y$. Suppose that X is (S)-conditionally compact and K is concave–convex-like. Then $\sup_X \inf_Y K(x, y) = \inf_Y \sup_X K(x, y)$.

Remark 5.3.1: Theorem 5.3.2 includes theorem 5.3.1 as a special case. It is possible to give a direct proof of theorem 5.3.2 without resorting to von Neumann's theorem, but we shall not be doing it.

PROOF OF THEOREM 5.3.2: Since X is conditionally compact in the S topology, given any $\varepsilon > 0$, there exists a finite set A_ε such that, for any x, one can find an $x_i \in A_\varepsilon$ with

$$K(x, y) \le K(x_i, y) + \varepsilon \text{ for all } y$$

Therefore we have

$$\inf_Y \sup_X K \le \inf_Y \max_{A_\varepsilon} K + \varepsilon$$

Now we observe the following. If A and B are any two finite subsets of X and Y, respectively, and if K is concave–convex-like, then there exists $x_0 \in X$, $y_0 \in Y$, such that $K(x_i, y_0) \le K(x_0, y_j)$ for all $x_i \in A$, $y_j \in B$. This observation is a consequence of von Neumann's theorem. In other words, if $A \subseteq X$, $B \subseteq Y$ are finite, then

$$\inf_Y \max_A K \le \sup_X \min_B K$$

Hence it follows that

$$\inf_Y \sup_X K \le \inf_{\mathscr{B}} \sup_X \min_B K + \varepsilon$$

where $\mathscr{B}$ denotes the class of all finite subsets of Y. Throughout we fix the ε that we have chosen already. Now we shall prove that

$$\inf_{\mathscr{B}} \sup_X \min_B K \le \sup_X \inf_Y K + 2\varepsilon$$

Let us write $v = \inf_{\mathscr{B}} \sup_x \min_B K$. If $v = -\infty$, then we are through; v can never be $+\infty$ because of the assumption that X is (S)-conditionally compact. Hence we shall assume v to be a finite real number. Suppose that the above inequality is not true; then we have

$$\sup_X \inf_Y K < v - 2\varepsilon$$

or

$$\inf_Y K(x, y) < v - 2\varepsilon \text{ for all } x$$

Therefore

$$\inf_Y K(x_i, y) < v - 2\varepsilon \text{ for all } x_i \in A_\varepsilon$$

It follows that there exists at least one y_i for every x_i such that

$$K(x, y_i) \le K(x_i, y_i) + \varepsilon < v - 2\varepsilon + \varepsilon = v - \varepsilon$$

In other words,

$$\sup_X \min_B K(x, y_i) \le v - \varepsilon$$

where B is the finite set $(y_1, y_2, \cdots)$. That is,

$$v \le v - \varepsilon$$

which is impossible. Hence we have

$$\inf_Y \sup_X K \le \sup_X \inf_Y K + 3\varepsilon$$

Since ε is arbitrary and $\inf_Y \sup_X K(x, y) \ge \sup_X \inf_Y K(x, y)$ holds good always, theorem follows, and thus the proof is complete. Now one can establish the following more general theorem, which includes theorem 5.3.2.

Theorem 5.3.3

Let $K(x, y)$ be a real valued function defined over $X \times Y$. Further, suppose that X is (S)-conditionally compact. Then the following conditions are equivalent:

(1) Given any $\varepsilon > 0$, and any finite subsets A and B of X and Y, respectively, there exists $x_0 \in X$, $y_0 \in Y$, such that $K(x_i, y_0) \le K(x_0, y_j') + \varepsilon$ for all $x_i \in A$, $y_j \in B$.

(2) $\sup_X \inf_Y K(x, y) = \inf_Y \sup_X K(x, y)$.

(3) $\inf_Y \max_A K(x, y) \le \sup_X \inf_Y K(x, y)$ for every finite set A of X.

(4) $\sup_X \min_B K(x, y) \ge \inf_Y \sup_X K(x, y)$ for every finite set B of Y.

PROOF: One can prove as before (proof of theorem 5.3.2) that $(1) \Rightarrow (2)$, and it is easy to check that $(2) \Rightarrow (1)$, (3), and (4). Using the (S)-conditional compactness of X, it can be shown that (3) as well as (4) implies (2). Hence the theorem follows. Theorem 5.3.3 includes the following theorem due to Teh-Tjoe Tie.

Theorem 5.3.4

Let K be defined over $X \times Y$ and let X be (S)-conditionally compact. Suppose that K is concave–convex-like and that $K(x, y)$ is bounded below in y for every fixed x. Then

$$\sup_X \inf_Y K(x, y) = \inf_Y \sup_X K(x, y)$$

Remark 5.3.2: Any function that is concave–convex-like will have property (1) of theorem 5.3.3. We shall now state a minimax theorem that is valid when one of the spaces is compact.

Theorem 5.3.5

Let X be a compact space and let $K(x, y)$ be upper semicontinuous in x for every fixed y. Then the following two statements are equivalent:

(1) $\inf_Y \sup_X K(x, y) = \sup_X \inf_Y K(x, y)$.

(2a) For any $\varepsilon > 0$, and any finite subsets A and B of X and Y, there exists $x_0 \in X$, $y_0 \in Y$, such that $K(x_i, y_0) \leq K(x_0, y_j) + \varepsilon$ for all $x_i \in A$, $y_j \in B$.

(2b) For any $C < \inf_Y \sup_X K(x, y)$, there exists a finite set A of X such that for every y there exists an $x \in A$ with $K(x, y) \geq C$.

PROOF: We shall prove that $(2) \Rightarrow (1)$, as it can be easily checked that $(1) \Rightarrow (2)$. From condition (2a) it follows that

$$\inf_Y \max_A K(x, y) \leq \sup_X \min_B K(x, y)$$

where A and B are any two finite sets of X and Y. Since X is compact and $K(x, y)$ is upper semicontinuous in x, it is not hard to prove the following: Given any $C > \sup_X \inf_Y K(x, y)$, there exists a finite set B of Y such that for every x there exists a $y \in B$ with $K(x, y) < C$. We shall denote this condition as (2b'). We shall now prove that

$$\sup_X \inf_Y K(x, y) \geq \inf_X \sup_Y K(x, y)$$

Suppose that it is not true; then

$$\sup_X \inf_Y K < \inf_Y \sup_X K$$

so we can find two real numbers, r_1 and r_2, such that

$$\sup_X \inf_Y K(x, y) < r_1 < r_2 < \inf_Y \sup_X K(x, y)$$

From (2b) and (2b') we have

$$\inf_Y \max_A K(x, y) \geq r_2$$

and

$$\sup_X \min_B K(x, y) \leq r_1$$

Since $r_1 < r_2$, we have $\sup_X \min_B < r_1 < r_2 < \inf_Y \max_A, K(x, y)$, which contradicts our assumption (2a). Hence the theorem follows.

In fact, one can prove the following for concave–convex-like functions. If K is concave–convex-like, and further if either (2b) or (2b') is satisfied (compactness assumption as well as upper semicontinuous assumption is omitted), then $\sup_X \inf_Y K = \inf_Y \sup_X K$. This observation is due to Sion [10]. One can easily construct examples to show that theorem 5.3.5 is logically independent of Sion's observation. If we omit the assumptions of compactness and upper semicontinuity in theorem 5.3.5, then the following question arises. Will (2a) and (2b) alone imply (1)? We suspect that the answer is no, but we do not have any counter example at present at our disposal.

For the sake of completeness, we shall state with proof the most general minimax theorem for quasi-concave–convex functions [10]. We shall assume X and Y to be convex subsets of some linear topological spaces. $K(x, y)$ is said to be quasi-concave in x if the set $\{x: K(x, y) > r\}$ is convex for every real number r and for every y, $K(x, y)$ is quasi-convex in y if the set $\{y: K(x, y) < r\}$ is convex for every real number r and for every x. $K(x, y)$ is quasi-concave–convex if it is quasi-concave in x and quasi-convex in y. The theorem of Sion can be stated as follows.

Theorem 5.3.6

Let X and Y be convex, and suppose that one of the spaces is compact. Further, assume that K is quasi-concave–convex and that $K(x, y)$ is upper semicontinuous in x and lower semicontinuous in y. Then

$$\sup_X \inf_Y K(x, y) = \inf_Y \sup_X K(x, y)$$

PROOF OF THEOREM 5.3.6: *Case (i)*: Both X and Y are compact convex. Suppose that $\sup_X \inf_Y K(x, y) < C < \inf_Y \sup_X K(x, y)$. Let

$$A_x = \{y: K(x, y) > C\} \text{ and } B_y = \{x: K(x, y) < C\}$$

Therefore we can choose finite subsets $A_1 \subseteq X$, $B_1 \subseteq Y$, such that for each $y \in Y$ and hence for each $y \in$ convex B_1 there is an $x \in A_1$ with $K(x, y) > C$, and for each $x \in X$ and hence for each $x \in$ convex A_1 there is a $y \in B_1$ with $K(x, y) < C$.

Let A_2 be a minimal subset of A_1 such that for each $y \in$ convex B_1 there exists an $x \in A_2$ with $K(x, y) > C$, and let B_2 be a minimal subset of B_1 such that for each $x \in$ convex A_2 there is a $y \in B_2$ with $K(x, y) < C$.

Thus, by repeating this process of alternately reducing the A_i and B_i after a finite number of steps, we can choose finite subsets $A \subseteq X$ and $B \subseteq Y$ such that A is minimal with respect to the property: For each $y \in$ convex B there is an $x \in A$ with $K(x, y) > C$, and B is minimal in the sense that for each $x \in$ convex A there is a $y \in B$ with $K(x, y) < C$. Hence, there exists $x_0 \in$ convex A with $K(x_0, y) < C$ for all $y \in B$, and hence $K(x_0, y) < C$ for all $y \in$ convex B.

Similarly, we have $y_0 \in$ convex B with $K(x, y_0) > C$ for all $x \in$ convex A. Hence $C < K(x_0, y_0) < C$, which is absurd, and therefore the theorem follows.

Case (*ii*): X and Y are convex, and one of them is compact; say that X is compact.

PROOF: If possible, let $\sup_{X} \inf_{Y} K < C < \inf_{Y} \sup_{X} K$. Then there exists a finite set $B \subseteq Y$ such that for any $x \in X$ there is a $y \in B$ with $K(x, y) < C$. Taking $K^1(x, y) = K(x, y) \,|\, X \times$ convex B, that is, K restricted to $X \times$ convex B, we get

$$\sup_{X} \inf_{\text{convex } B} K^1(x, y) < C < \inf_{\text{convex } B} \sup_{X} K^1(x, y)$$

contradicting case (*i*) with Y replaced by convex B and K by K^1. Thus the proof is complete.

5.4 GAMES ON THE UNIT SQUARE

In attempts to find methods of solving a fairly wide class of (two-person zero-sum) games over the unit square, real success has been attained in only two cases: (1) games in which the problem is essentially a finite dimensional one, notably those with polynomial or polynomial-like payoffs, and (2) games having solutions that are absolutely continuous and can be solved via differential or integral equations. These latter games have discontinuities in the payoffs or their derivatives. Besides, Karlin [3] has dealt with bell-shaped and Polya-type kernels successfully. Sion and Wolfe have given an example [11], thereby showing that the minimax theorem due to Glicksberg [2] cannot be extended in certain directions. Of particular interest in applications to problems of tactical games is the case where the payoff $K(x, y)$ is bounded and has discontinuities along the diagonal $x = y$. In general, the value of such games need not exist—Sion and Wolfe's example is one such. Our object is to show that such kernels have mixed value provided that one of the players' mixed strategies are restricted to absolutely continuous distributions with respect to the Lebesgue measure. We hope that this theorem will be of some theoretical interest.

Let X and Y stand for the unit interval $[0, 1]$; m_X and m_Y denote the class of probability distributions on $[0, 1]$; A_X or A_Y will denote the class of all absolutely continuous distributions in $[0, 1]$ with respect to the Lebesgue measure.

Definition 5.4.1: A sequence $\mu_n \in m_X$ is said to converge weakly to $\mu \in m_X$ if, for each bounded continuous function f on X,

$$\lim_{n \to \infty} \int_0^1 f \, d\mu_n = \int_0^1 f \, d\mu$$

We endow m_X and m_Y with weak topology. It is well known that m_X is metrizable and compact in the weak topology [13]. We shall state and prove the following theorem [8].

Theorem 5.4.1

Let $K(x, y)$ be bounded on the unit square $0 \le x, y \le 1$, and all points of discontinuity lie on a finite number of curves of the form $y = \varphi_K(x)$, $K = 1, 2, \cdots, n$, where $\varphi_K(x)$ are continuous functions. Further, suppose that

$$K(\mu, \lambda) = \int_0^1 \int_0^1 K(x, y) \, d\mu(x) \, d\lambda(y) = \int_0^1 \int_0^1 K(x, y) \, d\lambda(y) \, d\mu(x)$$

Then $\max_{m_X} \inf_{A_Y} K(\mu, \lambda) = \inf_{A_Y} \max_{m_X} K(\mu, \lambda)$.

PROOF OF THEOREM 5.4.1: First we shall prove that, for every $\lambda \in A_Y$, $\int_0^1 K(x, y) \, d\lambda(y)$ is continuous in x.

Let $\psi(x) = \int_0^1 K(x, y) \, d\lambda(y)$ and $M = \sup |K(x, y)|$ on the square $0 \le x$, $y \le 1$.

$G = \{(x, y): |y - \varphi_K(x)| < \varepsilon/(12Mn) \text{ holds for some } K = 1, 2, \cdots, n\}$

and $F = G' \cap (X \times Y)$, where G' denotes the complement of G.

Obviously F is closed, and $K(x, y)$ is continuous on F. Since F is compact, $K(x, y)$ is uniformly continuous on F. That is, there exists a δ such that for points $(x', y), (x'', y)$ in F for which $|x' - x''| < \delta$, the inequality

$|K(x', y) - K(x'', y)| < \varepsilon/3$ holds good. Now let x' and x'' be such that $|x' - x''| < \delta$. Then

$$| \psi(x') - \psi(x'') | \leq \int_0^1 | K(x', y) - K(x'', y) | \lambda'(y) \, dy$$

where $\lambda'(y)$ is the Radon–Nikodym derivative of λ with respect to y.

The right-hand side of the above can be evaluated by integrating over the sum of the intervals

$$| y - \varphi_K(x') | < \varepsilon/(12Mn)$$

$$K = 1, 2, \cdots, n$$

$$| y - \varphi_K(x'') | < \varepsilon/(12Mn)$$

(denote the sum of these intervals by S_1) and over the complement S_1' of S_1 with respect to the closed interval $[0, 1]$. It is clear that the length of S_1 does not exceed $\varepsilon/3M$. Hence

$$\int_{S_1'} | K(x', y) - K(x'', y) | \lambda'(y) \, dy \leq \frac{\varepsilon}{3}$$

Since λ is absolutely continuous with respect to the Lebesgue measure, it follows that

$$\int_{S_1} | K(x', y) - K(x'', y) | \lambda'(y) \, dy \leq 2Mr$$

where r is a small positive quantity (in fact, $\lambda(S_1) = r$) with the property that $r \to 0$ as $\varepsilon \to 0$. Therefore it follows that

$$\int_0^1 | K(x', y) - K(x'', y) | \lambda'(y) \, dy \leq \frac{\varepsilon}{3} + 2Mr$$

or

$$| \psi(x') - \psi(x'') | \leq \frac{\varepsilon}{3} + 2Mr$$

which in turn implies that $\psi(x)$ is continuous in x.

Since $\psi(x)$ is continuous in $[0, 1]$, it is bounded, and hence we can conclude that $K(\mu, \lambda)$ is continuous in μ for every fixed $\lambda \in A_Y$. Further, A_Y is convex, and m_X is convex and compact in the weak topology. The conclusion of the theorem can be seen via von Neumann's minimax theorem.

We shall now quote, without proof, certain results in this direction that are proved by Yanovskaya [18].

Theorem 5.4.2

Let $\eta_C = \left\{ F \mid \overline{\lim_{\varepsilon \to 0}} \left| \dfrac{F(x + \varepsilon) - F(x)}{\varepsilon} \right| \le C \right\}$, where $C > 1$, and F is a probability distribution function on $[0, 1]$. Suppose that $K(x, y)$ is measurable and $\int_0^1 K(x, y)\,dx$ and $\int_0^1 K(x, y)\,dy$ are uniformly convergent; then

$$\max_{\eta_C} \min_{\eta_D} K(\mu, \lambda) = \min_{\eta_D} \max_{\eta_C} K(\mu, \lambda)$$

Theorem 5.4.3

Let $K(x, y)$ be measurable, and suppose that $\int_S K(x, y)\,dy$ and $\int_S K(x, y)\,dx$ are continuous in x and y for any Lebesgue measurable set $S \subseteq [0, 1]$; then

$$\min_{\eta_D} \max_{m_X} K(\mu, \lambda) = \max_{m_X} \min_{\eta_D} K(\mu, \lambda)$$

Theorems 5.4.1 and 5.4.3 are similar in nature. A special case of theorem 5.4.2 is considered by Wald [16].

Example of Sion and Wolfe [11]

$$\text{Let } K(x, y) = \begin{cases} -1 & \text{if } x < y < x + \tfrac{1}{2} \\ 0 & \text{if } x = y \text{ or } y = x + \tfrac{1}{2} \\ +1 & \text{otherwise} \end{cases}$$

This function K has its discontinuities along $y = x$ and $y = x + \frac{1}{2}$. We shall now show that

$$\sup_{m_X} \inf_{m_Y} K(\mu, \lambda) = \tfrac{1}{3}$$

and

$$\inf_{m_Y} \sup_{m_X} K(\mu, \lambda) = \tfrac{3}{7}$$

Let μ be any probability measure on $[0, 1]$. If $\mu[0, \frac{1}{2}) \le \frac{1}{3}$, let $y_\mu = 1$. If $\mu[0, \frac{1}{2}) > \frac{1}{3}$, choose $\delta > 0$ such that $\mu[0, \frac{1}{2} - \delta) \ge \frac{1}{3}$, and let $y_\mu = \frac{1}{2} - \delta$. In either case it is quickly checked that,

$$\inf_{m_Y} \int_0^1 \int_0^1 K(x, y)\,d\mu(x)\,d\lambda(y) \le \int_0^1 K(x, y_\mu)\,d\mu(x) \le \frac{1}{3}$$

On the other hand, if μ is chosen so that

$$\mu(0) = \mu(\tfrac{1}{2}) = \mu(1) = \tfrac{1}{3}$$

then for all y,

$$\int_0^1 K(x, y)\,d\mu(x) = \frac{1}{3}\left[K(0, y) + K\left(\frac{1}{2}, y\right) + K(1, y) \right] \ge \frac{1}{3}$$

Hence

$$\sup_{m_X} \inf_{m_Y} K(\mu, \lambda) = \tfrac{1}{3}$$

Similarly, it can be shown that

$$\inf_{m_Y} \sup_{m_X} K(\mu, \lambda) = \tfrac{3}{7}$$

As the conditions of theorem 5.4.1 are satisfied in this example, it follows that

$$\max_{m_X} \inf_{A_Y} K(\mu, \lambda) = \inf_{m_Y} \max_{A_Y} K(\mu, \lambda)$$

and

$$\sup_{A_X} \min_{m_Y} K(\mu, \lambda) = \min_{m_Y} \sup_{A_X} K(\mu, \lambda)$$

This game can be considered as a continuous Blotto game as follows: Player I must assign a force x to the attack of one of the two mountain passes and $1 - x$ to the other. Player II must assign a force y to the defense of the first pass and $1 - y$ to the other, at which is also located an extra stationary defense force of $^1/_2$. A player receives from the other a payment of 1 at each pass if his force at that pass exceeds his opponent's, and he receives nothing if they are equal there. The payoff is thus

$$B(x, y) = \operatorname{sgn}(x - y) + \operatorname{sgn}[(1 - x) - (\tfrac{3}{2} - y)]$$

It is easily checked that $1 + B(x, y) = K(x, y)$, so that this game has the value

$$\left[-1 + \max_{m_X} \inf_{A_Y} K(\mu, \lambda) \right] \quad \text{and} \quad \left[-1 + \min_{m_Y} \sup_{A_X} K(\mu, \lambda) \right]$$

if player II (player I) restricts his attention only to absolutely continuous distributions.

5.5 CONVEX CONES AND FINITE OPTIMALS

Introduction

Among the continuous payoff kernels $K(x, y)$ on the unit square for an infinite zero-sum two-person game, the separable kernels, the generalized convex kernels, and certain analytic kernels are known to possess optimal mixed strategies for the two players with finite spectrum [3]. It is known that even among C^∞ kernels we can have very pathological optimal mixed strategies as their unique optimals [3]. Thus the problem of classifying kernels with finite optimals is unresolved. Here an attempt is made to look at this problem from the topological viewpoint. The binding geometric object between kernels and strategies could be chosen as the cone generated by functions $h_\alpha(x) = K(x, \alpha)$ where we fix α and view $K(x, \alpha)$ as a function of x. Some of the properties of the cones are reflected in the finiteness of the spectrum for an optimal strategy for a player. Similar versions could be stated for the other player by considering a related kernel where now the second player becomes the maximizer. Further, these cones in certain other topologies also characterize extreme optimals for a class of games. This characterization is given in the next section.

Preliminaries

Let X, Y be compact metric spaces and $K(x, y) > 0$ be a continuous payoff on $X \times Y$. Let E_X, E_Y be the Banach space of continuous functions on X and Y with their supremum norm ($\| \cdot \|$).

Let C be the closure of the convex cone in E_X generated by functions $h_\alpha(x)$ where $h_\alpha(x) = K(x, \alpha)$, $\alpha \in Y$. Let K be the cone of nonnegative functions in E_X. Let E_0 be the linear manifold $C - C$ and $\bar{E}_0$ its closure. By a positive operator we mean a linear operator A from $\bar{E}_0 \to E_X$ which maps the cone C into K. We would call the image of the cone C under A the range cone. The following is the main theorem.

Theorem 5.5.1

Let A be a positive operator from $\bar{E}_0 \to E_X$ continuous on the cone C. If A is isometric on C and if the range cone has a relative interior point in the closed linear manifold spanned by this cone, then player II has always an optimal mixed strategy whose spectrum is finite. If the cone C itself possesses an interior point relative to $\bar{E}_0$, then the conditions are trivially satisfied for the identity map and in this case the kernel is separable and both players have optimal mixed strategies with finite spectrum.

PROOF: Let P_Y denote the set of all probability measures on the Borel sets of Y. P_Y as a subset of $E_Y{}^*$ (the dual of E_Y) is compact metric in its weak topology [13]. Further, by Helly's theorem

$$\tau: v \to \int_Y K(x, y) \, dv(y)$$

is continuous from P_Y into E_X. Thus $\tau(P_Y) = B$ is compact in E_X. Trivially it is convex. Let T be the convex cone generated by B; that is,

$$T = \{\lambda f: \lambda \geq 0, f \in B\}$$

We will show that $T = C$. To see this let $p \in C$, $p \neq 0$. There exists $p_n \to p$, $p_n \neq 0$ where p_n is the form

$$p_n(x) = \sum_1^{\gamma_n} a_i K(x, a_i), \quad a_i \geq 0, \quad i = 1, 2, \cdots, \gamma_n$$

By assumption some $a_i > 0$ and that $\theta_n \cdot p_n \in B$ where $\theta_n = 1/\Sigma a_i$.

By the compactness of B, $\theta_n \cdot p_n$ has a convergent subsequence and without loss of generality $\theta_n \cdot p_n \to p_* \in B$. Since $K(x, y) > 0$, $\|p_*\| > 0$. We also have that $p_n \to p$ and that θ_n's are bounded. Without loss of generality $\theta_n \to \theta_0 > 0$. Thus $\theta_0 \cdot p = p^*$ and thus $p = (1/\theta_0) \cdot p_* \in T$. Thus $C \subset T$. Conversely, let $q \in T$, $q \neq 0$. For some $\lambda > 0$, $\lambda q \in B$. Thus $q = (1/\lambda)\tau(v)$ for some $v \in P_Y$. But in P_Y those measures which have finite spectrum are dense [13]. Therefore we have $(1/\lambda)\tau(v_n) = q_n$, $v_n \to v$, where v_n is a sequence in P_Y where each one has finite spectrum. Now q_n is of the form

$$q_n(x) = \frac{1}{\lambda} \cdot \sum_1^{\gamma_m} \mu_i K(x, \alpha_i), \quad \mu_i \geq 0, \quad \sum \mu_i = 1$$

Thus $q_n \in C$. The continuity of τ implies $q \in C$. Thus the two cones coincide.

Continuing with our proof, let $H = A(B)$ be the image of B under A and by the continuity of A, H is compact. Further, the range cone is given by $\bigcup_{\lambda \geq 0} \lambda H$. Let S be the compact convex null of 0 and H. Then the range cone is

$$\bigcup_{n=1}^{\infty} nS.$$

Since the range cone has nonnull interior in the closed linear span of its elements, by the Baire category theorem [3] some nS and hence S has nonnull interior. S being a compact subset of this Banach space, it is finite dimensional [1a]. Let k be the dimension of S. Then by Theorems 1.9.3 and 1.9.4 any point of S is the convex linear combination of at most $k + 1$ of its extreme points. Now since B is the closed convex closure of $h_\alpha(x)$, $\alpha \in Y$, the extreme points of B are contained in the compact set $\{h_\alpha(x), \alpha \in Y\}$ [1a]. The linearity and continuity of A implies that any extreme point of S is the A image of an extreme point of B. Consider the kernel $M(x, \alpha)$ where $M(x, \alpha) = A(h_\alpha(x))$. $M(x, \alpha)$ is continuous on $X \times Y$ and $M(x, \alpha)$'s as functions of x generate a finite–dimensional space.

Thus for this kernel, player II has an optimal strategy v_0 with finite spectrum. Let v_0 be the value of this game with payoff $M(x, \alpha)$. Thus

$$\int_Y M(x, \alpha)\, dv_0(\alpha) \leq v_0 \qquad \text{for all } x$$

But A maps C into the cone K and thus for any point $h \in B$ where

$$h(x) = \int_Y K(x, y)\, dv(y)$$

for some $v \in P_Y$ we have by linearity and isometry

$$Ah(x) = \int_Y M(x, y)\, dv(y) \quad \text{and} \quad \|Ah\| = \max_x \int_Y M(x, y)\, dv(y)$$

$$\|h\| = \max_x \int_Y K(x, y)\, dv(y)$$

This shows that

$$\min_v \max_x \int_Y K(x, y)\, dv(y) = \min_v \max_x \int_Y M(x, y)\, dv(y)$$

$$= \max_x \int_Y M(x, y)\, dv_0(y) = v_0$$

Thus both games have the same value v_0 and that v_0 is optimal also for the original payoff $K(x, y)$. This completes the proof of the main assertion in the theorem. If the cone C possesses an interior point of $\bar{E}_0$, then by the argument above $\bar{E}_0$ is finite dimensional. Let $h_{\alpha_1}, h_{\alpha_2}, \cdots, h_{\alpha_s}$ span $\bar{E}_0$. Let x, y be generic points and $x_1, x_2, \cdots, x_s, y_1, y_2, \cdots, y_s$ be any fixed set of points with the property (without loss of generality) that the determinant

$$\Delta = \begin{vmatrix} K(x_1, y_1), \cdots, K(x_1, y_s) \\ \cdots\cdots\cdots\cdots\cdots\cdots \\ K(x, y_1), \cdots, K(x_s, y_x) \end{vmatrix} \neq 0$$

Now $h_y(x) = K(x, y)$ and $h_{\alpha_1} \cdots h_{\alpha_s}$ form a dependent system and that

$$\begin{vmatrix} K(x, y), & K(x, y_1) & \cdots & K(x, y_s) \\ K(x_1, y) & K(x_1, y_1) & & K(x_1, y_s) \\ \cdots\cdots\cdots\cdots\cdots\cdots\cdots\cdots \\ K(x_s, y) & K(x_s, y_1) & & K(x_s, y_s) \end{vmatrix} = 0$$

Expanding the determinant, we have

$$\Delta K(x, y) + \sum_i \sum_j K(x, y_i) K(x_j, y) a_{i_j} = 0$$

Here a_{ij}'s are the suitable cofactors got by deleting the column $i + 1$ and row $j + 1$ in the foregoing determinant. Since $\Delta \neq 0$, we have

$$K(x, y) = -\sum_i \sum_j K(x, y_i) \cdot K(x_j, y) \frac{a_{ij}}{\Delta}$$

that is, $K(x, y)$ is separable. But in this case we know by symmetry that player I possesses an optimal strategy with finite spectrum [3].

The following two examples illustrate the existence of isometries for nonseparable kernels to which our main theorem is applicable.

Example 5.5.1: Let $K(x, y) > 0$ be continuous and convex in x for each y in $0 \leq x, y \leq 1$. For any α in $0 \leq \alpha \leq 1$ let the function $h_\alpha(x) = K(x, \alpha)$ be mapped to the function $g_\alpha(x) = (1 - x)K(0, \alpha) + xK(1, \alpha)$. Since the maximum for a convex function is either at 0 or 1, we have $\|h_\alpha\| = \|g_\alpha\|$. Further, for any finite $\lambda_1, \lambda_2, \cdots, \lambda_k \geq 0, \alpha_1, \alpha_{,2}, \cdots, \alpha_k \leq 1, \Sigma \lambda_i k_{\alpha_i}$ is convex and that if we define

$$A : \Sigma \lambda_i h_{\alpha_i} \rightarrow \Sigma \lambda_i g_{\alpha_i}$$

then we check that A is an isometry and it is linear and continuous in $\bar{C}$. The range of A consists of functions of the form $a + bx$ and that the range cone has relative interior. Thus continuous kernels on the unit square which are convex in x for each y have finite optimals for player II (of course, also for player I).

Example 5.5.2: Instead of demanding convexity in x we could weaken the condition by demanding $K(x, y) \leq (1 - x)K(0, y) + xK(1, y)$ for all x, y in the example above. The same argument works.

In these two examples we could see the finite optimals directly by domination arguments. It would be interesting to know whether we could construct such isometries of the theorem for some of the known kernels with finite optimals for player II, such as the generalized concave kernels of Karlin and certain Cauchy bell-shaped kernels.

Thus it seems that the cone C and its topological structure could give us some guidance in our problem of classification.

5.6 EXTREME OPTIMAL STRATEGIES IN INFINITE GAMES

Introduction

We have seen in Theorem 3.1.14 that the extreme optimal strategies for the players in a zero-sum two-person matrix game are the simple solution pairs of some square submatrices of the payoff matrix. A similar characterization is available also for the non-zero-sum bimatrix games (Theorem 7.6.1). In this section we shall give a characterization of the extreme optimal strategies of players for a special case of infinite zero-sum two-person games. This characterization is closely related to a more general problem of Douglas [1b].

Preliminaries

Let X, Y be compact metric spaces and let m_X^p, m_Y^p denote the space of all probability measures on the Borel subsets of X and Y, respectively. For any $\mu \in m_X^p$ let $\sigma(\mu)$ denote the spectrum of μ (i.e., the smallest closed set X_0 in X with $\mu(X_0) = 1$). Let $K(x, y)$ be a continuous payoff function on $X \times Y$; m_X^p and m_Y^p as subsets of $C^*(X)$ and $C^*(Y)$ (the dual space of the space of real continuous functions on X and Y) are convex and compact in their weak topologies [13]. Further

$$L(\mu, v) = \iint K(x, y) \, d\mu(x) \, dv(y)$$

is affine and continuous in each variable when the other is fixed. (Here μ ranges over m_X^p and v over m_Y^p.)

By general minimax theorems we know that there exists a $\mu_0 \in m_X^p$, $v_0 \in m_Y^p$ such that for some unique constant v (the value)

$$\int_X K(x, y) \, d\mu_0(x) \geq v \qquad \text{for all } y \in Y$$

$$\int_Y K(x, y) \, dv_0(y) \leq v \qquad \text{for all } x \in X$$

$$\int_X \int_Y K(x, y) \, d\mu(x) \, dv(y) = v$$

One easily checks that the set of all optimal strategies for a player, say player I, is a closed convex subset of m_X^p and hence compact convex. The weak topology being locally convex, Theorem 1.8.1 of Krein and Milman is applicable and we have extreme points for the set of optimal strategies for player I. Further, the closed convex closure of these extreme points is the set of all optimal strategies for the player. Our aim is to characterize these extreme points. We have the following theorem.

Theorem 5.6.1

Let $v \neq 0$ and let there exist an optimal strategy v_0 for player II with $v_0(G) > 0$ for all nonempty open subsets of Y. A necessary and sufficient condition for an optimal strategy μ_0 of player I to be extreme optimal is that the functions $h_\alpha(x) = k(x, \alpha)$, $\alpha \in Y$, generate a dense linear manifold in $L_1(X_0, \mathscr{B}, \mu_0)$ where $X_0 = \sigma(\mu)$ and $\mathscr{B}$ is the class of Borel sets on X_0.

PROOF: First we claim that

$$g(y) = \int_X K(x, y) \, d\mu_0(x) \equiv v \quad \text{on} \quad Y$$

For suppose $g(y_0) > v$ for some $y_0 \in Y$, then by the continuity of $g(y)$ $g(y) > v + \varepsilon$ for some $\varepsilon > 0$ and for all y in an open set G containing y_0. Thus

$$v = \int_Y \left(\int_X K(x, y) \, d\mu_0(x) \right) dv_0(y) = \int_G \left(\int_X K(x, y) \, d\mu_0(x) \right) dv_0(y)$$

$$+ \int_{(Y-G)} \left(\int_X K(x, y) \, d\mu_0(x) \right) dv_0(y)$$

$$\geq (v + \varepsilon)v_0(G) + v v_0(Y - G) = v + \varepsilon v_0(G)$$

Since $v_0(G) > 0$, $v > v$ a contradiction. Now we shall prove the theorem.

Necessity: Let μ_0 be extreme and let $\mathcal{M}$, the linear manifold spanned by $h_\alpha(x) = k(x, \alpha)$, $\alpha \in Y$, be not dense in $L_1(X_0, \mathcal{B}, \mu_0)$. Let $\bar{\mathcal{M}}$ be the closure of $\mathcal{M}$ in $L_1(X_0, \mathcal{B}, \mu_0)$. By assumption there exists an $f_0 \in L_1(X_0, \mathcal{B}, \mu_0)$, $f_0 \notin \bar{\mathcal{M}}$. By strong separation theorem (Theorem 1.6.2) we have a nontrivial continuous linear functional φ on $L_1(X_0, \mathcal{B}, \mu_0)$ and a constant C such that

$$\varphi(f_0) > C$$

$$\varphi(f) \leq C \qquad \text{for all } f \in \bar{\mathcal{M}}$$

If $f \in \bar{\mathcal{M}}$, so is $-f$ and that $\varphi(-f) = -\varphi(f) \leq C$. But $\varphi(f) \leq C$ implies $-\varphi(f) \geq -C$. Thus $-C \leq C$; that is, $C \geq 0$. But $\varphi(f)$ is never positive for any $f \in \bar{\mathcal{M}}$ for $\varphi(f_*) > 0$ for some $f_* \in \bar{\mathcal{M}}$, then $\varphi(\lambda f_*) > 0$ for any $\lambda > 0$ and that $\varphi(\lambda f_*) > C$ for some λ and $\lambda f_* \in \bar{\mathcal{M}}$. Thus $\varphi(f) \leq 0$ for all $f \in \bar{\mathcal{M}}$, which implies $\varphi(f) = 0$ for all $f \in \bar{\mathcal{M}}$. But any continuous linear functional φ on $L_1(X_0, \mathcal{B}, \mu_0)$ is of the form

$$\varphi(f) = \int_{X_0} f(x)g(x) \, d\mu_0(x), \qquad f \in L_1(X_0, \mathcal{B}, \mu_0)$$

where $g(x)$ is a μ_0 essentially bounded function [1a]. Let $|g(x)| \leq C$ almost everywhere μ_0. Thus

$$\int_{X_0} K(x, y)g(x) \, d\mu_0(x) = 0, \qquad y \in Y$$

(because $K(x, y)$ for fixed y belongs to $\mathcal{M}$. We have

$$0 = \int_Y \left(\int_{X_0} K(x, y)g(x) \, d\mu_0(x) \right) dv_0(y) = \int_{X_0} \left(\int_Y K(x, y) \, dv_0(y) \right) g(x) \, d\mu_0(x)$$

$$= v \int_{X_0} g(x) \, d\mu_0(x)$$

(The equality above follows by the fact that X_0 is the spectrum of μ_0 and that any open subset of X_0 has positive μ_0 mass.)

Since $v \neq 0$

$$\int_{X_0} g(x)\, d\mu_0(x) = 0$$

Let

$$\mu_1(E) = \int_E \left(1 + \frac{g(x)}{2C}\right) d\mu_0(x), \qquad E \in \mathscr{B}$$

$$\mu_2(E) = \int_E \left(1 - \frac{g(x)}{2C}\right) d\mu_0(x), \qquad E \in \mathscr{B}$$

Since φ is nontrivial, $\mu_1 \neq \mu_2$. Further, $1 \pm g(x)/2C \geq 0$ almost everywhere μ_0. Thus $\mu_1(E) \geq 0$, $\mu_2(E) \geq 0$ for $E \in \mathscr{B}$. Thus $\mu_1, \mu_2 \in m_X^p$. Further $\mu_0 = (\mu_1 + \mu_2)/2$. Lastly

$$\int_{X_0} K(x, y)\, d\mu_1(x) = \int_{X_0} K(x, y)\, d\mu_2(x) \equiv v, \qquad y \in Y$$

and they are also optimal for player I.

Thus $\mu_0 = (\mu_1 + \mu_2)/2$ contradicts the assumption that μ_0 is extreme optimal for player I. Hence the necessary part.

Sufficiency: Let μ_0 be optimal and let $h_\alpha(x) = K(x, \alpha)$, $\alpha \in Y$, generate a dense linear manifold $\mathscr{M}$ in $L_1(X_0, \mathscr{B}, \mu_0)$. Suppose $\mu_0 = (\mu_1 + \mu_2)/2$, μ_1, μ_2 optimal for player I with $\mu_1 \neq \mu_2$. We have

$$1 = \mu_0(X_0) = \frac{\mu_1(X_0) + \mu_2(X_0)}{2}, \qquad \mu_1(X_0) \leq 1, \qquad \mu_2(X_0) \leq 1$$

Thus $\mu_1(X_0) = \mu_2(X_0) = 1$. Therefore, $\mu_1 \neq \mu_2$ gives us a continuous function f_0 on X_0 with

$$\int_{X_0} f_0\, d\mu_1 < \int_{X_0} f_0\, d\mu_2$$

By assumption we have

$$h_n \in \mathscr{M} \qquad \text{and} \qquad \lim_{n \to \infty} \int |h_n - f_0|\, d\mu_0 = 0$$

that is,

$$\tfrac{1}{2} \int |h_n - f_0|\, d\mu_1 + \tfrac{1}{2} \int |h_n - f_0|\, d\mu_2 \to 0 \qquad \text{as} \qquad n \to \infty$$

But any $h \in \mathscr{M}$ is of the form

$$h(x) = \sum_{i=1}^{r} a_i K(x, \alpha_i)$$

and

$$\int_{X_0} h \, d\mu_1 = \sum_{i=1}^{r} a_i \int_{X_0} K(x, \alpha_i) \, d\mu_1 = v \sum_{i=1}^{r} a_i$$

(because of our initial assertion on μ_0, which applies for μ_1, μ_2 also). Similarly

$$\int_{X_0} h \, d\mu_2 = v \sum_{1}^{r} a_i$$

Thus

$$\int_{X_0} h_n \, d\mu_1 = \int_{X_0} h_n \, d\mu_2$$

and that $\int_{X_0} f_0 \, d\mu_1 = \int_{X_0} f_0 \, d\mu_2$. This contradicts our assumption about f_0. This establishes the sufficiency part. The proof of the theorem is complete.

Remark 5.6.1: Suppose we do not assume that there exists an optimal strategy for player II satisfying the condition of the theorem. Then we could consider the $h_\alpha(x) = K(x, \alpha)$, $\alpha \in Y_0$, where

$$\int K(x, \alpha) \, d\mu_0(x) = v, \qquad \alpha \in Y_0$$

$$> v, \qquad \alpha \notin Y_0$$

Clearly the spectrum of any optimal strategy for player II is contained in Y_0. One could possibly think that the h_α's for $\alpha \in Y_0$ generate a dense linear manifold in $L_1(X_0, \mathscr{B}, \mu_0)$ when μ_0 is extreme with $\sigma(\mu_0) = X_0$ and $v \neq 0$. Even though this statement is true in the case of matrix games, it fails to hold in general in infinite dimensions. The following classical example of Glicksberg and Gross [3] would suffice.

Let

$$K(x, y) = 1 + \sum_{1}^{\infty} (x^n - a_n) y^n \sin \frac{1}{y} + \exp^{-(1/y^2)}$$

$$0 \le x, y \le 1 \qquad \text{and} \qquad a_n = \int_0^1 x^n \, d_x$$

One checks [4] that the only optimal strategy for player I is the Lebesgue measure on $[0, 1]$ and the only optimal strategy for player II is the degenerate measure concentrated at 0. The value $v = 1$. Further, $Y_0 = \{0\}$, the set consisting of the single point 0. In this case $h_\alpha(x)$, $\alpha \in Y_0$, consists of just the 0 function, which certainly does not generate $L_1[0, 1]$.

Remark 5.6.2: That $v \neq 0$ is not a serious assumption, for when $v = 0$, we can add a constant to $K(x, y)$ and consider the new payoff. Both payoffs have the same set of optimals.

REFERENCES

[1] Berge, C. (1954). Sur une convexite regulier et ses applications a la theorie des jeux. *Bull. Soc. Math. France* **82**, 301–315.

[1a] Dunford, N., & Schwartz, J. T. (1958). "Linear operators", Interscience.

[1b] Douglas, R. G. (1964). On extremal measures and subspace density, *Michigan Math. J.* **11**, 243–246.

[2] Glicksberg, I. L. (1950). Minimax Theorem for Upper and Lower Semicontinuous Payoffs, Rand Corporation Research Memorandum No. 478.

[3] Karlin, S. (1959). Mathematical methods and the theory in games. "Programming Economics" **2**, Addison-Wesley Publishing Company, Massachusetts.

[4] Kelley, J. L. (1955). "General Topology," D. Van Nostrand Company, Princeton, New Jersey.

[5] Kneser, H. (1952). Sur un theorem fundamental de la theorie des jeux. *Compt. Rend. Acad. Sci.* **234**, 2418–2420.

[6] Ky Fan (1953). Minimax theorems. *Proc. Natl. Acad. Sci. U.S.* **39**, 42–47.

[7] Parthasarathy, T. (1965). A note on a minimax theorem of T-T. Tie. *Sankhya* **27**, 407–408.

[8] Parthasarathy, T. (1970). On games over the unit square. To appear in SIAM J. Appl. Math.

[9] Prohorov, Yu. V. (1956). Convergence of random processes and limit theorems in the theory of probability. In "Theory of Probability and Its Applications" (English translation), Vol. 1, pp. 157–214.

[9a] Raghavan, T. E. S. (1970). Convex cones and finite optimals, *Ann. Math. Stat.* **41**, 702–705.

[10] Sion, M. (1958). On general minimax theorem. *Pacific J. Math.* **8**, 171–176.

[11] Sion, M., and Wolfe, P. (1957). On a game without a value. In "Contributions to the Theory of Games," Vol. 3, pp. 299–306, *Ann. Math. Studies, No. 39,* edited by M. Dresher, A. W. Tucker and P. Wolfe, Princeton University Press, Princeton, New Jersey.

[12] Teh-Tjoe Tie (1963). Minimax theorems on conditionally compact sets. *Ann. Math. Stat.* **34**, 1536–1540.

[13] Varadarajan, V. S. (1958). Weak convergence of measures on separable metric spaces. *Sankhya* **20**, 15–22.

[14] von Neumann, J., and Morgenstern, O. (1944). "Theory of Games and Economic Behavior," Princeton University Press, Princeton, New Jersey.

[15] Wald, A. (1945). Generalization of a theorem by von Neumann concerning zero-sum two-person game. *Ann. Math.* **46**, 281–286.

[16] Wald, A. (1950). A note on zero-sum two-person game. *Ann. Math.,* **51**, 739–742.

[17] Wald, A. (1950). "Statistical Decision Functions," John Wiley & Sons, New York.

[18] Yanovskaya, E. B. (1964). Minimax theorems for games on the unit square. In "Theory of Probability and Its Applications" (English translation), Vol. 9, 500–502.

Chapter Six

APPLICATIONS OF MINIMAX THEOREMS

It is the purpose of this chapter to prove a few theorems regarding average distances of points in a finite collection from some point in a compact connected metric space. For each such space, in fact, there corresponds a unique constant. We shall also deduce a theorem in probability theory and threshold logic from minimax theorems. Finally, we shall prove a result on positive operators using minimax theorems and a result on Lusin's second separation principle for analytic sets using a theorem of Gale and Stewart on infinite games of perfect information.

A stronger motivation or raison d'être for this chapter, however, is furnished by the not-too-well exploited fact that game theory can be used as a tool in other branches of mathematics and physics, including geometry. In fact, well-known results in game theory render the proofs of the theorems that will be stated below almost trivial.

Theorem 6.1.1 [8]

Relative to a compact connected metric space there corresponds a unique constant K with the property that, given any finite collection of points, one can find a point y such that the average distance of the points in the collection from y is equal to K. Further, $d/2 \leq K < d$, where d is the diameter of the space. Conversely, if d and K are positive numbers such that $d/2 \leq K < d$, then there corresponds a compact connected metric space of diameter d whose associated constant given by the first part of the theorem will be K.

Theorem 6.1.3 [11]

Let X be any complete separable metric space, and let $\mathscr{A}$ be any family of real valued continuous functions on X. Suppose that (1) $\mathscr{A}$ is equicontinuous and (2) $\mathscr{A}$ is uniformly bounded; that is, there exists a constant c such that $|f(x)| \leq c$ for all $x \in X$ and $f \in \mathscr{A}$. Let μ_n and μ be probability measures on X, and let μ_n converge weakly to μ. Then

$$\lim_{n \to \infty} \sup_{\mathscr{A}} \left| \int f \, d\mu_n - \int f \, d\mu \right| = 0$$

Theorem 6.2.1 [2]

Given a switching function $F(x_1, x_2, \cdots, x_n)$ and $\| \alpha_{ik}^* \|$, then F is realizable with a single threshold gate if and only if the value of the game corresponding to $\| \alpha_{ik}^* \|$ is greater than $^1/_2$.

Theorem 6.3.1 [13]

Let C be a closed cone in real reflexive Banach space E. Let C and C^*, the conjugate cone, have non-null interior. Further, let A be a strongly positive operator. Then:

(1) $Az = \lambda_0 z$, $\lambda_0 > 0$, and $z \in$ interior of C.

(2) λ_0 is the spectral radius of A.

(3) The subspace $S_{\lambda_0} = \{y \mid A_y = \lambda_0 y\}$ is one-dimensional.

(4) A^* has an eigenvector ψ for λ_0 which is strictly positive on C^*.

(5) $S^*_{\lambda_0} = \{f \mid A^*f = \lambda_0 f\}$ is one-dimensional.

(6) No other vector linearly independent of z or ψ lies in C or C^*, respectively.

Theorem 6.4.1 [10]

Let Y be any complete separable metric space. If A and B are analytic subsets of Y, there are analytic sets A_1, B_1, such that

(1) $A_1 \supset A$, $B_1 \supset B$.

(2) $A_1 \cup B_1 = Y$.

(3) $A_1 \cap B_1 = A \cap B$.

6.1 PROOF OF THEOREM 6.1.1

Consider the two-person game with payoff $d(x, y)$. Pure strategy spaces for player 1 and player 2 are the given compact connected metric space X.
Let $v = \min_{\lambda} \max_{\mu} d(\mu, \lambda) = \max_{\mu} \min_{\lambda} d(\mu, \lambda)$, with

$$d(\mu, \lambda) = \iint d(x, y)\, d\mu(x)\, d\lambda(y)$$

where μ and λ are distributions on X. Let $x_1, x_2, \cdots, x_n$ be any finite collection of points. Then it follows that

$$\max_{\mu} d(\mu, \lambda) \geq v \text{ for every } \lambda$$

In particular, taking $\lambda = \left(\dfrac{1}{n}, \dfrac{1}{n}, \cdots, \dfrac{1}{n}\right)$ at $(x_1, x_2, \cdots, x_n)$, we have

$$\max_{\mu} \frac{1}{n} \sum_{1}^{n} d(\mu, x_i) \leq v$$

That is,

$$\max_{y} \frac{1}{n} \sum_{1}^{n} d(y, x_i) \geq v$$

From the fact that $d(x, y) = d(y, x)$, it is not hard to see that

$$\min_{y} \frac{1}{n} \sum_{1}^{n} d(y, x_i) \geq v$$

Let

$$g(y) = \frac{1}{n} \sum_{1}^{n} d(y, x_i)$$

Then $g(y)$ is continuous. Since X is connected, the range of $g(y)$ is also connected. Intervals are the only connected sets in the real line; it follows that there exists a point y^1 such that $g(y^1) = v$. Thus existence of a constant is proved.

Suppose that we have another constant K with the same property. Assume for the moment that $K > v$. Since X is compact metric, the space of probability measures is compact and metrizable in the weak topology [15]. Hence it follows that, given any $\varepsilon > 0$, there exists $x_1, x_2, \cdots, x_n$ and positive rational weights $\lambda_1, \lambda_2, \cdots, \lambda_n$, such that $\Sigma \lambda_i d(x_i, y) \leq v + \varepsilon$ for all y. If X consists of a singleton, the uniqueness is trivial. But if otherwise, since X is connected, there are infinitely many points in every neighborhood. Let $\lambda_i = n_i/N$ (N is the common denominator for all λ_i). Select N different points x_i^1 with n_i of them clustered around x_i to obtain an arithmetic average approximating an optimal strategy. This follows by the continuity of the metric. Hence

$$\frac{1}{N} \Sigma d(x_i^1, y) \leq v + \varepsilon \quad \text{for all} \quad y$$

This obviously leads to a contradiction to the assumption that K also has the desired property. Thus the uniqueness is established. Let

$$d = \max_{(x, y)} d(x, y) = d(x_1, x_2)$$

Then

$$K \geq \min_{y} \frac{1}{2}[d(x_1, y) + d(x_2, y)] \geq \frac{d}{2}$$

The last inequality follows by the triangle inequality of the metric d. Also it is clear that $K \leq d$. If possible, let $K = d > 0$. Without loss of generality we shall assume that $d = 1$. This assumption—namely, that $d = 1$—leads to the existence of an infinite number of points in the space all one unit from each other, contradicting the fact that X is compact metric.

Conversely, assume that there are positive constants d and K with $d/2 \leq K < d$. Without loss of generality, assume that $d = 1$. We have to construct a game on a compact connected metric space whose value K lies in $[{}^1/_2, 1)$. Consider $[0, 1]$. Let

$$D_\lambda(x, y) = \frac{(\lambda + 1)|x - y|}{\lambda |x - y| + 1}$$

where $0 \leq x, y \leq 1$, $\lambda \geq 0$. If $\lambda = 0$, the value of the game $v(\lambda)$ associated with the metric D_λ is ${}^1/_2$. Also $v(\lambda) \to 1$ as $\lambda \to +\infty$. Since $v(\lambda)$ is continuous, we can by a proper choice of λ obtain a game with the prescribed value. Thus proof is complete. Before proving theorem 2 we shall start with some preliminaries.

Definition 6.1.1: We say that $\mu_n \Rightarrow \mu$; that is, μ_n converges weakly to μ if $\lim_{n \to \infty} \int f \, d\mu_n = \int f \, d\mu$ for every continuous and bounded function on X.

Definition 6.1.2: We say that $f_n \to f$ in the uniform topology on compacta if $\lim_{n \to \infty} \sup_K |f_n(x) - f(x)| = 0$ for every compact subset K of X. From the Ascoli theorem it follows that the family of equicontinuous and uniformly bounded functions becomes conditionally compact in the uniform topology on compacta. We need the following theorem due to Prokhorov in the sequel.

Theorem 6.1.2

Let X be any complete separable metric space. Let Γ be any family of probability measures on X. Then Γ is conditionally compact if and only if, given $\varepsilon > 0$, there exists a compact set $K_\varepsilon \subset X$ such that $\mu(K_\varepsilon) > 1 - \varepsilon$ for all $\mu \in \Gamma$.

Define

$$\psi_n(f) = \left| \int f \, d\mu_n - \int f \, d\mu \right|$$

and

$$b_n(f) = \sup_{m \geq n} \psi_m(f)$$

The proof of theorem 6.1.3 is complete if we show the following, namely:

$$\sup_{\overline{\mathscr{A}}} \inf_{n} b_n(f) = \inf_{n} \sup_{\overline{\mathscr{A}}} b_n(f) \cdots *$$

where $\overline{\mathscr{A}}$ denotes the closure of $\mathscr{A}$ with respect to the uniform convergence on compacta. But this will follow easily if we show that $b_n(f)$ is upper semi-continuous in f for every fixed n. In fact, we are going to show that $b_n(f)$ is continuous in f. For the moment assume this. We shall now establish *. Write $v = \inf_{n} \sup_{\overline{\mathscr{A}}} b_n(f)$. If possible, let $v > \sup_{\overline{\mathscr{A}}} \inf_{n} b_n(f)$. Since $\overline{\mathscr{A}}$ is compact and $b_n(f)$ is continuous, it follows that there exist finite integers $n_1, n_2, \cdots, n_k$ such that, given any f, one can find an n_i with $b_{n_i}(f) < v$. Let

$$\max [n_1, n_2, \cdots, n_k] = n_0$$

Then we have $b_{n_0}(f) < v$ for all f. This means that $v < v$, which is impossible. Hence * follows.

We have yet to establish that $b_n(f)$ is continuous. Let $f_n \to f$ in the uniform convergence on compacta. Since $\mu_n \Rightarrow \mu$, $\{\mu, \mu_1, \mu_2, \cdots\}$ is compact in the weak topology. By Prokhorov's theorem, it follows that, given any $\varepsilon > 0$, there exists a compact set K_ε such that $\mu_n(K_\varepsilon) > 1 - \varepsilon$ for all n, and $\mu(K_\varepsilon) > 1 - \varepsilon$. Now it follows that

$$\left| \int f_m \, d\mu_n - \int f_m \, d\mu - \left(\int f \, d\mu_n - \int f \, d\mu \right) \right|$$

$$\leq \int |f_m - f| \, d\mu_n + \int |f_m - f| \, d\mu$$

$$\leq \int_{K_\varepsilon} |f_m - f| \, d\mu_n + \int_{K_\varepsilon'} |f_m - f| \, d\mu_n +$$

$$+ \int_{K_\varepsilon} |f_m - f| \, d\mu + \int_{K_\varepsilon'} |f_m - f| \, d\mu$$

$$< (2 + 4C)\varepsilon \quad \text{whenever} \quad m > m_0 \quad \text{and for all} \quad n$$

where $K_\varepsilon' = X - K_\varepsilon$. This step proves the stated assertion regarding $b_n(f)$ Thus the proof of theorem 6.1.3 is complete.

6.2 A THEOREM ON SWITCHING FUNCTION

In this section we shall give an application of minimax theorem to the testing and realization of switching function by a single threshold gate.

Definition 6.2.1: A switching function $F(x_1, x_2, \cdots, x_n)$ is defined to be a mapping from $A^n = A \times A \times \underset{(n \text{ times})}{\cdots A} \to A$, where A is the set consisting of two real numbers 0 and 1.

Definition 6.2.2: A switching function $F(x_1, x_2, \cdots, x_n)$ is said to be a threshold function (or linearly separable or 1 realizable or realizable by a single threshold gate, etc.) if there exist real numbers $w_0, w_1, w_2, \cdots, w_n$ such that

$$w_0 + \sum_{k=1}^{n} w_k x_k > 0 \quad \text{when} \quad F(x_1, x_2, \cdots, x_n) = 1$$

$$w_0 + \sum_{k=1}^{n} w_k x_k < 0 \quad \text{when} \quad F(x_1, x_2, \cdots, x_n) = 0$$

It is easy to see the following lemma.

LEMMA 6.2.1: A switching function $F(x_1, x_2, \cdots, x_n)$ is a threshold function if and only if there exists a real n-tuple $(w_1, w_2, \cdots, w_n)$ with $w_k \geq 0$ and $\Sigma w_k = 1$ and a threshold value $T (0 < T < 1)$ such that

$$\sum_{k=1}^{n} w_k x_k > T \quad \text{when} \quad F = 1$$

$$\sum_{k=1}^{n} w_k x_k < T \quad \text{when} \quad F = 0$$

A partially specified switching function is a function defined on a subset of A^n. Given a function $F(x_1, x_2, \cdots, x_n)$ having l input combinations for which F is 1 and m input combinations for which F is 0, let the $l \times n$ matrix of the input combinations which are mapped onto 1 be called the α-matrix and the $m \times n$ matrix of the input combinations which are mapped onto 0 be called the β-matrix.

Example: Let a function $F(x_1, x_2, x_3)$ be as shown in Table 1.

Table 1

x_1	x_2	x_3	$F(x_1, x_2, x_3)$	$\frac{1}{4}x_1 + \frac{1}{2}x_2 + \frac{1}{4}x_3$
1	1	0	1	$\frac{3}{4}$
0	1	1	1	$\frac{3}{4}$
0	1	0	1	$\frac{1}{2}$
1	0	1	1	$\frac{1}{2}$
1	0	0	0	$\frac{1}{4}$
0	0	1	0	$\frac{1}{4}$

For this F,

$$|\alpha_{iK}| = \begin{array}{c} \begin{array}{ccc} x_1 & x_2 & x_3 \end{array} \\ \begin{vmatrix} 1 & 1 & 0 \\ 0 & 1 & 1 \\ 0 & 1 & 0 \\ 1 & 0 & 1 \end{vmatrix} \end{array}$$

$$|\beta_{jh}| = \begin{array}{c} \begin{array}{ccc} x_1 & x_2 & x_3 \end{array} \\ \begin{vmatrix} 1 & 0 & 0 \\ 0 & 0 & 1 \end{vmatrix} \end{array}$$

This F is a threshold function and can be realized by the 3-tuple $(^1/_4, \, ^1/_2, \, ^1/_4)$ and a threshold value $T = \, ^3/_8$.

At this juncture we should like to make the following two observations: (1) If F is realizable with any threshold gate, it can also be realized with a gate having $T = \, ^1/_2$, provided that a properly weighted constant (0 or 1) is introduced as an additional input. (2) If F is realizable with a single threshold gate having $T = \, ^1/_2$, then for any input combination $(r_1, r_2, \cdots, r_n)$, $F(r_1, r_2, \cdots, r_n) = \overline{F}(\overline{r}_1, \overline{r}_2, \cdots, \overline{r}_n)$, where $\overline{r}_i = 1 - r_2$ $(i = 1$ to $n)$ and $\overline{F} = 1 - F$.

We shall now establish observation (1). Let F be realizable with weights $(w_1, w_2, \cdots, w_n)$ and a threshold value $T \neq \, ^1/_2$. If $T = \, ^1/_2 + \varepsilon \, (0 < \varepsilon < \, ^1/_2)$, introduce the constant 0 as an additional input with a weight of $2\varepsilon/(1 + 2\varepsilon)$, and multiply each w_k by $1/(1 + 2\varepsilon)$. If $T = \, ^1/_2 - \varepsilon$, add the constant 1 as an additional input with a weight of $2\varepsilon/(1 + 2\varepsilon)$, and multiply each w_k by $1/(1 + 2\varepsilon)$. Hence observation (1). This observation enables us to limit our search for a threshold gate to only those having $T = \, ^1/_2$, by introducing an additional column of zeros and a column of ones to any given table.

Proof of observation (2) is simple and hence omitted. This observation enables us to combine the α- and β-matrices into a single matrix. Thus our problem reduces to that of finding an n-tuple $(w_1, w_2, \cdots, w_n)$ with $w_i \geq 0$ and $\Sigma w_i = 1$ such that ⊛ $\cdots \Sigma w_k \alpha_{ik}^* > 1/2$ for all $i = 1, 2, \cdots, 1 + m$, where

$$\| \alpha_{ik}^* \| = \begin{vmatrix} \| \alpha_{ik} \| \\ \cdots \\ \| \overline{\beta}_{jk} \| \end{vmatrix} (1 + m) \times n$$

Thus to find the matrix $\| \alpha_{ik}^* \|$ for Table 1, we would first add columns of

6. APPLICATIONS OF MINIMAX THEOREMS

ones and zeros, yielding

$$\begin{Vmatrix} x_1 & x_2 & x_3 & 1 & 0 \\ 1 & 1 & 0 & 1 & 0 \\ 0 & 1 & 1 & 1 & 0 \\ 0 & 1 & 0 & 1 & 0 \\ 1 & 0 & 1 & 1 & 0 \end{Vmatrix} = \|\alpha_{ik}\|$$

$$\|\beta_{jk}\| = \begin{Vmatrix} x_1 & x_2 & x_3 & 1 & 0 \\ 1 & 0 & 0 & 1 & 0 \\ 0 & 0 & 1 & 1 & 0 \end{Vmatrix}$$

Then $\|\beta_{jk}\|$ would be complemented and appended to $\|\alpha_{ik}\|$ to give $\|\alpha_{ik}{}^*\|$.

$$\|\alpha_{ik}{}^*\| = \begin{Vmatrix} x_1 & x_2 & x_3 & 1 & 0 \\ 1 & 1 & 0 & 1 & 0 \\ 0 & 1 & 1 & 1 & 0 \\ 0 & 1 & 0 & 1 & 0 \\ 1 & 0 & 1 & 1 & 0 \\ 0 & 1 & 1 & 0 & 1 \\ 1 & 1 & 0 & 0 & 1 \end{Vmatrix}$$

Finally we shall give the following simple sufficient condition that an n-tuple $(w_1, w_2, \cdots, w_n)$ satisfying ⚠ does not exist. If there exists an $(1 + m)$-tuple $(v_1, v_2, \cdots, v_{1+m})$ with $\sum_{i=1}^{1+m} v_i = 1$ such that $\sum_i v_i \alpha_{ik}{}^* < {}^1/_2$ for all k, then there cannot exist any w satisfying *. Proof of this statement is simple and hence omitted.

PROOF OF THEOREM 6.2.1: From the minimax theorem it follows that we can always find an $(1 + m)$-tuple $(v_1{}^*, v_2{}^*, \cdots, v_{1+m}{}^*)$ and an n-tuple $w = (w_1{}^*, w_2{}^*, \cdots, w_n{}^*)$ and a value v such that $\sum_i v_i{}^* \alpha_{ik}{}^* < v$ for all k, and $\sum_k w_k{}^* \alpha_{ik}{}^* > v$ for all i. Now it is easily seen that, if $v > {}^1/_2$, then F is realizable with w and $T = {}^1/_2$. If $v \leq {}^1/_2$, then by the sufficient condition just stated above, it follows that F is not realizable. Thus the proof is complete.

6.3 PROOF OF THEOREM 6.3.1

For proving theorem 6.3.1 we need a few definitions and theorems of Krein [9].

Definition 6.3.1: A closed set, C, in a real Banach space, E, is called a cone if:

(1) $x, y \in C$ implies $x + y \in C$.

(2) $x \in C$, $\lambda \geq 0$, implies $\lambda x \in C$.

(3) $x, y \in C$, $x + y = 0$, implies $x = y = 0$.

Definition 6.3.2: A cone, C, in a real Banach space, E, is called normal if there exists a $\delta > 0$ such that, for any $x, y \in C$,

$$\| x + y \| \geq \delta \max \{ \| x \|, \| y \| \}$$

Definition 6.3.3: If a cone, C, in a real Banach space, E, satisfies $E = C - C$, then the set of continuous linear functionals on E that are nonnegative on C satisfies the axioms above for a cone and is called the dual cone C^* for C in the dual space E^*.

Definition 6.3.4: A linear operator, A, on a real Banach space, E, is called positive with respect to a cone C whenever A leaves the cone invariant. It is strongly positive if the cone has interior and if for any $x \in C$, $x \neq 0$, there exists a positive integer $n(x)$ such that $A^{n(x)}x$ is an interior point.

Theorem a (Krein [9])

If a normal cone, C, in a real Banach space, E, has an interior point, then the set $I_u = \{x : u \pm x \in C\}$ is bounded.

Theorem b (Krein [9])

Let C be a cone with interior in a real Banach space, E. If $x_0 \notin C$, then there exists an $f \in C^*$ with $f(x_0) < 0$.

PROOF OF THEOREM 6.3.1: By assumption, $E = E^{**}$ and both C and C^* have interior. Theorem b coupled with reflexivity of E gives $C = C^{**}$. In fact, C and C^* are normal cones. To see this, let u be interior to C, with $\| u \| = 1$ and $f_1, f_2 \in C^*$. For some $p > 0$, $u \pm p \dfrac{x}{\| x \|} \in C$ for all $x \in E$. Therefore

$$f_1(u) \geq p f_1 \left(\frac{x}{\| x \|} \right)$$

and

$$f_2(u) \geq p f_2 \left(\frac{x}{\| x \|} \right)$$

Thus $\| f_1 + f_2 \| \geq f_1(u) + f_2(u) \geq p \max (\| f_1 \|, \| f_2 \|)$. This proves that C^* is normal. Similarly, we can prove the normality of C.

Let $\Pi_1 = \{f : f(y_0) = 1\} \cap C^*$, where y_0 is an interior point of C with $\| y_0 \| = 1$.

Let $\Pi_2 = \{x : f_0(x) = 1\} \cap C$, where f_0 is an interior point of C^* with $\| f_0 \| = 1$.

Since y_0, f_0 are interior to C and C^*, respectively, it follows that, for some $p_1, p_2 > 0$,

$$f(y_0) \geq p_1 \| f \|$$

and

$$f(x) \geq p_2 \| x \|$$

Thus Π_1 and Π_2 are bounded; further, they are closed. Since Π_1 and Π_2 are convex sets, they are closed also in their respective weak topologies (for otherwise we can arrive at a contradiction by strong separation theorem). By their reflexivity they are weakly compact.

Consider the bilinear functional

$$K_\lambda(f, y) = (f, (A - \lambda) y) = f((A - \lambda) y)$$

on $E^* \times E$ for any real λ. The function $K_\lambda(f, y)$ is weakly continuous in each variable when the other is fixed.

We have by Sion's minimax theorem (see Chapter 5)

$$\min_{\Pi_2} \max_{\Pi_1} K_\lambda(f, y) = \max_{\Pi_1} \min_{\Pi_2} K_\lambda(f, y)$$

Let $v(\lambda) = \min_{\Pi_2} \max_{\Pi_1} K_\lambda(f, y)$. We observe the following:

(1) By the strong positivity of the operator A, $v(0) > 0$.

(2) $v(\lambda)$ is continuous and nonincreasing.

(3) $v(\lambda) \to -\infty$ as $\lambda \to +\infty$.

From the above three observations it follows that $v(\lambda_0) = 0$ for some $\lambda_0 > 0$. Further, since Π_1 and Π_2 generate the respective cones—that is:

$$\bigcup_{\alpha \geq 0, f \in \Pi_1} (\alpha f) = C^* \qquad \bigcup_{\alpha \geq 0, x \in \Pi_2} (\alpha x) = C$$

$$K_{\lambda_0}(g, y) \geq 0 \quad \text{for all} \quad y \in C$$

$$K_{\lambda_0}(f, z) \leq 0 \quad \text{for all} \quad f \in C^*$$

where g and z are the optimal solutions to the kernel $K_{\lambda_0}(f, y)$. By theorem b of Krein [9], we get

$$(A - \lambda I)^* g \in C^* \qquad \text{and} \qquad (\lambda_0 - A) z \in C$$

(Here $(A - \lambda I)^*$ is the conjugate operator of $(A - \lambda I)$, where I is the identity operator in the respective space. We shall denote by $(A - \lambda)$ the operator $(A - \lambda I)$ from now on.) Therefore

$$A(\lambda_0 - A)\,z + (\lambda_0 - A)\,\lambda_0 z \in C$$

That is,

$$\lambda_0^2 z - A^2 z \in C$$

Proceeding inductively, we have

$$\lambda_0^{n(z)} z - A^{n(z)} z \in C$$

For suitable $n(z)$ the strong positivity of A implies that $A^{n(z)}z$ belongs to the interior and that $\lambda_0^{n(z)}z$ and hence z belong to the interior of C. (Here we use the fact that the sum of an interior and a point of a cone is an interior point of the cone.)

Since z is optimal, $(A - \lambda_0)^*g = 0$, for otherwise

$$(g, (A - \lambda_0)\,z) = ((A - \lambda_0)^*g, z) > 0$$

contradicting the optimality of z. Thus $A^*g = \lambda_0 g$. Further, by strong positivity of A, for any $y \in C - \{0\}$,

$$\lambda_0^{n(y)} g(y) = (A^{*n(y)}g)\,(y) = g(A^{n(y)}y) > 0$$

for suitable $n(y)$. Thus g is strictly positive on $C - \{0\}$.

We claim that $(\lambda_0 - A)\,z = 0$, for otherwise, since $(\lambda_0 - A)\,z \in C$, for any $y \in C - \{0\}$, we have $g(y) > 0$, and thus $(g, (\lambda_0 - A)\,z) > 0$; that is, $(g, (A - \lambda_0)\,z) < 0$. This contradicts the optimality of g. Thus $Az = \lambda_0 z$, and z belongs to the interior of C.

Now let $Ax = \lambda_0 x$ for some $x \in E$ linearly independent of z in some complex extension $\bar{E}$ of E. Since λ_0 is real, we can even assume that x is real. (See [9] for a complex extension of E.) We can choose some α real with $x + \alpha z$ lying on the boundary of C. Thus $x + \alpha z$ is a boundary point of C and also an eigenvector. Since $\lambda_0 > 0$, the strong positivity would imply that $x + \alpha z$ is interior to C. This contradiction establishes the fact that

$$S_{\lambda_0} = \{y: Ay = \lambda_0 y, y \in E\}$$

is one-dimensional. Similarly

$$S_{\lambda_0}{}^* = \{f: A^*f = \lambda_0 f, f \in E^*\}$$

is one-dimensional.

Let $Av = \mu v$, $v \in C - \{0\}$, $\mu \neq \lambda_0$. Then

$$\mu(g, v) = (g, \mu v) = (g, Av) = (A^*g, v) = \lambda_0(g, v)$$

Since $\lambda_0 \neq \mu$, $(g, v) = 0$. But we know that $g(v) > 0$ for any $v \in C - \{0\}$. This proves the last assertion of the theorem.

To prove that λ_0 is the spectral radius of A, let $|\lambda| > \lambda_0$ and $y \in E$ with $z \pm y \in C$. By theorem a, $\|y\| \leq M$ for all such $y \in E$ for some M. Since $AC \subseteq C$,

$$\lambda_0^{-n} A^n (z \pm y) = z \pm \lambda_0^{-n} A^n y \in C$$

and

$$\| \lambda_0^{-n} A^n y \| \leq M$$

Therefore, if $|\lambda| > \lambda_0$, then

$$\sum_0^\infty \frac{\| A^n y \|}{|\lambda|^{n+1}} = \sum_0^\infty \frac{\| \lambda_0^{-n} A^n y \|}{|\lambda|^{n+1}} |\lambda_0^n| \leq M \sum_0^\infty \frac{|\lambda_0|^n}{|\lambda|^{n+1}} < \infty$$

Thus $R(\lambda, A) y = (A - \lambda)^{-1} y = -\sum_0^\infty \frac{A^n y}{\lambda^{n+1}}$ exists for $|\lambda| > \lambda_0$. Since

$z + p \dfrac{x}{\| x \|} \in C$ for some $p > 0$, for any $x \in E - \{0\}$ we can take

$y = p \dfrac{x}{\| x \|}$, and this would imply that $-\sum_0^\infty \dfrac{A^n x}{\lambda^{n+1}}$ exists for any x and that

$R(\lambda, A)$ exists. This shows that λ_0 is the spectral radius of A. The proof of the theorem is complete.

6.4 PROOF OF THEOREM 6.4.1

We develop the following preliminaries for the proof of theorem 6.4.1.

Definition: A subset, A, of a complete separable metric space, Y, is analytic if there is a continuous function f from Ω to Y with $A = f\Omega$, where Ω is the space of infinite sequences of positive integers.

Lusin's second separation principle asserts that, if A, B are analytic subsets of Y, there are analytic sets A_1, B_1 such that

$$A_1 \supset A, \ B_1 \supset B$$

$$A_1 \cup B_1 = Y$$

$$A_1 \cap B_1 = A \cap B$$

To obtain this theorem from the result of Gale and Stewart, for any finite sequence $x = (n_1, n_2, \cdots, n_k)$ of positive integers, denote by $\Omega(x)$ the set of all $w \in \Omega$ with x initial segment, and by $R(x)$, $S(x)$ the closures of $f\Omega(x)$, $g\Omega(x)$, respectively.

We associate with each $y \in Y$ a game, $G(y)$, with two players, α and β, played as follows. The players alternately choose positive integers, α choosing first, each choice made with complete information about previous choices. A play

$$\Pi = (m_1, n_1, m_2, n_2, \cdots)$$

will be called a win for α (in $G(y)$) if there is a positive integer k for which $y \in R(m_1, m_2, \cdots, m_i)$ for $1 < i < k$, $y \notin S(n_1, n_2, \cdots, n_k)$; a win for β if there is a positive integer k for which $y \in S(n_1, n_2, \cdots, n_i)$ for $1 < i < k$, $y \notin R(m_1, m_2, \cdots, m_k)$; and a draw if for every positive integer k, $y \in R(m_1, m_2, \cdots, m_k)$ and $y \in S(n_1, n_2, \cdots, n_k)$. Informally, α is trying to produce a $w = (m_1, m_2, \cdots)$ with $f(w) = y$, and β is trying to produce a $w = (n_1, n_2, \cdots)$ with $g(w) = y$. The first player to fail drastically, in the sense of producing a finite sequence x for which y is not even in the closure of the image of the set of w consistent with x, loses. If neither ever fails drastically, so that, as is easily checked, both succeed, the game is a draw.

Denote by A_1 the set of y such that α can force a draw or better in $G(y)$, and by B_1 the set of y such that β can force a draw or better in $G(y)$. Then A_1, B_1 have the properties asserted in the theorem. We sketch the proof.

First, A_1 is analytic. For, the set D of all pairs (y, φ), such that φ is a strategy for α that forces at least a draw in $G(y)$, is a Borel set, and A_1 is its projection on Y. Similarly A_2 is analytic.

Second, $A_1 \supset A$, for if $y \in A$, there is a $w = (m_1, m_2, \cdots)$ with $f(w) = y$. So α, by playing $m_1, m_2, \cdots$, never drastically fails, and thus secures at least a draw. Similarly $B_1 \supset B$.

Third, $A_1 \cup B_1 = Y$, for the set Q of plays Π that win for α in $G(y)$ is open, so that, according to Gale–Stewart theorem, either α can force $\Pi \in Q$, so that $y \in A_1$, or β can force $\Pi \in Q^c$, so that $y \in B_1$.

Finally, $A_1 \cap B_1 = A \cap B$, for, if $y \in A_1 \cap B_1$, either player can force a draw or better in $G(y)$. If both do this, the resulting play

$$\Pi = (m_1, n_1, m_2, n_2, \cdots)$$

is a draw. Then

$$y = f(m_1, m_2, \cdots) = g(n_1, n_2, \cdots)$$

That is,

$$y \in A \cap B$$

This completes the proof of theorem 6.4.1.

Examples of Borel sets X, Y, B such that (1) $B \subset X \times Y$, (2) the projection of B on X is X, but (3) for no Borel measurable d mapping X into Y is the graph of d, a subset of B, have been given by Sierpenski [14] and Addison [1]. Such examples are of interest in dynamic programming (see for instance [4]), since if we interpret X as the set of states of some system,

Y as the set of available acts, and $I_B(x, y)$, where I_B is the indicator of B, as your income if the system is in state x and you choose act y, you can earn 1 in every state, but there is no Borel measurable plan—that is, function d from X into Y, with $d(x)$ specifying the act to be chosen when the system is in state x—that earns 1 in every state. Blackwell [6] has produced a new example simpler than those previously given. The proof that it is an example uses ideas from Addison's construction and the theorem of Gale–Stewart [7]. Interested readers can refer to Blackwell [6].

REFERENCES

[1] Addison, J. W. (1958). Separation principles of classical and effective descriptive set theory. *Fundamenta Math.* **46**, 123–135.
[2] Akers, S. B., Jr. (1961). Threshold logic and two-person zero-sum games. In "Switching Circuit Theory and Logical Design," Conference paper read at Detroit, Michigan.
[3] Blackwell, D. (1961). Minimax and irreducible matrices. *J. Math. Anal. Appl.* **3**, 37–39.
[4] Blackwell, D. (1965). Discounted dynamic programming. *Ann. Math. Stat.* **36**, 226–235.
[5] Blackwell, D. (1967). "Infinite games and analytic sets," Preprint.
[6] Blackwell, D. (1967). "A Borel set not containing a graph," Preprint.
[7] Gale, D., and Stewart, F. M. (1953). Infinite games with perfect information. In "Contributions to the Theory of Games," Vol. 2, pp. 245–266 (*Ann. Math. Studies No. 28*), edited by H. W. Kuhn and A. W. Tucker. Princeton University Press, Princeton, New Jersey.
[8] Gross, O. (1964). The rendezvous value of a metric space. In "Advances in Game Theory." pp. 49–53 (*Ann. Math. Studies No. 52*), edited by M. Dresher, L. S. Shapley and A. W. Tucker, Princeton University Press Princeton, New Jersey.
[9] Krein, M. G., and Rutnam, M. A. (1950). Linear operators leaving invariant a cone in a Banach space. *Am. Math. Soc.* Translation No. 26, 1–128.
[10] Lusin, N. (1930). "Leçons sur les Ensembles Analytiques et leurs Applications," Gauthier-Villars, Paris.
[11] Parthasarathy, T. (1966). Minimax Theorems and Product Solutions, Ph.D. thesis submitted to Indian Statistical Institute, Calcutta.
[12] Prohorov, Yu. V. (1956). Convergence of random processes and limit theorems in the theory of probability. In "Theory of Probability and Its Applications" (English translation), Vol. 1, pp. 157–214.
[13] Raghavan, T. E. S. (1965). On linear operators leaving a convex set invariant in Banach spaces. *Sankhya* **27**, 293–302.
[14] Sierpenski, W. (1931). Sur deux complimentaires analytiques nonseparables B. *Fundamenta Math.* **17**, 296–297.
[15] Varadarajan, V. S. (1958). Weak convergence of measures on separable metric spaces. *Sankhya* **19**, 15–22.

Chapter Seven

NON-ZERO-SUM NONCOOPERATIVE TWO-PERSON GAMES

7.1 INTRODUCTION AND EXAMPLES

The elegant theory of zero-sum two-person games is hardly applicable in many realistic game theoretic problems in economics and in other social sciences. Even if we accept the noncooperation of players, the game is often non-zero-sum—that is, the game is not completely antagonistic. In this chapter we shall study the mathematical aspects of such games. Like the optimal strategies in the zero-sum two-person case, we have what are called equilibrium points for these more general cases. In the case with two players we shall also give an algorithm for finding one such equilibrium pair. We shall also discuss the relationship between equilibrium points and solutions of a certain nonlinear programming problem. We shall start with some classical examples of such games.

Example 7.1.1: Two suspects of a crime are taken into custody and are imprisoned in separate jails. The district attorney is certain that both of them have been responsible for the crime, but he can hardly prove it with the available evidence. The prisoners have the choice of confessing or not confessing. If both do not confess, the attorney can cook up a suitable case and punish them mildly. If both confess, then they are punished, but not severely. If one of them confesses, but not the other, then the one who confesses gets the mildest punishment, but the other gets the severest punishment. In terms of years of punishment, this has the following form:

		Prisoner II	
		Confess	Not confess
	Confess	3 years' imprisonment for both	3 months' imprisonment for prisoner I, and 8 years for prisoner II
Prisoner I	Not confess	3 months' imprisonment for prisoner II, and 8 years for prisoner I	2 years' imprisonment for both

How should they react, if they are criminals by profession?

Example 7.1.2: A certain commodity is produced by two producers. The price of the commodity in the market depends on their total production, and they also have restrictions in their production capacity. Both want to maximize their individual profits. Further, they are highly noncooperative. How best should they produce?

7.2 EQUILIBRIUM THEOREMS

The analog of minimax theorem for the non-zero-sum case is the existence of equilibrium pairs in mixed strategies. Let $K_1(x, y)$ and $K_2(x, y)$ be two real valued functions on $X \times Y$. A point $(x^0, y^0) \in X \times Y$ is called an equilibrium pair if

$$K_1(x^0, y^0) \geq K_1(x, y^0) \quad \text{for all} \quad x \in X$$
$$K_2(x^0, y^0) \geq K_2(x^0, y) \quad \text{for all} \quad y \in Y$$

We shall prove the following theorem of Nash [10] on the existence of equilibrium pairs in mixed strategies for finite non-zero-sum two-person games.

Theorem 7.2.1

Let $A = (a_{ij})$, $B = (b_{ij})$, $1 \leq i \leq m$, $1 \leq j \leq n$, be two payoff matrices for player I and player II, respectively. If the pure strategy i is chosen by player I and the pure strategy j is chosen by player II, then let player I receive a_{ij} units and let player II receive b_{ij} units. If they are noncooperative, then there exists a probability vector $x^0 = (\xi_1^0, \xi_2^0, \cdots, \xi_m^0)$ for player I, and there exists a probability vector $y^0 = (\eta_1^0, \eta_2^0, \cdots, \eta_n^0)$ for player II such that

$$K_1(x^0, y^0) = \sum_i \sum_j a_{ij}\xi_i^0\eta_j^0 \geq \sum_i \sum_j a_{ij}\xi_i\eta_j^0 = K_1(x, y^0)$$

$$\text{for all probability vectors} \quad x = (\xi_1, \xi_2, \cdots, \xi_m)$$

and

$$K_2(x^0, y^0) = \sum_i \sum_j b_{ij}\xi_i^0\eta_j^0 \geq \sum_i \sum_j b_{ij}\xi_i^0\eta_j = K_2(x^0, y)$$

$$\text{for all probability vectors} \quad y = (\eta_1, \eta_2, \cdots, \eta_n)$$

PROOF: Evidently $X \times Y$ is a compact convex set where X is the set of all probability vectors for player I and Y is the set of all probability vectors for player II. Let ψ be a point to set mapping of $X \times Y$ into itself defined by

$$\psi : (x_*, y_*) \rightarrow \{(x', y') : \begin{array}{l} \max_X K_1(x, y_*) = K_1(x', y_*) \\ \\ \max_Y K_2(x_*, y) = K_2(x_*, y') \end{array}\}$$

Since K_1, K_2 are bilinear in x, y, we have $\psi(x_*, y_*)$ a compact convex set in $X \times Y$. One can easily check that the conditions of Kakutani's theorem (1.11.5) are fulfilled for ψ. Thus from the theorem we have an $(x^0, y^0) \in \psi(x^0, y^0)$. That is,

$$K_1(x^0, y^0) \geq K_1(x, y^0) \quad \text{for all} \quad x \in X$$

$$K_2(x^0, y^0) \geq K_2(x^0, y) \quad \text{for all} \quad y \in Y$$

The theorem is generalized as in the general minimax theorems to the following theorem of Isoda and Nikaido [14].

Theorem 7.2.2

Let X and Y be two compact convex sets in R^m and R^n, respectively. Let $K_1(x, y)$ and $K_2(x, y)$ be two continuous functions on $X \times Y$, and further let $K_1(x, y)$ be concave in x for fixed y and $K_2(x, y)$ concave in y for fixed x. Then there exists an equilibrium pair (x^0, y^0).

PROOF: Let $p = (x, y) \in X \times Y$, and $q = (x_*, y_*) \in X \times Y$.
Define:

$$f(p, q) = K_1(x, y^*) + K_2(x^*, y)$$

We shall first show that there exists a q^0 for which

$$\max_{p \in X \times Y} f(p, q^0) = f(q^0, q^0)$$

If not, then for each $q \in X \times Y$ there exists a $p \in X \times Y$ such that $f(p, q) > f(q, q)$. Let

$$G_p = \{q: f(p, q) > f(q, q)\}$$

Since f is continuous, trivially G_p is open and clearly $X \times Y$ is covered by such G_p's. The compactness of $X \times Y$ implies that a finite of them will suffice to cover $X \times Y$. Let them be $G_{p_1}, G_{p_2}, \cdots, G_{p_k}$. Now define

$$\varphi_j(q) = \max\left(f(p_j, q) - f(q, q), 0\right)$$

The $\varphi_j(q)$'s are nonnegative, and by the definition of G_p's at least one $\varphi_j(q) > 0$ for each q.

Let ψ map the convex set $X \times Y$ into itself where

$$\psi: q \to \frac{1}{\varphi(q)} \sum_j \varphi_j(q) \cdot p_j$$

with $\varphi(q) = \sum_j \varphi_j(q)$. Since the φ_j's are continuous, ψ is a continuous map of $X \times Y$ into itself. Thus by Brouwer's fixed point theorem (1.11.4) we have $\psi(\tilde{q}) = \tilde{q}$ for some $\tilde{q} \in X \times Y$. That is,

$$\tilde{q} = \frac{1}{\varphi(\tilde{q})} \sum_j \varphi_j(\tilde{q}) \cdot p_j$$

Thus

$$f(\tilde{q}, \tilde{q}) = f\left(\frac{1}{\varphi(\tilde{q})} \Sigma \varphi_j(\tilde{q}) \cdot p_j, \tilde{q}\right)$$

But $f(p, q)$ is concave in p for fixed q, and thus

$$f(\tilde{q}, \tilde{q}) \geq \frac{1}{\varphi(\tilde{q})} \Sigma \varphi_j(\tilde{q}) f(p_j, \tilde{q}) \qquad (*)$$

But this is not true, for when $\varphi_j(\tilde{q}) > 0$, then $f(\tilde{q}, \tilde{q}) < f(p_j, \tilde{q})$, and further when $\varphi_j(\tilde{q}) = 0$,

$$\varphi_j(\tilde{q}) f(p_j, \tilde{q}) = \varphi_j(\tilde{q}) f(\tilde{q}, \tilde{q})$$

But, since some $\varphi_j(q) > 0$, we have

$$f(\tilde{q}, \tilde{q}) < \frac{1}{\varphi(\tilde{q})} \sum_j \varphi_j(\tilde{q}) f(p_j, \tilde{q})$$

Hence we have a contradiction to step (*) above.

Thus we always have a q^0 with

$$\max_{p \in X \times Y} f(p, q^0) = f(q^0, q^0)$$

That is, $K_1(x, y^0) + K_2(x^0, y) \leq K_1(x^0, y^0) + K_2(x^0, y^0)$ for all $x \in X$, $y \in Y$ and for $(x^0, y^0) = q^0$. Putting $x = x^0$ in the above inequality, we get

$$K_2(x^0, y) \leq K_2(x^0, y^0)$$

Similarly, by putting $y = y^0$ we get

$$K_1(x, y^0) \leq K_1(x^0, y^0)$$

Thus we have an equilibrium pair.

As in the zero-sum two-person games played on the unit square with a continuous payoff, we could prove the following theorem.

Theorem 7.2.3

Let $K_1(x, y)$ and $K_2(x, y)$ be two continuous functions on $X \times Y$, where X and Y are compact subsets of finite dimensional Euclidean spaces. If F and

G denote any probability distribution functions on X and Y, respectively, with

$$K_i(F, G) = \iint_{X \times Y} K_i(x, y) \, dF(x) \, dG(y), \qquad i = 1, 2$$

then $K_1(F, G)$ and $K_2(F, G)$ have an equilibrium pair in the space of distribution functions on X and Y.

PROOF: Essentially, if we check $K_i(F, G)$ to be bilinear and continuous in some sense on the space $\mathscr{F} \times \mathscr{G}$ ($\mathscr{F}$ and $\mathscr{G}$ are the space of probability distributions in X and Y), then we could repeat verbatim the proof of the previous theorem, except for the fact that we would apply Schauder's generalization of Brouwer's theorem 1.11.6 to this case. To see this, one invokes the following two theorems.

THEOREM (RIESZ [2]): If $C(X)$ denotes the space of continuous functions on a compact metric space X, then the dual space of $C(X)$ is the space of signed measures on the Borel subsets of X.

THEOREM (VARADARAJAN [16]): If X is a compact metric space, the space of probability measures on X is a compact metric space in the w^* topology of $C^*(X)$, the dual of $C(X)$.

Now to prove the theorem, what one has to see is the continuity of $K_i(F, G)$, $i = 1, 2$, in the w^* topology. That is, we assign the w^* topology to $\mathscr{F} \times \mathscr{G}$ viewed as a subset of $C^*(X) \times C^*(Y)$ and check that $K_i(F, G)$, $i = 1, 2$, are continuous in that topology. (This is highly nontrivial.) The rest of the proof is a repetition of the previous proof as indicated above. One could also prove this by Glicksberg's generalization [3] of Kakutani's fixed-point theorem (1.11.5). We shall apply our second theorem of this section to the following problem in economics which we vaguely stated in the beginning. We cannot claim that the model is realistic. From a mathematician's angle, perhaps it may be satisfactory. This was first given by Wald [18] in a more general form.

Example 7.2.1: Two sellers who are highly noncooperative are faced with the problem of selling a finite unit of a certain commodity. The consumer's demand function, $p = d(x)$, gives the price that the consumers are prepared to pay when the market has a total supply of x units of the commodity. The sellers have a cost $c(y)$ if they send y units for sales. Each seller just aims to maximize his profit without bothering about the other. How best should they sell their total stock? For a special case, this is solved in the following theorem [18].

Theorem 7.2.4

Let the demand function, $d(x)$, be a monotonically decreasing, twice-differentiable, continuous concave function in $0 \leq x \leq x_0$, with $d(0) > 0$ and $d(x) = 0$ for $x \geq x_0 > 0$. Let $a \geq x_0$, $b \geq x_0$, be the available stocks in the hands of sellers I and II, respectively. Let the cost function, $c(x)$, be a strictly increasing, twice-differentiable, non-negative convex function with $c'(0) < d(0)$. If they want to maximize their profits optimally (here profit is gross income — cost), then they both should sell the same amount, x^*, units of the commodity, where x^* is the unique solution of the equation

$$d(2x) + xd'(2x) - c'(x) = 0$$

on the interval $0 < x < x_0/2$.

PROOF: Let x_1, x_2 be the amount of units they send to the market. Their profits are

$$K_1(x_1, x_2) = x_1 d(x_1 + x_2) - c(x_1)$$

and

$$K_2(x_1, x_2) = x_2 d(x_1 + x_2) - c(x_2)$$

If they want to sell optimally, it is intuitively clear that the total amount, $x_1 + x_2$, they send should not exceed x_0; if $x_1 + x_2 > x_0$, their profits are $-c(x_1)$ and $-c(x_2)$. The nature of the cost function implies that $-c(x_1) < 0$, $-c(x_2) < 0$, when $x_1 > 0$, $x_2 > 0$, and $x_1 + x_2 > x_0$.

But we know by the second theorem of this section that there exists an equilibrium pair and certainly it cannot be (x_1, x_2) with $x_1 + x_2 > x_0$. Now let $x_1 + x_2 \leq x_0$. Let $(x_1{}^*, x_2{}^*)$ be an equilibrium pair. Clearly it is not $(0, 0)$, for if we take $(\varepsilon, 0)$ for any positive ε,

$$K_1(\varepsilon, 0) = \varepsilon d(\varepsilon) - c(\varepsilon)$$

Thus

$$K_1(\varepsilon, 0) - K_1(0, 0) = \varepsilon d(\varepsilon) - (c(\varepsilon) - c(0)) = \varepsilon d(\varepsilon) - \varepsilon c'(\zeta)$$

for some $0 < \zeta < \varepsilon$.

Since $c'(0) < d(0)$ by assumption, for some $\varepsilon > 0$, $K_1(\varepsilon, 0) > K_1(0, 0)$. Thus $(x_1{}^*, x_2{}^*) \neq (0, 0)$. Let us assume that $x_1{}^* > 0$. Clearly

$$0 < x_1{}^* < x_0 - x_2{}^*$$

Since $K_1(x_1, x_2{}^*)$ is a maximum at $x_1 = x_1{}^*$, we have

$$\frac{d}{dx_1} K_1(x_1, x_2{}^*) = 0 \quad \text{at} \quad x_1 = x_1{}^*$$

$$d(x_1{}^* + x_2{}^*) + x_1{}^* d'(x_1{}^* + x_2{}^*) - c'(x_1{}^*) = 0 \qquad (*)$$

From our assumption on d and c, we find that $x_2^* \neq 0$ and that

$$\frac{d}{dx_2} K_2(x_1^*, x_2) = 0 \quad \text{at} \quad x_2 = x_2^*$$

$$d(x_1^* + x_2^*) + x_2^* d'(x_1^* + x_2^*) - c'(x_2^*) = 0 \qquad (**)$$

Subtracting equation (**) from equation (*), we find

$$(x_1^* - x_2^*) d'(x_1^* + x_2^*) - (c'(x_1^*) - c'(x_2^*)) = 0 \qquad (***)$$

Since the function $c(x)$ is convex, $c''(x) \geq 0$ (theorem 1.10.3), and if $x_1^* < x_2^*$, then $-c'(x_1^*) + c'(x_2^*) \geq 0$. But in this case $d'(x_1^* + x_2^*) < 0$, and hence (x_1^*, x_2^*) does not satisfy equation (***).

By a similar arrangement we can prove that $x_1^* > x_2^*$ is not true. Thus we have $x_1^* = x_2^* = x^*$. Equation (*) becomes

$$d(2x^*) + x^* d'(2x^*) - c'(x^*) = 0$$

We know that $0 < x^* < x_0/2$.

Consider the equation:

$$h(x) = d(2x) + xd'(2x) - c'(x) = 0 \quad \text{in the interval} \quad 0 < x < \frac{x_0}{2}$$

$$h(0) = d(0) - c'(0) > 0 \quad \text{(by assumption)}$$

$$h\left(\frac{x_0}{2}\right) = \frac{x_0}{2} d'(x_0) - c'\left(\frac{x_0}{2}\right) < 0$$

$$h'(x) = 3d'(2x) + 2xd''(2x) - c''(x) < 0 \quad \text{for} \quad 0 \leq x \leq \frac{x_0}{2}$$

Thus $h(x)$ is strictly decreasing in $0 < x < x_0/2$, and it has a unique root in $0 < x < x_0/2$. Since x^* is a root when (x^*, x^*) is an equilibrium pair, it is unique and given by the solution of the equation

$$h(x) = 0 \text{ in the interval } 0 < x < \frac{x_0}{2}$$

Hence the theorem.

Remark 7.2.1: The minimax theorem for the zero-sum case is a sort of ultimate theorem of the subject, in the sense that all the behavior among two rational players with perfectly antagonistic interest is contained in the notion of mixed strategies. Under repetitions of the game, it makes sense to adopt the optimal mixed strategies. Further, there is a unique value for the game. In addition, if $K(x, y)$ is the expected income when (x, y) are the mixed strategies for the players, then for any optimal mixed strategies (x^*, y^*), (x^0, y^0) for the players,

$$K(x^*, y^*) = K(x^0, y^0) = K(x^*, y^0) = K(x^0, y^*) = \text{value of the game}$$

Further (x^*, y^*), (x^0, y^0), (x^*, y^0), and (x^0, y^*) are all equilibrium pairs for the payoffs $K_1(x, y) = K(x, y)$, $K_2(x, y) = -K(x, y)$.

But in the non-zero-sum case one or many of these properties may be violated for the equilibrium pairs. The equilibrium pairs (x^*, y^*), (x^0, y^0) may not be interchangeable. Even when the players are prohibited from cooperating in a long series of plays, they tend to cooperate indirectly. This is the usual situation when government tries to enforce rules preventing the formation of rings in a market with two sellers. In many of these problems each player can often control the income of the opponent, but not his own. Thus any realistic solution will be very much intermixed with the sociological aspects of the problem. For a critical analysis of these aspects for our first example of this chapter, see [6].

Thus one is led to the study of non-zero-sum cooperative games. Even in this case the theory does not possess the sort of completion of the various strategic aspects in the zero-sum case. Here one tries to defend a solution concept that possesses certain desirable properties. The desirable properties of a solution are rarely all met in practice in a cooperative case, and thus it looks as though one should be satisfied with the various definitions of solutions that are abstractions of concrete situations. We shall not deal with these concepts here. See [6] for a critical analysis of such cooperative two-person non-zero-sum games.

7.3 ALGORITHMS

Even though Nash's theorem asserts the existence of equilibrium points for bimatrix games, it does not tell us how to find one such pair. Although many algorithms have been proposed by Vorobev [17], Kuhn [4], and Mangasarian [7] to determine all equilibrium pairs, they are more of theoretical interest than for actual computation. The algorithm proposed recently by Lemke and Howson [5] seems to be the most effective one for finding an equilibrium pair. This is the subject of our study in this section.

Lemke and Howson's Algorithm

It is convenient to discuss the algorithm in vector notations; we shall use the following notations.

Let A and B denote $m \times n$ payoff matrices; B' denotes the transpose of B; x, x^*, and x^0 denote m vectors; and y, y^*, and y^0 denote n vectors. By $A > 0$ we mean that all entries in A are positive. By $x \geq 0$ we mean that each component of x is nonnegative. By e we denote an appropriate vector with all its entries unity. By E we denote a matrix with all its entries unity. The vectors are represented by a matrix with one column. If (x^0, y^0) is an equilibrium pair for a bimatrix game with payoff matrices A and B for player I and player II, then

$$(x^0, Ay^0) \geq (x, Ay^0) \quad \text{for all} \quad x \geq 0, \qquad (x, e) = 1$$
$$(x^0, By^0) \geq (x^0, By) \quad \text{for all} \quad y \geq 0, \qquad (y, e) = 1$$

Equivalently, by taking x, y as vectors with some entry unity and all other entries zero, we get

$$(x^0, Ay^0) \cdot e \geq Ay^0$$
$$(x^0, By^0) \cdot e \geq B'x^0$$

Instead of considering the given matrices A, B, if one considers $\alpha E - B'$, $\alpha E - A$, for a large positive constant α and appropriate matrices E with all entries unity, then

$$\alpha E - B' > 0, \alpha E - A > 0, \quad \text{for a suitable} \quad \alpha$$

Suppose that (x^*, y^*) satisfies

$$(\alpha E - B') x^* \geq e, x^* \geq 0, \qquad (y^*, ((\alpha E - B') x^* - e)) = 0$$

and

$$(\alpha E - A) y^* \geq e, y^* \geq 0, \qquad (x^*, ((\alpha E - A) y^* - e)) = 0$$

Then

$$x^0 = \frac{x^*}{(x^*, e)}, y^0 = \frac{y^*}{(y^*, e)}$$

is an equilibrium pair to A, B. Conversely, for any equilibrium pair (x^0, y^0) we have an x^*, y^* satisfying the above conditions for some α.

Thus there exists a one-to-one correspondence between equilibrium pairs (x^0, y^0) and (x^*, y^*) satisfying the above conditions.

We now assume that $A, B > 0$, and our problem is to find (x^*, y^*) with

$$B'x^* \geq e, \qquad x^* \geq 0, \qquad (y^*, B'x^* - e) = 0$$
$$Ay^* \geq e, \qquad y^* \geq 0, \qquad (x^*, Ay^* - e) = 0$$

We shall call any such (x^*, y^*) an equilibrium pair. Let X denote

$$X = \{x : x \geq 0, B'x - e \geq 0\}$$

Let $I = (e_1, e_2, \cdots, e_m)$ denote the $m \times m$ identity matrix. Let $B = (b_1, b_2, \cdots, b_n)$, where b_j is the jth column of B. Thus X is specified by points x with

$$(e_i, x) \geq 0, \qquad i = 1, 2, \cdots, m$$
$$(b_j, x) \geq 1, \qquad j = 1, 2, \cdots, n$$

Clearly x belongs to the boundary of X if at least one of these inequalities is an equality. Let $p(x)$, for any x, denote those e_i's or b_j's with

$$(e_i, x) = 0 \quad \text{or} \quad (b_j, x) = 1$$

If, for any x, $p(x)$ consists of e_i's and b_j's with exactly m of them linearly independent then x is unique for these e_i's and b_j's. We can write down $p(x)$ uniquely by the matrix

$$p(x) = (e_{\alpha_1}, e_{\alpha_2}, \cdots, e_{\alpha_k}, \quad b_{\beta_1}, b_{\beta_2}, \cdots, b_{\beta_l})$$

if it has $k + l$ of the e_i's and b_j's, chosen from the columns of I and B, with

$$\alpha_1 < \alpha_2 < \cdots < \alpha_k, \quad \beta_1 < \beta_2 < \cdots < \beta_l.$$

The x's that belong to X with $p(x)$ having rank m would be named as extreme points of X. We shall assume once and for all that the matrices A and B also fulfill the following nondegeneracy assumption.

Assumption: Let the columns of a matrix C be the columns of the matrix (B, I). Then the rank of C is the number of columns in C when $C = p(x)$ for some x. This is analogous to the nondegeneracy assumption that we imposed in linear programming.

As in linear programming, one can overcome this problem when the assumption fails [5]. But we shall give the algorithm only for the case when our assumption is satisfied. The algorithm is described in a series of theorems. But before that we shall see what implications follow by our assumption.

For any $x_0 \in X$, let $p(x_0)$ have rank r; $p(x_0)$ is an $m \times r$ matrix (r could be zero too).

$$p(x_0) = (p_1, p_2, \cdots, p_r)$$

Here p_j is the jth column of $p(x_0)$ for $j = 1, 2, \cdots, r$. Since $p_1, p_2, \cdots, p_r$ are linearly independent, we can extend it with vectors $p_{r+1}, \cdots, p_m$ so that

$$D = (p_1, p_2, \cdots, p_m)$$

is a nonsingular square matrix. Let

$$(D^{-1})' = (p^1, p^2, \cdots, p^m)$$

Thus

$$(p^i, p_j) = 1, \quad i = j$$
$$= 0, \quad \text{otherwise}$$

Consider any point $x = x_0 + \sum_{i=1}^{m} \lambda_i p^i$, where λ_i's are scalars.

Theorem 7.3.1

For a suitable constant k with $\sum_1^m \lambda_i^2 \leq k$, $\lambda_i \geq 0$, for $1 \leq i \leq r$, points x defined by

$$x = x_0 + \sum_1^m \lambda_i p^i$$

belong to X when $x_0 \in X$.

PROOF: Let p be any column of (I, B). Then

$$(p, x) = (p, x_0) + \sum_1^m \lambda_i(p, p^i)$$

If $p = p_i$, $1 \leq i \leq r$, then $(p, x) = (p, x_0) + \lambda_i$. Further, $(p, x_0) = 1$ if p is a column of B, and $(p, x_0) = 0$ if p is a column of I. Thus $(p, x_0) + \lambda_i \geq (p, x_0)$ if and only if $\lambda_i \geq 0$, $1 \leq i \leq r$. If p is any other column of (I, B), then $(p, x_0) > 0$ if $p(\neq p_1, p_2, \cdots, p_r)$ is a column of I, and $(p, x_0) > 1$ if $p(\neq p_1, p_2, \cdots, p_r)$ is a column of B. Thus, when $(p, x_0) > 0$, $(px) > 0$, and when $(p, x_0) > 1$, $(p, x) > 1$ for small values of λ_i's. Thus $x \in X$.

Remark 7.3.1: We can consider this theorem as the main effect of our assumption. As a consequence of our assumption, the above theorem tells us the following:

1. When x_0 is an extreme point of X, or equivalently when $p(x_0)$ is non-singular, the points $x = x_0 + \lambda_i p^i$ belong to X for small nonnegative values of λ_i's. When $\lambda_i > 0$, we get $p(x)$ from $p(x_0)$ by deleting the ith column p_i. Such points x constitute *an open edge* of X with *end point* x_0.

2. If $x_0 \in X$ with $p(x_0)$ having rank $m - 1$, then, for a suitable p^j, say p^m, and suitable $k > 0$ the point

$$x = x_0 + \lambda_m p^m, \qquad |\lambda_m| \leq k,$$

has $p(x) = p(x_0)$. Here $p_m \notin p(x_0)$. Any nonempty collection of such points x constitutes an *open edge* of x.

3. There are exactly m unbounded edges of X, each with one end point of the form $x = ke_i$ for suitable k. This is obvious, for

$$B'(ke_i) - e = k(B'e_i) - e > 0$$

for large k (here we need the property that $B > 0$). Any other open edge of X has two end points. We shall call them *adjacent extreme points*; x_0 and x_1 are adjacent extreme points if and only if $p(x_0)$ and $p(x_1)$ differ only in one column.

Similarly we can consider

$$Y = \{y: y \geq 0, Ay - e \geq 0\}$$

The boundary of Y consists of points y satisfying at least one of the $m + n$ equations:

$$(a_i, y) - 1 = 0, \qquad i = 1, 2, \cdots, m$$

$$(e_j, y) = 0, \qquad j = 1, 2, \cdots, n$$

where a_i is the ith column of A'; that is, $A' = (a_1, a_2, \cdots, a_m)$. Here e_j is the jth column of the $n \times n$ identity matrix. We shall assume that Y also satisfies a nondegeneracy assumption similar to the one we imposed on X. Let $q(y)$ be the matrix analogous to $p(x)$ that we considered for X. Let us consider $Z = X \times Y$. Any $z = (x, y)$ in Z is called an *extreme point* of Z if x is an extreme point of X and y is an extreme point of Y. We shall say that $z = (x, y)$ lies on an open edge of Z if x or y is an extreme point of X or Y and the other lies on an open edge. (x, y) is an equilibrium point with reference to $A, B > 0$ if and only if $(x, y) \in Z$ and

$$(e_i, x)((a_i, y) - 1) = 0, \qquad i = 1, 2, \cdots, m$$

$$(e_j, y)((b_j, x) - 1) = 0, \qquad j = 1, 2, \cdots, n$$

Theorem 7.3.2

Any equilibrium point of the nondegenerate problem is an extreme point of Z.

PROOF: If $z = (x, y)$ satisfies the above conditions, then either $(e_i, x) = 0$ or $(a_i, y) - 1 = 0$ for $i = 1, 2, \cdots, m$, and either $(e_j, y) = 0$ or $(b_j, x) - 1 = 0$ for $j = 1, 2, \cdots, n$ should hold good. But by our nondegeneracy assumption any $x \in X$ can satisfy at most m conditions of the form $(e_i, x) = 0, (b_j, x) - 1 = 0$, and any $y \in Y$ can satisfy at most n conditions of the form $(e_j, y) = 0$, $(a_i, y) - 1 = 0$. The fact that x, y fulfill in total at least $m + n$ conditions of these types implies that they satisfy exactly $m + n$ of these conditions. Thus $x \in X$ satisfies exactly m of the conditions

$$(e_i, x) = 0, \qquad i = 1, 2, \cdots, m$$

$$(b_j, x) - 1 = 0, \qquad j = 1, 2, \cdots, n$$

and x is an extreme point of X. Similarly y is an extreme point of Y. Thus $z = (x, y)$ is an extreme point of Z. Thus any equilibrium point $z = (x, y) \in Z$ is recognized by the following argument. For each s in $1 \leq s \leq m + n$, either the sth column of (I, B) is in $p(x)$, or the sth column of (A', I) belongs to $q(y)$, but not both.

Now let us consider $z = (x, y) \in Z$ satisfying at least $m + n - 1$ of the conditions

$$(e_i, x)((a_i, y) - 1) = 0, \qquad i = 1, 2, \cdots, m$$

$$(e_j, y)((b_j, x) - 1) = 0, \qquad j = 1, 2, \cdots, n$$

Equivalently, let H_s denote the set of $z \in Z$ which satisfy all but, possibly, $(e_s, y)((b_s, x) - 1) = 0$.

Theorem 7.3.3

Each point of H_s is either an extreme point of Z or a point on an open edge of Z.

PROOF: If $z \in H_s$ and satisfies the $m + n$ conditions above, then z is an extreme point. If it satisfies $m + n - 1$, then it satisfies at least $(m + n - 1)$ of

$$(a_i, y) - 1 = 0, \qquad i = 1, 2, \cdots, m$$

$$(e_j, y) = 0, \qquad j = 1, 2, \cdots, n$$

and

$$(e_i, x) = 0, \qquad i = 1, 2, \cdots, m$$

$$(b_j, x) - 1 = 0, \qquad j = 1, 2, \cdots, n$$

Hence it is an open edge of Z.

Theorem 7.3.4

There is a unique unbounded open edge of Z composed of points of H_s.

PROOF: Let $y_0 = k_0 e_s$. For a suitable k_0, $k_0 e_s$ is an extreme point of Y. Since $(e_j, y_0) = 0$ for $j \neq s$, and since they are $n - 1$ in number, the fact that y_0 is an extreme point of Y demands that $(a_i, y_0) - 1 = 0$ be satisfied exactly for one i—say, $i = r$. Thus

$$(a_r, y_0) - 1 = 0$$

$$(e_j, y_0) = 0, \qquad j \neq s$$

For any point $x = k e_r$, $x \in X$ when k is sufficiently large. The points (x, y_0) forming an unbounded edge of Z are points of H_s. If k_1 is the least value of k for which this is true (that is, $k e_r \in X$), then for $x_0 = k_1 e_r$, (x_0, y_0) is the end point of this edge. We shall call this the edge F. Let $s \neq s'$. Let z be an arbitrary point on an edge of Z. Clearly z may be in at most one of H_s or H_s'. This proves our assertion in the theorem.

Theorem 7.3.5

Let z be an extreme point of Z and a point of H_s. There are then one or two open edges of Z, consisting wholly of points of H_s which have z as end point; z is an equilibrium point if and only if there is one such edge.

PROOF: Let z be an extreme point of Z and a point of H_s. We have two cases to discuss.

Case 1: $(e_s, y)((b_s, x) - 1) = 0$. Here z is an equilibrium point. Conversely, any equilibrium point is a point of H_s. Since any equilibrium point is specified by the usual $m + n$ equations, either $(e_s, y) = 0$ or $(b_s, x) - 1 = 0$, but not both.

Suppose that $(e_s, y) = 0$, $((b_s, x) - 1 = 0$ is similarly treated). Since (x, y) is an extreme point of Z, m of the equations

$$(e_i, x) = 0, \qquad i = 1, 2, \cdots, m$$
$$(b_j, x) - 1 = 0, \qquad j = 1, 2, \cdots, n$$

are satisfied. Further, n of the equations

$$(e_j, y) = 0, \qquad j = 1, 2, \cdots, n$$
$$(a_i, y) - 1 = 0, \qquad i = 1, 2, \cdots, m$$

are satisfied. We know that $(e_s, y) = 0$ is one of them. Therefore there are $m + n$ edges of Z with z as a common end point. Along any one of these, precisely one of the $m + n$ conditions satisfied by z is violated. Only that edge along which $(e_s, y) = 0$ is violated will consist of points of H_s.

Case 2: $(e_s, y)((b_s, x) - 1) > 0$. Here $(e_s, y) > 0$ and $(b_s, x) - 1 > 0$. Since (x, y) is an extreme point, it must satisfy as usual the $m + n$ conditions above. Thus we need precisely one q for which, say,

$$(e_q, x) = 0, \qquad ((a_q, y) - 1) = 0$$

Then the two edges, one violating only $(e_q, y) = 0$ and the other violating only $((a_q, y) - 1) = 0$, consist wholly of points of H_s. Along any other edge with end point z, some one equilibrium condition defining H_s must be violated.

We shall call two open edges of Z *adjacent* if they have a common end point. Consider sequences of adjacent open edges of H_s together with their end points defined as s-paths.

The above theorems furnish us an extreme point of Z that lies in H_s. Starting from such an extreme point z, one may move along an edge of Z consisting of points of H_s. Either that edge ends up in an extreme point, say z_1, or it is the single unbounded edge, F, of points of H_s. In the former case, either z_1 is an equilibrium point, in which case one cannot proceed, or else z_1 is not an equilibrium point, in which case there is another edge with points wholly in H_s, with end point z_1 along which one may continue. We shall terminate the process starting at z when either (1) we enter the unbounded edge F of H_s, (2) we reach an equilibrium point other than z itself, or (3) we return to some point previously traveled.

We notice the following: (1) No edge of H_s is traversed twice, for otherwise we shall have an extreme point which is the end point of three edges of the path. (2) The number of extreme points is finite. (3) Only when z is not an equilibrium point may one return to z.

Thus, starting from z one returns to z or not. If one returns to z we call such a path a closed path. If one does not, then he ends up either in an equilibrium point or in the edge F. This leads to two cases: Either z is an equilibrium point or z is not an equilibrium point, in which case one can repeat the process along the other edge of points of H_s which has z as end point. We thus have the next theorem.

Theorem 7.3.6

The set H_s is nonempty, and H_s is the union of a finite number of disjoint s-paths.

Theorem 7.3.7

Let P be that s-path that contains the unbounded edge F. Then P has precisely one equilibrium point. This may be computed by traversing P starting with the unbounded edge P. The number of equilibrium points is finite and odd.

PROOF: Starting with F, the unique path through the s-path, F must terminate in an extreme point which is then an equilibrium point. Any path other than P which is not a closed s-path will have two end points each of which is an equilibrium point. These must be distinct. Hence the theorem. For the degenerate case, see [5].

7.4. NONLINEAR PROGRAMMING AND BIMATRIX GAMES

In this section we shall see how the problem of finding an equilibrium pair is closely related to nonlinear programming problems. We shall use the matrix notations of the previous section.

Theorem 7.4.1

A pair (x^0, y^0) is an equilibrium pair for a bimatrix game with payoff matrices A, B if and only if for some scalars p, q

$$Ay^0 \leq pe, \qquad B'x^0 \leq qe, \qquad (x^0, (A + B) y^0) = p + q$$

PROOF: If (x^0, y^0) is an equilibrium pair, then for $p = (x^0, Ay^0)$, $q = (x^0, By^0)$, the assertion holds trivially. Conversely, let (x^0, y^0, p, q) satisfy the conditions of the theorem:

$$Ay^0 \le pe$$

$$B'x^0 \le qe$$

That is,

$$(x^0, (A + B) y^0) = p + q$$

For any $x \ge 0$, $(x, e) = 1$, $y \ge 0$, $(y, e) = 1$, we have

$$(x, Ay^0) \le p$$

$$(B', x^0, y) \le q$$

and $(x, Ay^0) + (x^0, By) \le (p + q) = (x^0, Ay^0) + (x^0, By^0)$. In particular, $(x^0, Ay^0) \le p$ and $(x^0, By^0) \le q$, but with sum $p + q$. Thus $(x^0, Ay^0) = p$ and $(x^0, By^0) = q$. Therefore

$$(x, Ay^0) \le (x^0, Ay^0) \quad \text{and} \quad (x^0, By) \le (x^0, By^0)$$

This proves that (x^0, y^0) is an equilibrium pair.

Theorem 7.4.2

A pair (x^0, y^0) is an equilibrium pair to a bimatrix game with payoff matrices A, B if and only if (x^0, y^0, p, q) is a solution to the problem

maximize $[(x, Ay) + (x, By) - p - q]$

subject to $Ay \le pe$, $B'x \le qe$, $x, y \ge 0$, $(x, e) = 1$, $(y, e) = 1$

PROOF: The constraints evidently imply that

$$(x, Ay) + (x, By) - p - q \le 0$$

Thus the optimal value of the objective function is nonpositive.

Let (x^0, y^0) be an equilibrium pair. Clearly

$$\{x^0, y^0, p = (x^0, Ay^0), q = (x^0, By^0)\}$$

is feasible, and at (x^0, y^0, p, q), the value of the objective function is zero. Thus it is an optimal solution to the nonlinear problem. Conversely, let (x^0, y^0, p, q) be an optimal solution to the problem. Since by Nash's theorem we always have an equilibrium pair, and since at such a pair the optimal value is zero, so it is at (x^0, y^0, p, q). Thus

$$(x^0, Ay^0) + (x^0, By^0) - p - q = 0$$

An elementary argument shows that (x^0, y^0) is actually an equilibrium pair.

Let us consider the convex sets

$$S = \{(x, q): B'x - qe \leq 0, x \geq 0, (e, x) = 1\}$$

$$T = \{(y, p): Ay - pe \leq 0, y \geq 0, (e, y) = 1\}$$

(Here e's are chosen as usual in appropriate dimensions.) Any $(x^0, q) \in S$ and $(y^0, p) \in T$ that satisfies $(x^0, Ay^0) + (x^0, By^0) - p - q = 0$ is an equilibrium point.

Definition 7.4.1: (x^0, y^0, p, q) is called an extreme equilibrium point if (x^0, q) is an extreme point of S and (y^0, p) is an extreme point of T, and further (x^0, y^0) is an equilibrium point for A, B.

Theorem 7.4.3

Any equilibrium point of a bimatrix game with payoffs A, B is a convex combination of some extreme equilibrium points.

PROOF: Let (x^*, y^*, p^*, q^*) be an equilibrium point. Thus for the linear programming problem,

$$\text{maximize } \{(x, Ay^*) + (x, By^*) - p^* - q\}$$

$$\text{over } S = \{(x, q) : B'x - qe \leq 0, x \geq 0, (x, e) = 1\}$$

any solution and in particular (x^*, q^*) is a convex combination of the extreme points of S. Similarly (y^*, p^*) is a convex combination of the extreme points of T. Let U be the collection of such extreme points in S, and let V be the collection of such extreme points in T. Then

$$((x^*, Ay^*) - p^*e) + ((y^*, B'x) - qe) = 0, \quad \text{for} \quad (x, q) \in U$$

Since $\quad x \geq 0, \quad Ay^* - p^*e \geq 0, \quad y^* \geq 0, \quad B'x - qe \leq 0,$ we have

$$(y^*B'x - qe) = 0, \quad \text{for} \quad (x, q) \in U$$

Since (y^*, p^*) is a convex combination of some vertices $(y, p) \in V$, y^* is a convex combination of some nonnegative y's. Thus from the above equation,

$$(y, B'x - qe) = 0 \quad \text{for} \quad (x, q) \in U \quad \text{and for some} \quad (y, p) \in V$$

Similarly

$$(x, Ay - pe) = 0 \quad \text{for} \quad (y, p) \in V \quad \text{and for some} \quad (x, q) \in U$$

Thus $(xAy - pe) + (yB'x - qe) = 0$ for some $(x, q) \in U$ and some $(y, p) \in V$, or $(x, (A + B) y) - p - q = 0$ for some $(x, q) \in U$ and some $(y, p) \in V$. Hence some $(x, q) \in U$ and some $(y, p) \in V$ are extreme equilibrium points, and (x^*, q^*) and (y^*, p^*) are convex combinations of them.

Thus we could enumerate all the extreme points of S and T. Extreme equilibrium points are those vertices of S and T that satisfy

$$(x, (A + B) y) - p - q = 0$$

There are algorithms available for finding the extreme points of a polyhedra. Any such algorithm could be used. For an example see [1].

7.5 COMPUTATION OF EQUILIBRIUM POINTS AND AN EXAMPLE

In this section we shall compute a pair of Nash equilibrium points for two payoff matrices A_1, B_1. By the algorithm of the previous section it is seen that the problem is equivalent to finding a suitable $\alpha > 0$, such that

(1) $A = \alpha E - A_1 > 0$

(2) $B = \alpha E - B_1 > 0$

(3) $x \geq 0, y \geq 0$

(4) $Ay \geq f$

(5) $B'x \geq e$

(6) $(x, Ay - f) = 0$

(7) $(y, B'x - e) = 0$

Here, as usual, E is the matrix with unity in all entries, and e, f are vectors with unity in all entries.

To start, we shall assume the nondegeneracy hypothesis to be true.

1. Find A, B for some suitable $\alpha > 0$. Let $a_1, a_2, \cdots, a_m$ be the rows of A, and let $b_1, b_2, \cdots, b_n$ be the columns of B. Let $e_1, e_2, \cdots, e_n$ be the basis of unit vectors in R^n. That is, e_j has all but the jth coordinate zero, and the jth coordinate is 1 for $j = 1, 2, \cdots, n$. Let $f_1, f_2, \ldots, f_m$ be unit vectors in R^m.

2. Write down $a_1, a_2, \cdots, a_m$ and $e_1, e_2, \cdots, e_n$ as columns as in Table A.

Table A

	a_1	a_2	$a_3 \cdots a_m$	e_1	$e_2 \cdots e_n$
e_1	a_{11}	a_{21}	$a_{31} \cdots a_{m1}$	1	$0 \cdots 0$
e_2	a_{12}	a_{22}	$a_{32} \cdots a_{m2}$	0	$1 \cdots 0$
$\vdots$					
e_n	a_{1n}	a_{2n}	$a_{3n} \cdots a_{mn}$	0	$0 \cdots 1$

The entries $e_1, e_2, \cdots, e_n$ in the first column indicate that this is the present basis with respect to which the vectors $a_1, a_2, \cdots, a_m$ and $e_1, e_2, \cdots, e_n$ have the above matrix representation (the same as in the simplex tables).

3. Similarly, write down Table B as shown.

Table B

	b_1	b_2	$\cdots$ b_n	f_1	f_2	$f_3 \cdots f_m$
f_1	b_{11}	b_{12}	$\cdots b_{1n}$	1	0	$0 \cdots 0$
f_2	b_{21}	b_{22}	$\cdots b_{2n}$	0	1	$0 \cdots 0$
	b_{31}	b_{32}	$\cdots b_{3n}$	0	0	$1 \cdots 0$
$\vdots$						
f_m	b_{m1}	b_{m2}	b_{mn}	0	0	$0 \cdots 0$

4. Start with $y^0 = (1/l, 0, \cdots, 0)$ as an extreme point of Y where $l = \min(a_{11}, a_{21}, \cdots, a_{m1})$. Clearly $y^0 \in Y$ is an extreme point by theorem 7.3.4. Let $l = a_{31}$.

5. Choose $x^0 = (0, 0, 1/k, 0, \cdots, 0)$ so that x^0 is an extreme point of X. This is so for $k = \min(b_{31}, b_{32}, \cdots, b_{3n})$. Let $k = b_{32}$. By theorem 7.3.4, $(x^0, y^0) \in H_1$, where H_1 consists of points of $X \times Y$ satisfying all the equations:

$$(e_j, y^0)((b_j, x^0) - 1) = 0, \qquad j = 2, 3, \cdots, n$$

$$(f_i, x^0)((a_i, y^0) - 1) = 0, \qquad i = 1, 2, \cdots, m$$

and perhaps $(e_1, y^0)((b_1, x^0) - 1) = 0$.

If this is also satisfied for our (x^0, y^0), then (x^0, y^0) is an equilibrium pair in the sense of Lemke and Howson. By normalizing them we find the Nash equilibrium pair. If it is not satisfied, then proceed to the next instruction.

6. Determine $p(x^0)$, $q(y^0)$ in the notations of the previous section. That is,

$$p(x^0) = \{f_i, b_j: (b_j x^0) - 1 = 0, (f_i x^0) = 0\}$$

$$q(y^0) = \{e_j, a_i: (a_i y^0) - 1 = 0, (e_j y^0) = 0\}$$

In our case

$$p(x^0) = \{f_1, f_2, b_2, f_4, \cdots, f_m\}$$

$$q(y^0) = \{a_3, e_2, e_3, e_4, \cdots, e_n\}$$

7. Change the basis to $p(x^0)$ and $q(y^0)$, and write down Table A* and Table B*. As in the simplex method we throw out e_1 from the basis and bring a_3 into the basis to get Table A*. This is done as follows. Subtract from each row of

Table A suitable multiples of row 1 so that the third entry in the second, third, and nth rows is zero. Divide the first row by a_{31}. We thus get Table A*.

Table A*

a_1	a_2	a_3	$\cdots$	a_k	$\cdots$	a_m	e_1	e_2 $\cdots$ e_j	e_n	λ
α_{11}	α_{21}	1	$\cdots$	α_{k1}	$\cdots$	α_{m1}	q^{11}	q^{12}	q^{1n}	λ_1
α_{12}	α_{22}	0	$\cdots$	α_{k2}	$\cdots$	α_{m2}	q^{21}	q^{22}	q^{2n}	λ_2
α_{1j}	α_{2j}	0	$\cdots$	α_{kj}	$\cdots$	α_{mj}	q^{j1}	$q^{j2}\ \cdots$	q^{jn}	λ_j
α_{1n}	α_{2n}	0	$\cdots$	α_{kn}	$\cdots$	α_{mn}	q^{n1}	$q^{n2}\ldots\ldots$	q^{nn}	λ_n
ξ_1-1	ξ_2-1	ξ_3-1		ξ_k-1		ξ_m-1	y_1^0	y_2^0	y_n^0	

(row labels at left: a_3, e_2, $\vdots$, e_j, e_n)

8. Write down $y^0 = (y_1^0, y_2^0, \cdots, y_n^0)$ in the last row as shown in Table A*.

9. Find $\xi_i = (a_i, y^0)$ and enter in Table A*, $\xi_i - 1$, $i = 1, 2, \cdots, m$, as shown above.

10. Consider each row in the table. For example, for the jth row find

$$\min_{\substack{\alpha_{kj}<0,\ q^{jr}<0 \\ 1\leq k\leq m \\ 1\leq r\leq n}} \left\{ -\frac{(\xi_k - 1)}{\alpha_{kj}}, \ -\frac{y_r^0}{q^{jr}} \right\} = \lambda^*$$

Also find

$$\max_{\alpha_{sj}>0,\ q^{jt}>0} \left\{ -\frac{(\xi_k - 1)}{\alpha_{sj}}, \ -\frac{y_t^0}{q^{jt}} \right\} = \lambda^{**}$$

One can show by the extreme point nature of (x^0, y^0) that, when both λ^* and λ^{**} exist, then one of them is 0. Enter λ^* or λ^{**}, the one that is non-zero, in the λ column as λ_j. Otherwise enter zero.

We claim that $y = y^0 + \lambda_j q^j$ is an extreme point of Y, where q^j is the jth row of the (q^{ij}) matrix in our table. To show this, we shall first prove that

$$(a_k, y^0) + \lambda_j(a_k, q^j) \geq 1$$

Since (q^{ij}) is the inverse of the basis matrix with columns $a_3, e_2, \cdots, e_n$ appearing in the same order, we have

$$(a_k, q^j) = \alpha_{kj}, \qquad k = 1, 2, \cdots, m, \qquad j = 1, 2, \cdots, n$$

Thus $(a_k, y^0) + \lambda_j(a_k, q^j) = \xi_k + \lambda_j\alpha_{kj}$. That is,

$$(\xi_k - 1) + \lambda_j\alpha_{kj} = -\alpha_{kj}\left(-\frac{\xi_k - 1}{\alpha_{kj}} - \lambda_j \right)$$

When $\lambda_j > 0$, the choice of λ_j implies that $-(\xi_k - 1)/\alpha_{kj} - \lambda_j \geq 0$, when $-\alpha_{kj} > 0$, and we then have $(\xi_k - 1) + \lambda_j\alpha_{kj} \geq 0$. On the other hand, when $\lambda_j > 0$ and $\alpha_{kj} > 0$, clearly, by the fact that $y^0 \in Y$,

$$(\xi_k - 1) + \lambda_j\alpha_{kj} \geq 0, \qquad k = 1, 2, \cdots, m$$

A similar proof applies when $\lambda_j < 0$.

The way y is defined, it is also an extreme point of Y. Since y is the other end of an edge with one end at the extreme point, y^0, of Y, we cannot have $|\lambda_j|$ taking two positive values, for otherwise

$$y^0 + \lambda_j^{(1)}q^j \in Y, \qquad y^0 + \lambda_j^{(2)}q^j \in Y, \qquad \lambda_j^{(1)} > 0, \qquad \lambda_j^{(2)} < 0$$

is valid, which contradicts the fact that y^0 is an extreme point of Y.

11. Similarly, form Table B* with similar interpretations for $\eta_1 - 1, \eta_2 - 1, \cdots, \eta_n, \mu_i$, and p^{ij}'s.

Table B*

	b_1	b_2	$\cdots$	b_n	f_1	f_2	$\cdots$	f_m	μ
f_1	β_{11}	β_{12}	$\cdots$	β_{1n}	p^{11}	p^{12}	$\cdots$	p^{1m}	μ_1
f_2	β_{21}	β_{22}	$\cdots$	β_{2n}	p^{21}	p^{22}	$\cdots$	p^{2m}	μ_2
b_2	β_{31}	β_{32}	$\cdots$	β_{3n}	p^{31}	p^{32}	$\cdots$	p^{3m}	μ_3
$\vdots$									$\vdots$
f_m	β_{m1}	β_{m2}	$\cdots$	β_{mn}	p^{m1}	p^{m2}	$\cdots$	p^{mm}	μ_m
	η_1-1	η_2-1	$\cdots$	η_m-1	x_1^0	x_2^0		x_m^0	

12. Find $x_i = x^0 + \mu_i p^i$, $\quad i = 1, 2, \cdots, m$

$\qquad\quad y_j = y^0 + \lambda_j q^j$, $\quad j = 1, 2, \cdots, n$

13. Find $q(y_j)$, $p(x_i)$ for $i = 1, 2, \cdots, m$ and $j = 1, 2, \cdots, n$. If some $\lambda_j > 0$, then the minimum of the ratios that gave $\lambda^* = \lambda_j$ would be attained for some

$$-\frac{(\xi_k - 1)}{\alpha_{kj}} \quad \text{or} \quad -\frac{y_r^0}{q^{jr}}, \quad \alpha_{kj} < 0, \quad q^{jr} < 0$$

Say that it is $-(\xi_m - 1)/\alpha_{mj}$. Put $*$ at α_{mj}, and we have

$$q(y_j) = \{q(y_0) \cup a_m\} - \{e_j\}$$

as a set of columns. In our case it is $\{a_3, e_2, \cdots, e_{j-1}, a_m, e_{j+1}, \cdots, e_n\}$. Similarly, we find $p(x_i)$.

14. Define for each pair (x_i, y_j),

$$M(x_i, y_j) = \{e_r, f_s \colon e_r \in M(x_i, y_j) \;\; \text{if} \;\; e_r \in q(y_j)$$

$$\text{or} \quad b_r \in p(x_i); f_s \in M(x_i, y_j) \;\; \text{if}$$

$$f_s \in p(x_i) \quad \text{or} \quad a_s \in q(y_j)\}$$

15. Check whether, for any pair (x_i, y_j),

$$M(x_i, y_j) = \{e_1, e_2, \cdots, e_n, f_1, f_2, \cdots, f_m\}$$

Then stop at that stage. We can check that (x_i, y_j) forms an equilibrium pair.

16. If no such equilibrium pair exists, find that pair (x, y) for which e_1 alone is missing in $M(x, y)$. Equivalently, $(x, y) \in H_1$. By theorem 7.3.5, we have two such pairs that are other ends of two edges meeting at (x^0, y^0).

17. Choose any one of the two pairs (x, y) in H_1, change the basis to $p(x)$, $q(y)$, and proceed as before. Precisely we end up with an equilibrium point after a finite number of steps by theorem 7.3.7.

Geometrically the theorems of the algorithm tell us the following. If $z_0 = (x_0, y_0) \in H_1$, then we either have (x_0, y_0), an equilibrium point, or we have a graph like the one below:

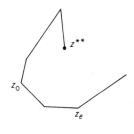

Graph A.

where $z_0, z_1, z_2, \cdots, z_e$ are the adjacent vertices that we find, and z_e is the end of the unbounded edge to which we may reach (all belonging to H_1), in which case we travel along the other edge and reach an equilibrium point z^{**} in a finite number of steps. (Everywhere we take only adjacent vertices in H_1.)

Or else we may be circling around and around by coming back to a vertex traveled without ever crossing an equilibrium point as in the graph below:

Graph B.

Last of all, if we neither reach F from any edge of z in H_1, nor do we cycle as above in adjacent vertices of H_1, then we reach an equilibrium point, z^*, by traveling along an edge. Since we cannot have two edges emanating from such an equilibrium point, we start traveling through the other edge of z. Since, by assumption, we do not end up in F, we reach another equilibrium point z^{**}. This looks like the graph below:

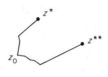

Graph C.

Our algorithm corresponds to the first graph, when we start with z_e, the end point of the unique unbounded edge in H_1. It is no wonder that z_e itself is an equilibrium point, in which case we have a graph like the one below:

Graph D.

Example 7.5.1: Find a pair of Nash equilibrium points to the following matrices A_1, B_1:

$$A_1 = \begin{bmatrix} 6 & 3 & 2 & 8 \\ 4 & 9 & 7 & 2 \\ 8 & 2 & 3 & 6 \end{bmatrix} \qquad B_1 = \begin{bmatrix} 2 & 7 & 8 & 1 \\ 9 & 2 & 4 & 4 \\ 4 & 8 & 3 & 5 \end{bmatrix}$$

For $\alpha = 10$ we have $A = \alpha E - A_1$, $B = \alpha E - B_1$, given by

$$A = \begin{bmatrix} 4 & 7 & 8 & 2 \\ 6 & 1 & 3 & 8 \\ 2 & 8 & 7 & 4 \end{bmatrix} \qquad B = \begin{bmatrix} 8 & 3 & 2 & 9 \\ 1 & 8 & 6 & 6 \\ 6 & 2 & 7 & 5 \end{bmatrix}$$

Let $y_0 = (1/2, 0, 0, 0)$, $x_0 = (0, 0, 1/2)$, be the first extreme pair to start with, as in instructions 4 and 5.

We have $(x_0, y_0) \in H_1$ and $(e_1, y)\,((b_1, x_0) - 1) \neq 0$. We get

<table>
<tr><td colspan="7" align="center">Table A</td><td colspan="7" align="center">Table B</td></tr>
<tr><td></td><td>a_1</td><td>a_2</td><td>a_3</td><td>e_1</td><td>e_2</td><td>e_3</td><td>e_4</td><td></td><td>b_1</td><td>b_2</td><td>b_3</td><td>b_4</td><td>f_1</td><td>f_2</td><td>f_3</td></tr>
<tr><td>e_1</td><td>4</td><td>6</td><td>2</td><td>1</td><td>0</td><td>0</td><td>0</td><td>f_1</td><td>8</td><td>3</td><td>2</td><td>9</td><td>1</td><td>0</td><td>0</td></tr>
<tr><td>e_2</td><td>7</td><td>1</td><td>8</td><td>0</td><td>1</td><td>0</td><td>0</td><td>f_2</td><td>1</td><td>8</td><td>6</td><td>6</td><td>0</td><td>1</td><td>0</td></tr>
<tr><td>e_3</td><td>8</td><td>3</td><td>7</td><td>0</td><td>0</td><td>1</td><td>0</td><td>f_3</td><td>6</td><td>2</td><td>7</td><td>5</td><td>0</td><td>0</td><td>1</td></tr>
<tr><td>e_4</td><td>2</td><td>8</td><td>4</td><td>0</td><td>0</td><td>0</td><td>1</td><td></td><td></td><td></td><td></td><td></td><td></td><td></td><td></td></tr>
</table>

$$q(y_0) = (a_1, e_2, e_3, e_4), \qquad p(x_0) = (b_2, f_1, f_2)$$

We get the new Tables A* and B* by changing the basis to $p(x_0)$ and $q(y_0)$ and filling $\xi_i - 1$ and $\eta_j - 1$ for $i = 1, 2, 3, j = 1, 2, 3, 4$, and the λ's and μ's.

Table A*

	a_1	a_2	a_3	e_1	e_2	e_3	e_4	λ
a_3	2	3	1*	$\frac{1}{2}$	0	0	0	0
e_2	-9	-23*	0	-4	1	0	0	$\frac{2}{23}$
e_3	-6	-18*	0	$-\frac{7}{2}$	0	1	0	$\frac{1}{9}$
e_4	-6*	-4	0	-2	0	0	1	$\frac{1}{6}$
	1	2	0	$\frac{1}{2}$	0	0	0	

Table B*

	b_1	b_2	b_3	b_4	f_1	f_2	f_3	μ
f_1	-1	0	$-\frac{17}{2}$*	$\frac{3}{2}$	1	0	$-\frac{3}{2}$	$\frac{5}{17}$
f_2	-23*	0	-22	-14	0	1	-4	$\frac{2}{23}$
b_2	3	1*	$\frac{7}{2}$	$\frac{5}{2}$	0	0	$\frac{1}{2}$	0
	2	0	$\frac{5}{2}$	$\frac{3}{2}$	0	0	$\frac{1}{2}$	

We thus get

$$y_1 = (\tfrac{1}{2}, 0, 0, 0) \qquad x_1 = (0, 0, \tfrac{1}{2})$$

$$y_2 = (\tfrac{7}{46}, \tfrac{4}{46}, 0, 0) \qquad x_2 = (0, \tfrac{4}{46}, \tfrac{7}{46})$$

$$y_3 = (\tfrac{1}{9}, 0, \tfrac{1}{9}, 0) \qquad x_3 = (\tfrac{5}{17}, 0, \tfrac{1}{17})$$

$$y_4 = (\tfrac{1}{6}, 0, 0, \tfrac{1}{6})$$

$$e_1 \notin M(x_0, y_1)$$

We check that $M(x_2, y_2) = \{e_1, e_2, e_3, e_4, f_1, f_2, f_3\}$; that is, (x_2, y_2) is an equilibrium pair. Normalizing them, we have

$$y^* = (\tfrac{7}{11}, \tfrac{4}{11}, 0, 0), \qquad x^* = (0, \tfrac{4}{11}, \tfrac{7}{11})$$

as an equilibrium pair to the payoffs A_1, B_1.

7.6 A THEOREM OF VOROBEV AND KUHN

In the case of zero-sum two-person matrix games, the theorem (3.1.14) of Shapley and Snow characterizes all extreme optimal strategies for the players. In this section we shall see the analogous characterization for the case of bimatrix games.

Let A, B be $m \times n$ payoff pair for players I and II. Let a_i be the ith row of A, and let b_j be the jth column of B. Let e be an appropriate vector with all entries unity. The x's would denote mixed strategies for player I, and the y's would denote mixed strategies for player II. Let $\mathscr{E}$ be the set of all Nash equilibrium points for A, B. For any finite set $x = (x_1, x_2, \cdots, x_s)$ of mixed strategies of player I, let

$$S(x_1, x_2, \cdots, x_s) = \{y: (x_i, y) \in \mathscr{E}, i = 1, 2, \cdots, s\}$$

One observes that the set $S(x_1, x_2, \cdots, x_s)$ is compact and convex. (This could possibly be empty.)

Definition 7.6.1: The mixed strategy y^0 is a V-extreme equilibrium point for player II, if y^0 is an extreme point of $S(x_1, x_2, \cdots, x_s)$ for some finite number of mixed strategies $x_1, x_2, \cdots, x_s$ of player I.

With the help of the next few lemmas we shall characterize the V-extreme equilibrium points of player II. A similar characterization holds good also for player I.

LEMMA 7.6.1: Let y^0 be a V-extreme equilibrium point with respect to $S(x_1, x_2, \cdots, x_s)$. Let $\lambda^0 = \max_k (x_k A y^0)$. Then (y^0, λ^0) is an extreme point of the compact convex set

$$K = \left\{ (y, \lambda): \begin{array}{l} (a_i, y) \leq \lambda, i = 1, 2, \cdots, m \\ (x_k, b_j) \leq (x_k, By), k = 1, 2, \cdots, s, j = 1, 2, \cdots, n \\ (e, y) = 1 \\ y \geq 0 \end{array} \right\}$$

PROOF: By assumption, $(x_k, y^0) \in \mathscr{E}$, $k = 1, 2, \cdots, s$, and $(x_k, Ay^0) \leq \lambda^0$, $k = 1, 2, \cdots, s$. Evidently $(y^0, \lambda^0) \in K$. Suppose that it is not an extreme point of K; then there exists (y', λ'), $(y'', \lambda'') \in K$, $(y', \lambda') \neq (y'', \lambda'')$, $\frac{1}{2}((y', \lambda') + (y'', \lambda'')) = (y^0, \lambda^0)$. We have

$$\lambda^0 = \max_i (a_i, y^0) \leq \frac{1}{2}(\max_i (a_i, y') + \max_i (a_i, y''))$$
$$\leq \frac{1}{2}(\lambda' + \lambda'') = \lambda^0$$

Hence $\max_i (a_i, y') = \lambda'$, $\max_i (a_i, y'') = \lambda''$. Further,

$$\lambda^0 = (x_k, Ay^0) = \frac{1}{2}(x_k, Ay') + \frac{1}{2}(x_k, Ay'') \leq \frac{1}{2}(\lambda' + \lambda'') = \lambda^0$$
$$k = 1, 2, \cdots, s$$

Hence $(x_k, Ay') = \lambda'$, $(x_k, Ay'') = \lambda''$, $k = 1, 2, \cdots, s$. Thus

$$y', y'' \in S(x_1, x_2, \cdots, x_s)$$

with $y^0 = (y' + y'')/2$. This contradicts our assumption that y^0 is a V-extreme equilibrium point. Hence the lemma.

LEMMA 7.6.2: Let y^0 be a V-extreme equilibrium point with respect to $S(x_1, x_2, \cdots, x_s)$. Let $\lambda^0 = \max_i (a_i y^0)$. Then (y^0, λ^0) is an extreme point of the set

$$T = \left\{ (y, \lambda): \begin{array}{l} (a_i, y) \leq \lambda, i = 1, 2, \cdots, m \\ (e, y) = 1 \\ y \geq 0 \end{array} \right\}$$

PROOF: By the previous lemma (y^0, λ^0) is extreme to K. Thus $(x_k, b_j) \leq (x_k, By^0)$ attains the equality sign for some j. This implies that, when $(x_k, b_l) < \max_j (x_k, b_j)$, then y_l^0, the lth coordinate of y^0, is zero. Let $N = \{l: (x_k, b_l) < \max_j (x_k, b_j)\}$. Then $y_l^0 = 0$ for $l \in N$. To prove the lemma, we have to show that (y^0, λ^0) is extreme to T. Evidently it is in T. Suppose that it is not extreme to T. Then we have (y', λ'), $(y'', \lambda'') \in T$, $(y', \lambda') \neq (y'', \lambda'')$, $\frac{1}{2}(y', \lambda') + \frac{1}{2}(y'', \lambda'') = (y^0, \lambda^0)$. Since $y', y'' \geq 0$, $y_l' = y_l'' = 0$ for $l \in N$. Thus when $y_r' > 0$ for some component r, then $y^0 > 0$, and $(x_k, b_r) = \max_j (x_k, b_j)$. Thus y', y'' also satisfy the constraints:

$$(x_k, b_j) \leq (x_k, By'), \qquad k = 1, 2, \cdots, s, \qquad j = 1, 2, \cdots, n$$

$$(x_k, b_j) \leq (x_k, By'')$$

Thus (y', λ'), $(y'', \lambda'') \in K$, with $\frac{1}{2}(y' + y'') = y^0$, $\frac{1}{2}(\lambda' + \lambda'') = \lambda^0$; this contradicts the assertion in lemma 7.5.1 above. Hence the lemma.

LEMMA 7.6.3: Let (y^0, λ^0) be an extreme point of the compact convex set

$$T = \left\{ (y, \lambda): \begin{array}{l} (a_i, y) \leq \lambda, i = 1, 2, \cdots, m \\ (e, y) = 1 \\ y \geq 0 \end{array} \right\}$$

Then there exists a $p \times p$ submatrix P of A such that

$$P_1 = \begin{bmatrix} P & -e' \\ e & 0 \end{bmatrix}$$

is nonsingular. (Here e' is the column vector $(1, 1, \cdots, 1)'$.) Further renumbering if necessary the rows and columns of A such that P is the principal $p \times p$ minor of A, we have

$$y_j^0 = \frac{\sum_{i=1}^{p} P_{ij}}{|P_1|} = \frac{\sum_{i=1}^{p} P_{ij}}{\sum_{i=1}^{p}\sum_{j=1}^{p} P_{ij}}, \qquad j = 1, 2, \cdots, p$$

$$y_j^0 = 0, \qquad j = (p+1), \cdots, n$$

$$\lambda^0 = \frac{|P|}{|P_1|} = \frac{|P|}{\sum_{i=1}^{p}\sum_{j=1}^{p} P_{ij}}$$

(Here P_{ij} as usual denotes the cofactor of the i, jth entry in P; $|P|$ and $|P_1|$ are as usual the determinantal values of P and P_1, respectively.)

PROOF: Without loss of generality we could assume

$$(a_i, y^0) = \lambda^0, \qquad i = 1, 2, \cdots, r$$
$$(a_i, y^0) < \lambda^0, \qquad i = (r+1), \cdots, m$$
$$y_j^0 > 0, \qquad j = 1, 2, \cdots, p$$
$$y_j^0 = 0, \qquad j = (p+1), \cdots, n$$

We claim that the matrix

$$L = \begin{bmatrix} a_{11} \cdots a_{1p} & -1 \\ a_{r1} \cdots a_{rp} & -1 \\ 1 \quad 1 & 0 \end{bmatrix}$$

has rank $(p+1)$. If it is not true, then there exists an n-vector

$$\pi = (\pi_1, \pi_2, \cdots, \pi_p, 0, 0, \cdots, 0)$$

and a constant λ such that $(\pi_1, \pi_2, \cdots, \pi_p, \lambda) \neq 0$, with

$$\sum_{j=1}^{p} a_{ij}\pi_j - \lambda = 0$$

$$\sum_{j=1}^{p} \pi_j = 0$$

Thus $y^0 \pm \varepsilon\pi$ are mixed strategies for player II when ε is sufficiently small.

Let $u = y^0 + \varepsilon\pi$, $v = y^0 - \varepsilon\pi$. We shall choose the arbitrarily small ε suitably.

$$(a_i, u) = \lambda^0 + \varepsilon\lambda, \qquad i = 1, 2, \cdots, r$$

Since $(a_i, y^0) < \lambda^0$, $i = (r + 1), \cdots, m$, we can choose ε in such a way that

$$(a_i, u) < \lambda^0 + \varepsilon\lambda, \qquad i = (r + 1), \cdots, m$$

$$(a_i, v) < \lambda^0 - \varepsilon\lambda, \qquad i = (r + 1), \cdots, m$$

Since $(a_i, v) = \lambda^0 - \varepsilon\lambda$, $i = 1, 2, \cdots, r$, then $(u, \lambda^0 + \varepsilon\lambda)$, $(v, \lambda^0 - \varepsilon\lambda)$ belong to T with

$$\left(\frac{u + v}{2}, \frac{(\lambda^0 + \varepsilon\lambda) + (\lambda^0 - \varepsilon\lambda)}{2}\right) = (y^0, \lambda^0)$$

This contradicts the fact that (y^0, λ^0) is extreme to T. Thus the rank of L is $(p + 1)$, and $r + 1 \geq p + 1$. Thus we can choose some $(p + 1)$ of the rows of L (including the last row), such that

$$P_1 = \begin{vmatrix} a_{11} \cdots a_{1p} & -1 \\ a_{p1} \cdots a_{pp}, & -1 \\ 1 \cdots \quad 1 & 0 \end{vmatrix},$$

is nonsingular, and (y^0, λ^0) is the unique solution of

$$\sum_{j=1}^{p} a_{ij}y_j = \lambda, \qquad i = 1, 2, \cdots, p$$

$$\sum_j y_j = 1$$

The rest of the assertions of the theorem follow by Cramer's rule.

Thus we have for each V-extreme equilibrium point y^0 of player II, a unique submatrix of A. Since there are only a finite number of square submatrices of A we have the following theorem.

Theorem 7.6.1

It is possible to effectively enumerate the finite set Y_0 of V-extreme equilibrium points of player II. We notice that not all square submatrices need correspond to a V-extreme equilibrium point of player II.

Theorem 7.6.2

For any finite set $X = \{x_1, \cdots, x_s\}$, it is possible to describe effectively the set $S(x_1, \cdots, x_s)$.

PROOF: Since $S(x_1, x_2, \cdots, x_s)$ is effectively described by its extreme points, it is sufficient to enumerate them. But the set of extreme points of $S(x_1, x_2, \cdots, x_s)$ is a subset of Y_0. Thus when we check whether $y_0 \in Y_0$ is an extreme point of $S(x_1, x_2, \cdots, x_s)$, then we have effectively enumerated $S(x_1, x_2, \cdots, x_s)$. But this is equivalent to checking the finite number of inequalities

$$(x_k, Ay^0) \geq (a_i, y^0), (x_k, By^0) \geq (x_k, b_j), k = 1, 2, \cdots, s, j = 1, 2, \cdots, n.$$

Hence the theorem.

Theorem 7.6.3

$$\mathscr{E} = \bigcup_{Y \subseteq Y_0} \mathrm{con}\,\{Y\} \times S(Y)$$

Here con $\{Y\}$ is the convex set generated by Y; $S(Y)$ is the set $S(y_1, y_2, \cdots, y_\alpha)$, where $Y = \{y_1, y_2, \cdots, y_\alpha\}$; and Y_0 is the set of V-extreme equilibrium points of player II.

PROOF: Let $(x^*, y^*) \in \mathscr{E}$. Then the extreme points of $S(x^*) \subseteq Y_0$. Let $Y \subseteq Y_0$ be this set; that is, $Y = Y_0 \cap S(x^*)$. Since Y generates $S(x^*)$, and since $y^* \in S(x^*)$, $y^* \in \mathrm{con}\,\{Y\}$. Since Y generates $S(x^*)$, $x^* \in S(y)$ for all $y \in Y$, and $x^* \in S(Y)$. Thus $\mathscr{E}$ is a subset of the right-hand side. Conversely, let $y^* \in \mathrm{con}\,\{Y\}$, $x^* \in S(Y)$ for some finite set $Y \subseteq Y_0$. Then

$$y_1, y_2, \cdots, y_\alpha \in S(x^*)$$

where $Y = \{y_1, y_2, \cdots, y_\alpha\}$. Therefore con $\{Y\} \subseteq S(x^*)$. Since $y^* \in \mathrm{con}\,\{Y\}$, $y^* \in S(x^*)$; that is, $(x^*, y^*) \in \mathscr{E}$. Hence the theorem.

Remark 7.6.1: The corresponding statements are valid also for player I.

Remark 7.6.2: One can prove the following result—if every pair of equilibrium strategies is completely mixed for (A, B) then (A, B) are square matrices and the equilibrium pair is unique [15]. (See the next section.)

7.7 COMPLETELY MIXED STRATEGIES IN BIMATRIX GAMES

As in the previous section, let A, B be payoffs for two players in a non-zero-sum game. The mixed strategy x is completely mixed for player I (as in zero-sum games) if $x > 0$. Consider the sets

$$S(x^0) = \{y: (x^0, y) \in \mathscr{E}, \text{ the Nash equilibrium pairs}\}$$
$$T(y^0) = \{x: (x, y^0) \in \mathscr{E}\}$$

We say that a nonempty set $S(x^0)$ is completely mixed if every element of the set is completely mixed. Similar definition holds for $T(y^0)$. We have the following generalizations of the theorems of Kaplansky (Chapter 3) for bimatrix games.

Theorem 7.7.1

Let $(x^0, y^0) \in \mathscr{E}$ and $(x^0, By^0) = 0$. If $T(y^0)$ is completely mixed, then the rank $\rho(B)$ of B is m or $m - 1$. If it is $(m - 1)$, then x^0 is the only element in $T(y^0)$.

PROOF: Let $e = (1, 1, \cdots, 1)'$ and $v_1 = (x^0, Ay^0)$. By assumption $Ay^0 \leq v_1 e$ and $B'x^0 \leq 0$. Since $T(y^0)$ is completely mixed, $x^0 > 0$ and that $Ay^0 = v_1 e$, for otherwise $v_1 = (x^0, Ay^0) < (x^0, v_1 e) = v_1$. Thus

$$Ay^0 = v_1 e, \qquad B'x^0 \leq 0$$

The rank $\rho(B)$ of B could be m or less than m. In case it is less than m we will show that it is $(m - 1)$. Suppose if $\rho(B) \leq m - 2$, then there exist two linearly independent vectors, at least one of which is independent of x^0 and satisfying the equations $B'z = 0$.

Let $B'\pi = 0$ where $\pi = (\pi_1, \pi_2, \cdots, \pi_m)$ be independent of x^0. Without loss of generality $\Sigma\pi_i = 0$ or $\Sigma\pi_i = 1$. In case $\Sigma\pi_i = 0$, then $x^* = x^0 - \lambda\pi$ with $1/\lambda = \max_i (\pi_i/x_i^0)$ defines a mixed strategy with some one of its components zero (it corresponds to an index i for which the value $1/\lambda$ is attained in π_i/x_i^0). Consider the pair (x^*, y^0).

$$(x^*, Ay^0) = v_1 \geq (x, Ay^0)$$
$$(x^*, By^0) = (B'x^*, y^0) = (B'(x^0 - \lambda\pi), y^0) = (x^0, By^0) = \lambda(B'\pi, y^0)$$
$$= (x^0, By^0) \geq (x^0, By) = (B'x^0, y) = (B'(x^0 - \lambda\pi), y) = (x^*, By)$$

Thus $(x^*, y^0) \in \mathscr{E}$ and $x^* \in T(y^0)$. But x^* is not completely mixed. Thus when $\Sigma\pi_i = 0$, we have established that $\rho(B)$ is not less than $(m - 1)$. In case $\Sigma\pi_i = 1$ again the mixed strategy

$$\tilde{x} = (1 + \lambda)x^0 - \lambda\pi \qquad \text{with} \qquad \frac{(1 + \lambda)}{\lambda} = \max_i \frac{\pi_i}{x_i^0}$$

can be considered and $(\tilde{x}, y^0) \in \mathscr{E}$ with $\tilde{x}$, not completely mixed. Thus when $\rho(B)$ is not m, then it is $m - 1$. In this case we have a unique (to within scalar multiple) solution to $B'z = 0$. But $B'x^0 \leq 0$. If x^0 were to be independent of the solution π, then we could repeat the proof verbatim to contradict the fact that $T(y^0)$ is completely mixed. Thus $B'x^0 = 0$. Lastly, the same type of arguments we gave above eliminates the possibility of any other x in $T(y^0)$ when $B'x^0 = 0$. Thus x^0 is the only element of $T(y^0)$.

Theorem 7.7.2

Let $m > n$ and $(x^0, y^0) \in \mathscr{E}$. Then there exists an x^* that is not completely mixed and $(x^*, y^0) \in \mathscr{E}$.

PROOF: Let E denote the $m \times n$ matrix with all entries unity. Clearly for the matrix pair $(A, B - v_2E)$ with $v_2 = (x^0, By^0)$, the equilibrium set $\mathscr{E}$ is the same. Thus without loss of generality $v_2 = 0$ for the pair (x^0, y^0). If no $(x^*, y^0) \in \mathscr{E}$ with x^* not completely mixed, then $T(y^0)$ is completely mixed and by the previous theorem $m - 1 \le \rho(B) \le n < m$.

This inequality implies $m - 1 = n$. Since $\rho(B) \le \min(m, n)$ we have $\rho(B) = m - 1 = n$. By the last assertion of the previous theorem we have $B'x^0 = 0$. Without loss of generality let the first n columns be linearly independent in B'. Thus we have a solution $\tilde{\pi} = (\pi_1, \pi_2, \cdots, \pi_n)'$ for the equations

$$\begin{bmatrix} b_{11} & b_{21} & \cdots & b_{m-1,1} \\ b_{12} & b_{22} & \cdots & b_{m-1,2} \\ \cdot & \cdot & \cdots & \cdot \\ b_{1n} & b_{2n} & \cdots & b_{m-1,n} \end{bmatrix} \begin{bmatrix} \pi_1 \\ \pi_2 \\ \cdot \\ \pi_n \end{bmatrix} = \begin{bmatrix} 1 \\ 1 \\ \cdot \\ 1 \end{bmatrix}$$

Here the matrix above is the submatrix of B' with the first $n = m - 1$ columns.

Let $\pi = (\pi_1, \pi_2, \cdots, \pi_n, 0)'$. Clearly $B'\pi = e$. Since $x^0 > 0$ and since the last component of π is zero, x^0 and π are linearly independent. As in the previous theorem we can get a mixed strategy x^* with some component zero and

$$B'x^* = \alpha e, \qquad \alpha \ne 0$$
$$Ay^0 = v_1 e$$

Thus $(x^*, y^0) \in \mathscr{E}$. This contradicts our assumption that $T(y^0)$ is completely mixed. Hence the theorem.

Theorem 7.7.3

Let for $(x^0, y^0) \in \mathscr{E}$, y^0 be not completely mixed. Then so is $T(y^0)$.

PROOF: Without loss of generality let $(x^0, By^0) = 0$ and $y_1^0 = 0$ where $y^0 = (y_1^0, y_2^0, \cdots, y_n^0)$. Consider the matrices A, B by deleting the first columns in A and B. Let $\underset{\sim}{y}^0 = (y_2^0, \cdots, y_n^0)$. Let $T(\underset{\sim}{y}^0)$ be completely mixed. Thus $x^0 > 0$ and

$$\underset{\sim}{A}\underset{\sim}{y}^0 = v_1 e, \qquad B'x^0 \le 0$$

Further if the jth component $(\underset{\sim}{B}', x^0)_j$ of $\underset{\sim}{B}'x^0$ is <0 for some $j = 2, 3, \cdots, n$, then we have $y_j^0 = y_j^0 = 0$. Thus $(x^0, \underset{\sim}{y}^0)$ is an equilibrium pair for $(\underset{\sim}{A}, \underset{\sim}{B})$. Further, $(x^0 \underset{\sim}{B}y^0) = 0$. Thus by the previous theorem we have an $(x^*, \underset{\sim}{y}^0)$ a new equilibrium pair for $(\underset{\sim}{A}, \underset{\sim}{B})$, where x^* is not completely mixed. For $j = 2, 3, \cdots, n$ we have

$$(B'x^*)_j = (\underset{\sim}{B}'x^*)_j$$

Thus if $(B'x^*)_j < 0$ for some $j = 2, 3, \cdots, n$, then we have $(B'x^*)_j < 0$ and that $y_j^0 = 0$. Clearly, whether $(B'x^*)_1 < 0$ or not $y_1^0 = 0$. Thus we have $(x^*, y^0) \in \mathscr{E}$ for the matrices (A, B). This contradicts the assumption that $T(y^0)$ is completely mixed. Hence the theorem.

Theorem 7.7.4

If every pair $(x, y) \in \mathscr{E}$ is completely mixed, then the matrices (A, B) should be square matrices and the equilibrium pair is unique.

PROOF: By Theorem 7.7.2 it is easily seen that A and B are square matrices. Let $(x^0, y^0) \in \mathscr{E}$. Without loss of generality let $v_2 = (x^0, By^0) = 0$. By Theorem 7.7.1, $\rho(B) = n$ or $n - 1$. Since $y^0 > 0$, we have $B'x^0 = 0$. Thus $\rho(B) \neq n$. By the last part of Theorem 7.7.1, x^0 is unique for y^0. Let $(x^*, y^*) \in \mathscr{E}$. Since $x^*, y^* > 0$ we have

$$Ay^* = v_1^*e \quad \text{where} \quad v_1^* = (x^*, Ay^*)$$

$$B'x^* = v_2^*e \quad \text{where} \quad v_2^* = (x^*, By^*)$$

Thus (x^*, y^0) and $(x^0, y^*) \in \mathscr{E}$. By the argument above $x^* = x^0$ and $y^* = y^0$, as $Y(y^0)$ has x^0 as its unique element. Thus (x^0, y^0) is the unique equilibrium pair for (A, B). Hence the theorem.

REFERENCES

[1] Balinski, M. L. (1961). An algorithm for finding all vertices of convex polyhedral sets. *J. Soc. Indust. Appl. Math.* **9**, 72–88.
[2] Dunford, N., and Schwartz, J. J. (1958). "Linear Operators," Vol. I, Interscience Publishers, New York.
[3] Glicksberg, I. L. (1952). A further generalization of the Kakutani fixed point theorem, with application to Nash equilibrium points. *Proc. Am. Math. Soc.* **3**, 170–174.
[4] Kuhn, H. W. (1961). An algorithm for equilibrium points in bimatrix games. *Proc. Natl. Acad. Sci. U.S.* **47**, 1657–1662.
[5] Lemke, C. E., and Howson, Jr., J. J. (1964). Equilibrium points of bimatrix games. *J. Soc. Indust. Appl. Math.* **12**, 413–423.
[6] Luce, R. D., and Raiffa, H. (1957). "Games and Decisions," John Wiley & Sons, New York.
[7] Mangasarian, O. L. (1964). Equilibrium points of bimatrix games. *J. Soc. Indust. Appl. Math.* **12**, 778–780.
[8] Mayberry, J. P., Nash, J. F. and Shubik, M. (1953). A comparison of treatments of a duopoly situation. *Econometrica* **21**, 141–154.
[9] Mills, H. (1960). Equilibrium points in finite games. *J. Soc. Indust. Appl. Math.* **8**, 397–402.
[10] Nash, J. F. (1950). Equilibrium points in n-person games. *Proc. Natl. Acad. Sci. U.S.* **36**, 48–49.
[11] Nash, J. F. (1950). The bargaining problem. *Econometrica* **18**, 155–162.
[12] Nash, J. F. (1951). Non-cooperative games. *Ann. Math.* **54**, 286–295.
[13] Nash, J. F. (1953). Two-person cooperative games. *Econometrica* **21**, 128–140.
[14] Nikaido, H., and Isoda, K. (1955). Note on non-cooperative convex games. *Pacific J. Math.* **5**, 807–815.
[15] Raghavan, T. E. S. (1970). Completely mixed strategies in bimatrix games. To appear in *Jour. London Math Soc.*
[16] Varadarajan, V. S. (1958). Weak convergence of measures over separable metric spaces. *Sankhya* **19**, 15–22.
[17] Vorobev, N. N. (1958). Equilibrium points in bimatrix games. In "Theory of Probability and Its Applications" (English translation), Vol. 3, pp. 297–309.
[18] Wald, A. (1951). "On some systems of equations of mathematical economics" (translation). *Econometrica* **19**, 368–403.

Chapter Eight

DIFFERENTIAL GAMES

The subject of differential games was begun by Isaacs, who was concerned with game theoretic problems where players have to take decisions continuously in time. In a formal way he analyzed many such multistage models and obtained explicit solutions to a class of such problems. Independently Berkovitz and Fleming started rigorizing the technical aspects of a wide class of problems and a variety of the formal concepts and proofs of Isaacs. When only one player is involved in the games, then the games turn out to be problems in dynamic programming and optimal control processes. Rather than go into the full details of the subject, we shall be content with a few examples and some theory.

8.1 EXAMPLES OF DIFFERENTIAL GAMES

In this section we shall present a few examples that come under the purview of differential games.

Example 8.1.1: A thief runs in the plane with simple motion and uniform speed w_2. A police constable chases him on a bicycle with speed w_1, but the radius of curvature of the cycle is bounded below by ρ. He steers the hand bar by selecting the value of the curvature at each instant. The capture occurs when their distance is $\leq l$. We also assume that $w_1 > w_2$ (natural!).

Example 8.1.2: A pursuer, P, and an evader, E, have the following problem. The pursuer, P, has to guard a certain compact connected set S (it may be a garden full of oranges and apples) from the evader (perhaps a naughty child). They both run in simple motion and have the same speed. The payoff is the distance from S to the point where capture occurs.

Example 8.1.3: Two nations, A and B, engaged in a war estimate the period of war to be at most N days. Nation A has p planes, and nation B has q planes. The war consists in a series of strikes or moves each day, and every day the nations have to allocate their available planes for three important tasks—namely, the task of air attacks, which destroy the enemy's aircraft and air base complex; the task of air defense, which is used to defend from the enemy's air attacks; and the task of ground support, which helps the infantry to advance and occupy enemy's territory. In any strike only a proportion of the allotted planes for air attack can penetrate through the enemy's air defense. Further, both sides get constant supplies of planes between any two air strikes. (The constants may be different for the two.) Occupation of the enemy's territory and the area captured would be the aim of the nations.

Example 8.1.4: Businessmen from n countries who are highly noncooperative compete to get a certain commodity which is sold in lots in an auction place over a period of weeks. The raw material comes in three stages, and the available quantity varies randomly from year to year. The material itself could be graded in three different qualities, and these are mixed in a certain proportion in any lot (the producer plays the market). The buyers have their own requirements of the different qualities of the commodity and have fixed amounts for purchases. The buyers cannot quote very low prices, because the producers would incur heavy losses and hence stop producing the material. The bidding is done in a secret way, in the sense that each buyer quotes his price for any lot, and this price is not revealed to others. Finally the organizer of the auction compares their prices, and the one who quotes the highest price gets the lot. If many quote the same highest price, they share the lot equally. Every lot is auctioned like this. The available commodity is sold in twenty such auctions over a period of three months.

8.2 MODEL BUILDING IN DIFFERENTIAL GAMES

The problem of building up models for many of these games may be a difficult task in the following sense. If every aspect of the real problem is to be incorporated in a mathematical model of the game, then the analysis may be very complex, and perhaps we shall not arrive at any solution. If the model is drastically simple, the solution from this may not be useful in any sense. But when one compares models of varying complexity, one often finds a striking similarity in the solutions. In fact, this similarity in the solutions would be the useful information that one could hope to utilize in any actual problem. For example, we shall build up a suitable mathematical model for the first and third examples above. We shall take the example of air war first.

Let us assume that nation A allots x planes for air attack, u planes for air defense, and m planes for ground support, Let y, v, n be the corresponding numbers for nation B. We assume that they allot every plane for one of these purposes. Hence

$$x + u + m = p, \qquad y + v + n = q$$

Let us assume that the two nations know only the total number of their enemy's planes before the first strike. Let the kill potential of nation B or the capacity to destroy the attacking planes of nation A be a constant, c. We assume that this job can be done only by the number, v, of planes allotted for air defense. Thus only $x - cv$ planes penetrate nation B's air defense, when $x > cv$. Thus the number of planes of nation A that penetrate the air defense of nation B is

$$\max (x - cv, 0)$$

Let each plane of nation A that penetrates the enemy's air defense destroy b planes. Thus

$$b \max (x - cv, 0)$$

planes of nation B are destroyed by nation A in the first strike. The constant b depends on the nature of the aircraft, the nature of the air base, and a host of such complex phenomena. Further, let s planes be constantly added in between each strike by nation B. (These may be produced in a factory of their own, or they may be bought from their allies who are not directly involved in the combat.) We shall assume that any plane of nation A that is used for air attack and that survives the enemy's air defense can be used again in the next strike. Thus the initial strike of nation A destroys

$$\min (q + s, b \max (0, x - cv))$$

planes of nation B. Therefore, after the first strike nation B is left with

$$q_1 = q + s - \min (q + s, b \max (0, x - cv))$$
$$= \max (0, q + s - b \max (0, x - cv))$$

planes. Similarly, for nation A it is

$$p_1 = \max (0, p + r - e \max (0, y - ku))$$

where e, k, r have similar interpretations as in the above model. The two nations after the first day's battle are left with p_1 and q_1 planes. Let $s(m)$ denote the payoff to nation A, where $s(m)$ is a positive strictly increasing function of m. Let $T(n)$ be the corresponding function for nation B. (Here $s(m)$ measures the area advanced by the infantry of nation A with the ground support of m planes in the initial strike.) Thus we have the payoff

$$\sum_{1}^{N} (s(m) - T(n))$$

after N days of battle for nation A. The aim of the nations in terms of this payoff is apparent. The problem was completely solved by Dresher and Berkovitz [10] for the case

$$s(m) = m, \qquad T(n) = n, \qquad b = e = c = k = 1$$

The following is a verbal instruction of the optimal strategies, when nation A has more planes than nation B (that is, $p > q$).

The war ends with a series of strikes on ground support; that is, during the closing period of the war both nations concentrate all their available forces on ground-support missions. In the earlier periods of the war their strategies are very different. During the earlier strikes, nation A with its greater force employs in specific proportions all of its forces to the three tasks. This proportion may vary from strike to strike. In this connection, if r_j denotes their force ratio $p_j : q_j$ at the end of the jth strike, then there exists a constant r such that, for those strikes j for which $r_j < r$, nation A employs its full force only for air attack and air defense and completely neglects ground support. But when $r_j \geq r$, nation A has to allocate its force to all three tasks. For nation B there is no optimal allocation of its forces for each of its strikes (that is, no single allocation is best). If it is very weak (namely, when the force ratio is larger than r), then nation B should allocate its entire air force to any one of the three tasks chosen according to a random experiment.

If nation B is not very weak (that is, when the force ratio is less than r), then it should concentrate its entire force on one of air attack or air defense according to a chance experiment. Thus, to whichever tasks the stronger side allocates its force, the weaker side concentrates its entire force on one of those tasks.

One could hardly guess the nature of the optimal strategies by any simulation process.

Now let us formulate the model for the first example.

We could reasonably formulate this game by taking into consideration the position coordinates in the plane at time t—the direction of the wheel bar at time t for the policeman, the running direction for the thief, and the curvature that the policeman takes at time t. To start, let the policeman be at (x_1, y_1) and the thief at (x_2, y_2) in the plane at a particular time $t = t_0$. Let their travel directions be φ and ψ, respectively, as in Fig. 1. After a small time interval, Δt_0, the policeman is at $(x_1(t_0 + \Delta t_0), y_1(t_0 + \Delta t_0))$, as shown in Fig. 2. Thus the distance traveled is

$$\sqrt{(x_1(t_0 + \Delta t_0) - x_1(t_0))^2 + (y_1(t_0 + \Delta t_0) - y_1(t_0))^2}$$

His speed being w_1, this is the same as $w_1 \cdot \Delta t_0$. Thus

$$\frac{y_1(t_0 + \Delta t_0) - y_1(t_0)}{w_1 \cdot \Delta t_0} = \cos \varphi$$

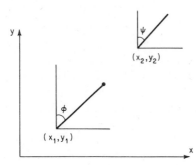

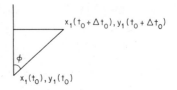

Fig. 1.   Fig. 2.

We have

$$\frac{dy_1}{dt} = w_1 \cos \varphi$$

Similarly we have

$$\frac{dx_1}{dt} = w_1 \sin \varphi$$

For the thief they are

$$\frac{dx_2}{dt} = w_2 \sin \psi \qquad \text{and} \qquad \frac{dy_2}{dt} = w_2 \cos \psi$$

Lastly, we should take into consideration the curvature restriction for the policeman and the curvature he has at time t. The curvature at a point $(t, u(t))$ of a curve $u = u(t)$ is the rate of change of the angle the tangent makes with the t-axis with respect to the length of the curve. If α is the angle the tangent makes with the t-axis, then

$$\text{Curvature} = \frac{d\alpha}{ds}$$

Here s is the length of the curve.

If $c(t)$ is the curvature at $(t, u(t))$ of a curve, then, by assumption, $|1/c(t)| \geq \rho$. If we define $\rho c(t) = \theta$, then $|\theta| \leq 1$. But

$$\frac{d\varphi}{dt} = \frac{d\varphi}{ds} \cdot \frac{ds}{dt} = w_1 c(t)$$

$$= \frac{w_1}{\rho} \times \rho c(t) = \frac{w_1}{\rho} \theta$$

Thus we have the differential equation

$$\frac{d\varphi}{dt} = \frac{w_1}{\rho}\,\theta, \qquad |\theta| \leq 1$$

describing the curvature restriction.

The payoff can be described in terms of the time of capture if the thief is sure that the policeman would ultimately be at a distance d less than l from him. In this case the thief aims to delay the time of capture as much as he can. The policeman wants to minimize this. The capture occurs when

$$(x_1 - x_2)^2 + (y_1 - y_2)^2 \leq l^2$$

Isaacs [20] explicitly finds the condition under which the thief can continue indefinitely without being caught. Naturally this depends on the nature of their speeds and the curvature restriction. In fact, his solution is the following: The thief would be ultimately caught (in the sense of the policeman nearing him at a distance of at most l) if

$$\frac{l}{\rho} > \sqrt{\left(1 - \frac{w_2^2}{w_1^2}\right)} + \sin^{-1}\frac{w_2}{w_1} - 1$$

If the inequality is reversed, the thief can indefinitely be at a distance $\geq l$ and hence never be caught.

Suppose that in the above problem one assumes that the thief does not run; then the problem would be one of minimizing the time of capture, and this comes under the purview of optimal control processes.

8.3 MATHEMATICAL FOUNDATIONS OF DIFFERENTIAL GAMES

Taking into consideration various such problems, we can put them into this general format.

Let $x = (x_1, x_2, \cdots, x_n)$ be a vector in n-space and let t denote time. Let $\mathscr{R}$ be a fixed region in (t, x) space. At each instant (time) t, let $y = (y_1, y_2, \cdots, y_\sigma)$ be chosen from a set Y by player I. Let $z = (z_1, z_2, \cdots, z_s)$ be chosen from a set Z by player II. The choice of the functions $y = y(t, x)$ and $z = z(t, x)$ are governed by the position $x = x(t)$ which satisfies the differential equations

$$\frac{dx_i}{dt} = g_i(x, y, z, t), \qquad i = 1, 2, \cdots, n$$

Here x is called the state vector, and y, z are called control variables or pure strategies for the players. Each player has to choose $y(t, x)$ and $z(t, x)$ from a set of permissible functions such that, for each time t and x satisfying the differential equations above, $y(t, x) \in Y$ and $z(t, x) \in Z$. The game is one of perfect information; they know how the game proceeds, and they know the complete past history. The play begins at $t = t_0$ and at the state $x(t_0) = x_0$. The game terminates when $(t, x(t)) \in \mathscr{C}$, where $\mathscr{C}$ is a predetermined surface on the boundary of $\mathscr{R}$. If (T, x_T) denotes the point of termination of the game starting at (t_0, x_0), then player I receives

$$V^0(t, x_T) + \int_{t_0}^{T} f(t, x(t), y(t), z(t)) \, dt$$

Player I aims to maximize his payoff, and player II wants to minimize it.

The differential equations are often called the kinematic equations of the system. In our game of policeman and thief, the angle ψ is under the control of the thief, and the angle φ and the parameter θ are under the control of the policeman. Here we notice that θ is restricted to the interval $-1 \le \theta \le 1$. The (φ, θ) space constitutes the control space for the policeman.

One cannot always hope for optimal pure strategies even when the game is of the discrete type as in example 8.1.3. Even when perfect information exists in these games, as long as the choices are infinite one cannot always hope for a theorem like theorem 2.7.1. The following is a counter-example by Berkovitz [3].

Example 8.3.1: Blue chooses $y = y(t, x)$, where $y(t, x)$ has the required continuity and differentiability properties and $0 \le y \le 1$. Red chooses $z = z(t, x)$, where $z(t, x)$ has the required continuity and differentiability properties and $0 \le z(t, x) \le 1$. The kinematic equation is

$$\frac{dx}{dt} = (y - z)^2, \qquad x(0) = x_0$$

Play terminates at a fixed time $t = T$. Blue wishes to maximize

$$\int_{0}^{T} x \, dt$$

and red wishes to minimize the above integral.

Here for any given strategy, y, of Blue, Red can choose the same, and thus $dx/dt = 0$ on $0 \le t \le T$. Hence

$$\int_{0}^{T} x \, dt = x_0 T$$

Therefore

$$\max_{y} \min_{z} \int_{0}^{T} x \, dt \le x_0 T$$

But since $dx/dt = (y - z)^2 \geq 0$, $x(t)$ is nondecreasing, and $\int_0^T x \, dt \geq x_0 T$ for any pair of strategies (y, z). Thus

$$\max_y \min_z \int_0^T x \, dt = x_0 T$$

On the other hand, if Red chooses a z, then Blue can choose a y defined by

$$y(t, x) = 1 \quad \text{if} \quad z(t, x) \leq \tfrac{1}{2}$$
$$= 0 \quad \text{if} \quad z(t, x) > \tfrac{1}{2}$$

This guarantees that $dx/dt \geq {}^1/_4$ for $0 \leq t \leq T$. Hence

$$\min_z \max_y \int_0^T x \, dt \geq x_0 T + \frac{1}{8}T^2$$

For the strategy $z(t, x) \equiv {}^1/_2$ of Red and for any strategy y of Blue, $dx/dt \leq {}^1/_4$ on $0 \leq t \leq T$. Thus

$$\min_z \max_y \int_0^T x \, dt = x_0 T + \frac{1}{8}T^2$$

Comparing the minimax and max min of the payoff, we find that there can be no pure optimals for the players.

Thus one is tempted to ask the following question. What are the necessary and sufficient conditions for a differential game to have pure strategy optimals? This is too difficult a question, and we can split it up into two questions, one asking the necessary condition and the other asking for sufficient condition.

The major contribution of Pontryagin and Bellman is to give necessary conditions for optimal solutions for one-player games, under very general assumptions on the kinematic equations and the payoff.

For our game theoretic case of at least two players it is much more difficult, and the main difficulty lies in thinking of solutions that anticipate every possible position of the opponent. In the next section we shall develop a method of considering the discrete case and interpreting the value of games in the continuous case to be the limit value of suitable discrete games.

8.4 THE CONVERGENCE PROBLEM

Let t be a time point in the interval $(0, T)$, x a point in n-space, and y, z points of Y and Z, respectively. We assume Y and Z to be compact convex sets in R^σ and R^s, respectively. Let $g = (g_1, g_2, \cdots, g_n)$, where $g_i = g_i(x, y, z)$ is a continuous function in its arguments. Let $f = f(x, y, z)$ be a continuous function. Further, let f and g satisfy a uniform Lipschitz condition in x, namely:

$$| f(x, y, z) - f(x^*, y, z) | \leq K \| x - x^* \|$$

$$\| g(x, y, z) - g(x^*, y, z) \| \leq K \| x - x^* \|$$

We define the time-discrete version of the game as follows. For a fixed T_0 let $T \leq T_0$, and let

$$\Delta_n = 2^{-n} T_0, \qquad n = 1, 2, \cdots,$$

$$T = j\Delta_n, \qquad j = 1, 2, \cdots, 2^n$$

The game starts at an initial position, $x = x_1$. It is played in j moves. In any move $i \leq j$, player I chooses a $y_i \in Y$, and player II chooses an $z_i \in Z$ simultaneously. Both have full knowledge of all the previous moves and in particular the position x_i of the game. The new position is given by the difference equation:

$$x_{i+1} = x_i + \Delta_i g(x_i, y_i, z_i)$$

At the end of the jth move, player I receives from player II

$$\sum_{i=1}^{j} \Delta_i f(x_i, y_i, z_i) + V^0(x_{j+1})$$

where V^0 is a given function giving the worth of the terminal position for player I. We assume for some L,

$$| V^0(x) - V^0(x^*) | \leq L \| x - x^* \|$$

We also make the following additional assumptions:

(1) $f(x, y, z) = p(x, y) + q(x, z)$.

(2) $g(x, y, z) = a(x, y) + b(x, z)$.

(3) p is concave in y for fixed x, and q is convex in z for fixed x.

(4) a and b are linear in y and z for fixed x.

Theorem 8.4.1

The discrete game with j moves for a fixed n has value $V_n(x, T)$ depending on the initial position, x, and the time length, T, satisfying the functional equation.

$$V_n(x, T) = \underset{y, z}{\text{Value}} \, [\Delta_n f(x, y, z) + V_n(x + \Delta_n g(x, y, z), T - \Delta_n)]$$

and $V_n(x, 0) = V^0(x)$. Here value denotes the game value of the bracketed expression as a function of y, z.

PROOF: First we shall explain what we mean by the pure strategy and mixed strategies for these multimove games, and then we shall prove the theorem by an induction argument on the number of moves. We shall define the pure strategies too, inductively on the number of moves. For example, player I may decide about his complete course of action in the following way. He chooses $y_1 \in Y$ in the first move; $y_2(z_1) \in Y$ in the second move as a function of z_1 when z_1 happens to be the choice of player II in his first move; $y_3(z_1, z_2) \in Y$ as his choice in his third move when z_1, z_2 happen to be player II's choices in his two moves; and so on. Thus a pure strategy σ_j in a j-move game is specified by a pair

$$\sigma_j = (\sigma_{j-1}, c(z_1, z_2, \cdots, z_{j-1}))$$

where σ_{j-1} is a pure strategy in the truncated $(j-1)$-move game which indicates completely the strategy to be taken in the first $(j-1)$ moves and the choice $c(z_1, z_2, \cdots, z_{j-1}) \in Y$ to be taken in the jth move, when $z_1, z_2, \cdots, z_{j-1}$ were the choice of the opponent in his first $(j-1)$ moves.

Often there is an advantage in considering moves counting from the last. This is very helpful in continuous cases, for which we should define our pure strategies differently.

For example, by move j we mean j moves to end the game. Thus by move 1 we actually mean the last move. Let σ_{j-1} be a pure strategy for player I for a game with $(j-1)$ moves to finish. Then a pure strategy σ_j for player I for a game with j moves to finish is specified by a pair (y, φ), $y \in Y$, where φ is a function on z with image in the pure strategy space of the game with $(j-1)$ moves to finish. Here y specifies the first move and, depending on the choice z in the first move of player II, player I has to handle a game with $(j-1)$ moves. Naturally any procedure he would adopt could take care of player II's choice, z. But any such procedure is a pure strategy σ_{j-1} for the new game with $(j-1)$ moves. Since it may vary depending on the choice of z, we have $\varphi(z)$ as the actual pure strategy selected.

To define mixed strategies for j-move games, we naturally demand, say for player I, a probability distribution g on Y and a function φ on $Y \times Z$ which maps every specific outcome (y, z) in the first move (in the usual sense) to a mixed strategy in the game with $(j-1)$ moves to finish. Even earlier we could have taken φ to be defined on $Y \times Z$ rather than on Z. Being a pure strategy, in the previous case y is determined by player I completely, and his choice is known. Since it is not known in advance, here it is reasonable to expect him to select his mixed strategy in the $(j-1)$-move game depending on his choice y and his opponents' choice z at the start.

Now coming to the proof of the theorem, we can easily check the truth for games with $j = 1$, and by induction we assume the following.

1. For any given initial state, the game with $(j-1)$ moves to finish has

optimal mixed strategies G^0, H^0 for players I and II. Evidently this depends on the initial state. If the initial state is $x + \Delta_n g(x, y, z)$, then the value is $E(G^0, H^0) = V_n(x + \Delta_n g(x, y, z), T - \Delta_n)$. We assume this to be continuous in y and z for fixed x. (This, too, is an induction assumption.)

2. For any pure strategy σ_{j-1} of player I and τ_{j-1} of player II,

$$E(G^0, H^0) \geq E(\sigma_{j-1}, H^0)$$

$$E(G^0, H^0) \leq E(G^0, \tau_{j-1})$$

Now let (g^0, φ) and (h^0, ψ) be any two mixed strategies for the j-move game with initial state x. Further let us assume $G^0 = \varphi(y, z)$, $H^0 = \psi(y, z)$ to be optimal for the $(j-1)$-move game with initial state $x + \Delta_n g(x, y, z)$. Then the expected value for the j-move game with respect to (g^0, φ), (h^0, ψ) is given by

$E((g^0, \varphi), (h^0, \psi))$

$$= \iint \Delta_n f(x, y, z)\, dg^0\, dh^0 + \iint E(\varphi(y, z), \psi(y, z))\, dg^0\, dh^0$$

$$= \iint \Delta_n f(x, y, z)\, dg^0\, dh^0 + \iint E(G^0, H^0)\, dg^0\, dh^0$$

$$= \iint \Delta_n f(x, y, z)\, dg^0\, dh^0 + \iint V_n(x + \Delta_n g(x, y, z), T - \Delta_n)\, dg^0\, dh^0$$

$$\iint [\Delta_n f(x, y, z) + V_n(x + \Delta_n g(x, y, z), T - \Delta_n)]\, dg^0\, dh^0$$

Further, if $\sigma_j = (y^0, \varphi(z))$, with $\varphi(z) = \sigma_{j-1}$, then

$E(\sigma_j, (h^0, \psi))$

$$= \int \Delta_n f(x, y^0, z)\, dh^0 + \int E(\varphi(z), \psi(y^0, z))\, dh^0$$

$$= \int \Delta_n f(x, y^0, z)\, dh^0 + \int E(\sigma_{j-1}, H^0)\, dh^0$$

$$\leq \int \Delta_n f(x, y^0, z)\, dh^0 + \int E(G^0, H^0)\, dh^0$$

$$\leq \int \Delta_n f(x, y^0, z)\, dh^0 + \int V_n(x + \Delta_n g(x, y^0, z), T - \Delta_n)\, dh^0$$

$$\leq \int [\Delta_n f(x, y^0, z) + V_n(x + \Delta_n g(x, y^0, z), T - \Delta_n)]\, dh^0$$

Similarly, for any pure strategy

$$\tau_j = (z^0, \psi(y) = \tau_{j-1}) \text{ for player II}$$

$$E((g^0, \varphi), \tau_j) \geq \int [\Delta_n f(x, y, z^0) + V_n(x + \Delta_n g(x, y, z^0), T - \Delta_n)] \, dg^0$$

Thus we can claim (g^0, φ) and (h^0, ψ) to be optimal for the j-move game if we can show that

$$E(\sigma_j, (h^0, \psi)) \leq E((g^0, \varphi), (h^0, \psi)) \leq E((g^0, \varphi), \tau_j)$$

for any pure strategy σ_j of player I and τ_j of player II. Let

$$H(y, z) = \Delta_n f(x, y, z) + V_n(x + \Delta_n g(x, y, z), T - \Delta_n)$$

Then what we demand on g^0 and h^0 so that (g^0, φ) and (H^0, v) are optimal for the j-move game is that

$$\int H(y, z) \, dg^0 \geq \iint H(y, z) \, dg^0 \, dh^0$$

so that

$$E((g^0, \varphi), \tau_j) \geq \int H(y, z) \, dg^0 \geq \int H(y, z) \, dg^0 \, dh^0 = E((g^0, \varphi), (h^0, \psi)$$

Similarly we demand that

$$\int H(y, z) \, dh^0 \leq \iint H(y, z) \, dg^0 \, dh^0$$

so that

$$E(\sigma_j, (h^0, \psi)) \leq \int H(y, z) \, dh^0 \leq \iint H(y, z) \, dg^0 \, dh^0 = E((g^0, \varphi), (h^0, \psi))$$

Thus $(g^0, \varphi), (h^0, \psi)$ are optimal to the j-move game when g^0, h^0 are optimal for $H(y, z)$. By assumption, $H(y, z)$ is continuous, and hence g^0, h^0 can be chosen optimally. Hence $(g^0, \varphi), (h^0, \psi)$ can be chosen optimally. In fact the theorem is proved by induction.

Our aim is to observe the behavior of $V_n(x, T)$ as $n \to \infty$. This will help us in considering the continuous analog of our problem and viewing it as the limit of such discrete games as above (by discrete we mean discreteness in time t).

Let $V_n^-(x, T)$ denote the value of the above game where now player II is first informed of the choice of player I in the ith move, $i = 1, 2, \cdots, j$, before he makes his choice in the ith move. Now we can dispense with mixed strategies, and we easily find

$$V_n^-(x, T) = \max_y \min_z [\Delta_n f(x, y, z) + V_n^-(x + \Delta_n g(x, y, z), T - \Delta_n)]$$

$$V_n^-(x, 0) = V^0(x), \qquad n = 1, 2, \cdots$$

Similarly, when player II has to commit himself and inform player I of his choice in each move before player II makes his choice, the value of this game is

$$V_n^+(x, T) = \min_z \max_y [\Delta_n f(x, y, z) + V_n^+(x + \Delta_n g(x, y, z), T - \Delta_n)]$$

$$V_n^+(x, 0) = V^0(x), \qquad n = 1, 2, \cdots$$

As in finite matrix games one can easily see that

$$V_n^-(x, T) \le V_n(x, T) \le V_n^+(x, T)$$

Suppose that, under our assumptions on f, g, we are able to prove that

$$\lim_n V_n^-(x, T) = \lim_n V_n(x, T) = \lim_n V_n^+(x, t)$$

then in a sense we can say that the players have optimal pure strategies for the time-continuous case; this we shall discuss at the end.

Consider the following two functions, U_{nk}^- and U_{nk}^+, defined inductively as follows.

$$U_{nk}^-(x, T) = \max_{y_1} \min_{z_1} \max_{y_2} \cdots \cdots \cdots \min_{z_m} H^- \qquad \text{(A)}$$

$$U_{nk}^+(x, T) = \min_{z_1} \max_{y_1} \min_{z_2} \cdots \cdots \cdots \max_{y_m} H^+ \qquad \text{(B)}$$

$$U_{nk}^-(x, 0) = U_{nk}^+(x, 0) = V^0(x)$$

where

$$k \le n, \qquad m = 2^{n-k}$$

$$T = j\Delta_k, \qquad j = 1, 2, \cdots, 2^k$$

$$H^\pm = \sum_{i=1}^m \Delta_n f(x, y_i, z_i) + U_{nk}^\pm\left(x + \sum_{i=1}^m \Delta_n g(x, y_i, z_i), T - \Delta_k\right)$$

Here each $y_i[z_i]$ might depend on all the preceding $z_p[y_p]$ for the two equations A and B above.

Theorem 8.4.2

The inequality $U_{nk}^- \le U_{nk}^+$ holds for all n, k.

The statement is true for any n, k and for $t = 0$; for in this case

$$U_{nk}^-(x, 0) = U_{nk}^+(x, 0) = V^0(x)$$

Assume the truth of the proposition for all x, n, k and for all

$$t = \Delta_k, 2\Delta_k, \cdots, (j - 1)\Delta_k$$

We shall prove for $t = j\Delta_k$.

Clearly, by definition,

$$U_{nk}^-(x, T) = \max_{y_1} \min_{z_1} \cdots \min_{z_m} \left[\sum_1^m \Delta_n f(x, y_i, z_i) + \right.$$

$$\left. + U_{nk}^-\left(x + \sum \Delta_n g(x, y_i, z_i), T - \Delta_k\right)\right]$$

By assumption,

$$U_{nk}^-(x + \Sigma\Delta_n g(x, y_i, z_i), T - \Delta_k) \le U_{nk}^+(x + \Sigma\Delta_n g(x, y_i, z_i), T - \Delta_k)$$

Hence by adding $\Sigma\Delta_n f(x, y_i, z_i)$ to both sides and taking minimum over z_m on the left, we have

$$\min_{z_m} \left[\sum_1^m \Delta_n f(x, y_i, z_i) + U_{nk}^-(x + \sum \Delta_n g(x, y_i, z_i), T - \Delta_k)\right]$$

$$\le \left[\sum_1^m \Delta_n f(x, y_i, z_i) + U_{nk}^+(x + \sum \Delta_n g(x, y_i, z_i), T - \Delta_k)\right]$$

Taking maximum with respect to y_m on both sides for fixed $z_1, z_2, \cdots, z_{m-1}$, $y_1, y_2, \cdots, y_{m-1}$, we get

$$\max_{y_m} \min_{z_m} \le \min_{z_m} \max_{y_m}$$

Repeating this $(m - 1)$ times, we end up with

$$U_{nk}^-(x, T) \le U_{nk}^+(x, T)$$

Theorem 8.4.3

For any two initial positions x_1, x_1^* and for fixed values $y_1, y_2, \cdots, y_j$, $z_1, z_2, \cdots, z_j$, the state x_i, x_i^* of the system at the ith stage satisfies

$$\|x_i - x_i^*\| \le \|x_1 - x_1^*\|e^{kT}, \qquad i = 1, 2, \cdots, j + 1$$

PROOF: Since

$$x_2^* = x_1^* + \Delta_n g(x_1^*, y_1, z_1)$$

$$x_2 = x_1 + \Delta_n g(x_1, y_1, z_1)$$

we have

$$\|x_2 - x_2^*\| = \|(x_1 - x_1^*) + \Delta_n g(x_1, y_1, z_1) - \Delta_n g(x_1^*, y_1, z_1)\|$$

$$\le \|x_1 - x_1^*\| + \Delta_n \|g(x_1, y_1, z_1) - g(x_1^*, y_1, z_1)\|$$

$$\le \|x_1 - x_1^*\| + \Delta_n K \|x_1 - x_1^*\|$$

$$\le (1 + \Delta_n K)\|x_1 - x_1^*\|$$

By an easy induction argument,

$$\|x_i - x_i{}^*\| \leq (1 + \Delta_n K)^{i-1} \|x_1 - x_1{}^*\|$$

$$\leq \left(1 + \frac{TK}{j}\right)^{i-1} \|x_1 - x_1{}^*\|$$

$$\leq \left(1 + \frac{TK}{j}\right)^{j} \|x_1 - x_1{}^*\|$$

Since $(1 + TK/j)^j$ is nondecreasing in j, by taking limit $j \to \infty$ we have

$$\|x_i - x_i{}^*\| \leq e^{KT}\|x_1 - x_1{}^*\|$$

Theorem 8.4.4

On any bounded set in (x, T) space, each of the functions V_n^-, V_n^+, U_{nk}^-, U_{nk}^+ satisfies a Lipschitz condition with the Lipschitz constants independent of k and n.

PROOF: We shall prove the result only for the case of V_n^+, and the proof is similar for the rest. The previous theorem is valid even if we assumed any y_i to depend on $z_1, z_2, \cdots, z_i$ and any z_i on $y_1, y_2, \cdots, y_{i-1}$. Therefore, for any two initial positions $x_1, x_1{}^*$ and for any $y_1, y_2, \cdots, y_j, z_1, z_2, \cdots, z_j$, of the required type, we have

$$\left| \left[\sum_{i=1}^{j} \Delta_n f(x_i, y_i, z_i) + V^0(x_{j+1}) \right] - \left[\sum_{i=1}^{j} \Delta_n f(x_i{}^*, y_i, z_i) + V^0(x_{j+1}{}^*) \right] \right|$$

$$\leq \sum_{i=1}^{j} \Delta_n \, |f(x_i, y_i, z_i) - f(x_i{}^*, y_i, z_i)| + |V^0(x_{j+1}) - V^0(x_{j+1}{}^*)|$$

$$\leq \sum_{i=1}^{j} \Delta_n K \|x_i - x_i{}^*\| + L e^{KT}\|x_1 - x_1{}^*\|$$

$$\leq \sum_{i=1}^{j} \Delta_n K e^{KT}\|x_1 - x_1{}^*\| + L e^{KT}\|x_1 - x_1{}^*\|$$

$$\leq (TK + L)e^{KT}\|x_1 - x_1{}^*\|$$

Since this is true for any sets of y_i's and z_i's of the above type, we have

$$|V_n^+(x_1, T) - V_n^+(x_1, T)| \leq (TK + L)e^{KT}\|x_1 - x_1{}^*\|$$

By assumption the (x, T) is bounded, and by the continuity of f and g and by the compactness of Y and Z, there exists an upperbound M for $|f(w, y, z)|$, $\|g(w, y, z)\|$ for w ranging over the bounded set (x, T) and

$(y, z) \in Y \times Z$. If $T^* < T$, then for any given initial position x and any pair of strategies for a game of duration T^*, the final positions are no more distant than $M(T - T^*)$. Thus

$$|V_n^+(x, T) - V_n^+(x, T^*)| \leq M(1 + L)(T - T^*)$$

This shows the Lipschitz property of V^+ in x, T. A similar proof applies for the rest of the functions stated in the theorem.

Theorem 8.4.5

On any bounded set in (x, T) space there exists a constant N such that

$$|U_{nk}^\pm(x, T) - V_n^\pm(x, T)| \leq NT\Delta_k$$

for any $k = 1, 2, \cdots, n \geq k$, and

$$T = j\Delta_k, \qquad j = 1, 2, \cdots, 2^k$$

PROOF: This follows by an induction on T.

Theorem 8.4.6

$$U_{nk}^- \geq V_k^- \quad \text{and} \quad U_{nk}^+ \leq V_k^+$$

By our concavity–convexity assumption of f in y and z and by the continuity of f and the compactness of z, we have

$$V_k^-(x, T) = \min_z [\Delta_k f(x, y^0, z) + V_k^-(x + \Delta_k g(x, y^0, z), T - \Delta_k)]$$

for some $y^0 \in Y$; we also have $z_1, z_2, \cdots, z_m$ such that z_i is a function of $y_1, y_2, \cdots, y_i$, and

$$\sum_1^m \Delta_n f(x, y_i, z_i) + U_{nk}^- \left(x + \sum_1^m \Delta_n g(x, y_i, z_i), T - \Delta_k\right) \leq U_{nk}^-(x, T)$$

for any, admissible $y_1, y_2, \cdots, y_m$. Put $y_i = y^0$ and $z_i^0 = z_i(y^0, \cdots, y^0)$, $i = 1, 2, \cdots, m$. Further, put $z^0 = (1/m)(z_1^0 + \cdots + z_m^0)$.

For $T = 0$, the theorem is valid. We shall assume the truth for $T - \Delta_k$ and all initial x. We have

$$V_k^-(x, T) \leq \Delta_k f(x, y^0, z^0) + V_k^-(x + \Delta_k g(x, y^0, z^0), T - \Delta_k)$$

Since $2^{n-k} = m$, $\Delta_n = (1/m)\Delta_k$, and since f is convex in z for fixed x, y, the right side of the above inequality is at most

$$\Delta_n \sum_1^m f(x, y^0, z_i^0) + V_k^- \left(x + \Delta_n \sum_1^m g(x, y^0, z_i^0), T - \Delta_k\right)$$

By induction hypothesis, we have,

$$V_k^-(x, T) \le \Delta_n \sum_1^m f(x, y^0, z_i^0) + U_{nk}^- \left(x + \Delta_n \sum_1^m g(x, y^0, z_i^0), T - \Delta_k \right)$$

Similarly one can prove that $U_{nk}^+ \le V_k^+$.

Remark 8.4.1: We notice that the continuous analog of our discrete time game is the game discussed in Section 8.3. We then hope to see whether the analog of the functional equation for $V(x, T)$ of theorem 8.4.1, namely

$$\frac{\partial V}{\partial T} = \operatorname*{val}_{(y, z)} [f(x, y, z) + (\nabla V, g(x, y, z))]$$

$$V(x, 0) = V^0(x)$$

has a solution V, where ∇V is the gradient vector

$$\left(\frac{\partial V}{\partial x_1}, \frac{\partial V}{\partial x_2}, \dots, \frac{\partial V}{\partial x_n} \right)$$

Since f satisfies concavity–convexity properties in y and z, and since g is bilinear in y, z, the continuity properties give us optimal pure strategies for the game with payoff

$$f(x, y, z) + \langle \nabla V, g(x, y, z) \rangle$$

and the discrepancies in the values occasioned by neglecting mixed strategies in the discrete version seems to approach zero as $n \to \infty$.

There is no theorem to guarantee that the above differential equation has a solution. We shall prove from now on, the fundamental theorems of this section.

Theorem 8.4.7

$$V_n^+(x, T) \to V^+(x, T), \qquad V_n^-(x, T) \to V^-(x, T)$$

as $n \to \infty$. Further, $V^-(x, T) \le V^+(x, T)$. The convergence is uniform on bounded sets.

From theorems 8.4.5 and 8.4.6, we know that

$$V_n^-(x, T) \ge V_k^-(x, T) - NT\Delta_k \quad \text{for} \quad n \ge k \quad \text{and} \quad T = j\Delta_k$$

Suppose that for some (x, T) the sequence $\{V_n^-\}$ has two limit points, say a, b, with $a < b$. For some k^*

$$2NT\Delta_k^* < b - a$$

We can find $a, k \ge k_0$, such that $V_k^-(x, T)$ is near b and $n > k$ with $V_n^-(x, T)$ near a. This is a contradiction. Thus $V_n^-(x, T)$ tends to a limit. The family

$\{V_n^-(x, T)\}$ is equicontinuous and uniformly bounded by theorem 8.4.4, and hence they form a conditionally compact set (see [22]). Thus the convergence of $V_n^-(x, T)$ pointwise to $V^-(x, T)$ is uniform on bounded sets.

Theorem 8.4.8

From our assumptions on f and g, we have

$$V^-(x, T) = V^+(x, T) = \lim_n V_n(x, T)$$

PROOF: As a result of theorem 8.4.5, it is sufficient to prove that

$$\lim_{n \to \infty} [U_{nk}^-(x, T) - U_{nk}^+(x, T)] = 0$$

uniformly for x in any bounded set. Again we proceed by an induction argument.

For $T = 0$, this is true by the definition of U_{nk}^+ and U_{nk}^-. Assuming the result for $T - \Delta_k$, let

$$z_1^0(y_1), z_2^0(y_1, y_2), \cdots, z_m^0(y_1, \cdots, y_m)$$

be an optimal strategy for player II for the game where player I has to commit first. Let

$$y_1^0(z_1), y_2^0(z_1, z_2), \cdots, y_m^0(z_1, \cdots, z_m)$$

be an optimal strategy for player I for the game where player II has to commit first. For an arbitrary $z_1{}^*$, let

$$z_i{}^*(y_1, \cdots, y_{i-1}) = z_{i-1}^0(y_1, y_2, \cdots, y_{i-1}), \qquad i = 2, 3, \cdots, m$$

In the game where player I is in a favorable position (in the sense that he knows in advance the opponent's choice in any move before he commits), if he chooses $y_1^0, y_2^0, \cdots, y_m^0$ and player II chooses $z_1{}^*, z_2{}^*, \cdots, z_m{}^*$, then

$$U_{nk}^+(x, T) \le \Delta_n P + U_{nk}^+(x + \Delta_n Q, T - \Delta_k)$$

where

$$P = p(x, y_1^0) + q(x, z_1{}^*) + \cdots + p(x, y_m^0) + q(x, z_m{}^*)$$
$$Q = a(x, y_1^0) + b(x, z_1{}^*) + \cdots + a(x, y_m^0) + b(x, z_m{}^*)$$

For the game with player II in a favorable position, let

$$y_i{}^*(z_1, \cdots, z_{i-1}) = y_i^0(z_1{}^*, z_1, \cdots, z_{i-1}), \qquad i = 1, 2, \cdots, m$$

Let

$$P' = p(x, y_1{}^*) + q(x, z_1^0) + \cdots + p(x, y_m{}^*) + q(x, z_m^0)$$
$$Q' = a(x, y_1{}^*) + b(x, z_1^0) + \cdots + a(x, y_m{}^*) + b(x, z_m^0)$$

Then

$$\Delta_n P' + U_{nk}^-(x + \Delta_n Q', T - \Delta_k) \le U_{nk}^-(x, T)$$

But

$$y_1^* = y_1^0(z_1^*), \qquad z_1^0(y_1^*) = z_2^*(y_1^0), \qquad y_2^*(z_1^0) = y_2^0(z_1^*, z_2^*), \text{ etc.}$$

$$P - P' = q(x, z_1^*) - q(x, z_m^0)$$

$$Q - Q' = b(x, z_1^*) - b(x, z_m^0)$$

Let A be a bound for $|q|$, $|b|$, and let C be a Lipschitz constant for U_{nk}^+. Therefore, from the above steps

$$U_{nk}^+(x, T) \le \Delta_n P' + U_{nk}^+(x + \Delta_n Q', T - \Delta_k) + 2A\Delta_n + 2CA\Delta_n$$

and

$$0 \le U_{nk}^+(x, T) - U_{nk}^-(x, T) \le U_{nk}^+(x + \Delta_n Q', T - \Delta_k) -$$
$$- U_{nk}^-(x + \Delta_n Q', T - \Delta_k) + 2A(1 + C)\Delta_n$$

Since the right side of the inequality tends to zero uniformly in x on any bounded set as $n \to \infty$, the result follows.

8.5 CONTINUOUS DIFFERENTIAL GAMES

The theory of differential games for the discrete case is well established in the sense that we can talk about pure strategy sets Y and Z for the players and we have the value satisfying a functional equation (stated in theorem 8.4.1) for them. For any choice y_i, z_i of the players in the ith move, depending on their choices in the previous moves and therefore depending on the state of the system, they can think of pure strategies. Further, the state of the system after any move is uniquely determined once the choices in that move are selected by the players. But we have formidable difficulties in the continuous analog defined by a system of differential equations:

$$\frac{dx_i}{dt} = g_i(x, y, z), \qquad i = 1, 2, \cdots, n$$

$$x(0) = x_0$$

First it is not clear what we mean by pure strategies. Since pure strategies in any move in the discrete case for a player depend on the state of the system at that move and the choices of the opponent in all his previous moves, and since the current state itself is a result of the opponent's choices, one can think of pure strategies as functions $y = y(x)$ and $z = z(x)$ for the players.

But in that case we do not know whether there exists a solution to the above system of differential equations for any possible function $y(x)$ and $z(x)$ with ranges y, z in Y and Z, respectively. In fact, what we need in many of our physical problems is not a solution to the above differential equation but a solution for a slightly less restrictive system. Since the controls are needed only to modify the future course of action, essentially this requirement corresponds to demanding solutions only to

$$\lim_{\Delta t \to 0^+} \frac{x_i(t + \Delta t) - x_i(t)}{\Delta t} = g_i(x, y, z), \qquad i = 1, 2, \cdots, n$$

Although the existence criteria for such equations are broader, still they have their own limitations, and we start with the following notion of k-strategies as pursued by Isaacs. We assume that players have to restrict themselves to piecewise constant functions for their pure strategies; that is, they divide the time interval into a finite number of parts, and in each interval player I chooses the pure strategy $y = y(x, t) = y(x_i, t_i)$ if $t_i \leq t < t_{i+1}$, where $t_1, t_2, \cdots, t_N$ is a time partition of the total time period of the game and $x(t_i) = x_i$. A similar strategy is chosen by player II, and they should find optimal strategies only from these collections in case they exist. Since this is essentially discretizing the situation, we have no problem about the differential equations and solutions. They always exist with this assumption. We study the behavior of the strategies and the value as the intervals of the time partitions shrink to zero.

But at this stage one can raise a question as to our reason for involving only the state x of the system in the definition of the pure strategies—namely, our reason for assuming $y = y(x)$ and not $y = y(x, dx/dt)$ or any other form. We can always make our models more and more complex, but in all physical problems one feels that the players can make rational judgments at time t purely on the state of the system at t.

If we assume without loss of generality that the time partitions are dyadic —that is, for player I it is some $t_1, t_2, \cdots, t_s$ where $s = 2^{k_1}$ for some k_1, and for player II it is $t_1, t_2, \cdots, t_r$ where $r = 2^{k_2}$ for some k_2—then player I chooses s functions $u_1(x), u_2(x), \cdots, u_s(x)$ with values in Y, and player II chooses r functions $v_1(x), v_2(x), \cdots, v_r(x)$ with values in Z.

Let $n = \max(k_1, k_2)$. This induces a finer time partition. The payoff is calculated as in the nth time-discrete game, using the difference equation of that case for these specific strategies. If $k_1 = n$, then player I chooses $y_i = u_i(x_i)$, $i = 1, 2, \cdots$. If $k_1 < n$, then this strategy is defined in the following way.

For $i = 1, 2, \cdots$, define $y_i = u_p(x_p)$, where $p \leq i$ is the largest integer such that $p\Delta_n = j\Delta_{k_1}$. Similarly for player II. With this notion we have the following theorem.

Theorem 8.5.1

From our assumptions on f and g, $\lim\limits_{n \to \infty} V_n(x, T)$ is the value of the time-continuous game.

PROOF: Define $u_i(x)$ such that $T = i\Delta_k$, $i = 1, 2, \cdots, s$, and

$$V_k^-(x, T) = \min_z \, [\Delta_k f(x, u_i(x), z) + V_k^-(x + \Delta_k g(x, u_i(x), z), T - \Delta_k)]$$

where $k = k_1$ will be chosen presently.

By theorem 8.4.5 and 8.4.6 this strategy for player I yields against any strategy for player II at least $V_k^-(x, T) - NT\Delta_k$. But by theorem 8.4.8, for any $\varepsilon > 0$ and (x, T) we have a k_1 such that $T = j\Delta_{k_1}$ for some positive integer j, and

$$V(x, T) - \varepsilon < V_{k_1}^-(x, T) - N\Delta_{k_1}T$$

Similarly, player II has a strategy against which player I can never get more than $V(x, T) + \varepsilon$. Since ε is arbitrary, $V(x, T)$ is the value of the time-continuous game.

Remark 8.5.1: We constrained the functions f and $g_1, \cdots, g_n$ to a special form to prove the uniform convergence of $V_n(x, T)$ to a limit function on bounded sets in the (x, T) space. Fleming proves [16] this result for more general $f, g_1, g_2, \cdots, g_n$. The proof in that article depends on the existence of solutions for a certain nonlinear parabolic partial differential equation [16, 17], closely related to the formal partial differential equation of remark 8.4.1 from which we can form the value function in the continuous case.

8.6 SOLUTION TO AN EXAMPLE IN DIFFERENTIAL GAMES

Often, we have theorems that only prove the existence of a solution, but are not powerful enough to give an algorithm for finding solutions, if they exist. For example, the proof of the minimax theorem we give in Chapter Two is nonconstructive. But a constructive proof could be given by the simplex method and at the same time lead to an efficient algorithm. Here we are faced with the same problem. We shall assume that a certain differential game in continuous time has an optimal pure strategy for the players (an optimal control). We shall build up a solution to the game by assuming the existence of a solution to a certain partial differential equation. In spite of the simplicity of the model we find many new concepts emerging out of the solution. This leads to a more general mathematical theory of differential games as discussed in a series of papers [3–5, 7] by Berkovitz. We have not stressed that approach, for the simple reason that the prerequisite for that development is a fair amount of the calculus of variations, and space does not permit us to venture more.

Now let us turn to our example 8.1.3 on tactical air war. Let us assume that the players have to allocate for only two tasks, "attrition" and "attack." By attrition we mean depleting the enemy's rate of weapon supply, and attack means entering into an outright battle. Suppose that, for a mathematical treatment of the problem in its continuous version, we assume that the time variable is continuous. Let us build up a much simpler model for this case. At any time t, nation A has to decide what proportion, φ, of his total force he would allocate for attrition. We shall assume φp to be meaningful for any $0 \le \varphi \le 1$, where p denotes the number of planes for nation A. We can build up a similar proportion, ψ, for nation B. Let λ be the constant rate at which nation A produces planes, and let μ be the rate for nation B. Let us assume that their loss rate is directly proportional to the total planes allotted by the enemy for attrition. If $p(t)$ denotes the number of planes at time t for nation A and $q(t)$ denotes the number of planes at time t for nation B, then in small time interval $(t, t + \Delta t)$, the loss for nation A is approximately $\alpha \psi(t) q(t) \cdot \Delta t$ for some constant α, and the approximate production is $\lambda \cdot \Delta t$. Thus the number of planes at time $t + \Delta t$ for nation A is given by

$$p(t + \Delta t) \doteq p(t) + \lambda \cdot \Delta t - \alpha \psi(t) q(t) \cdot \Delta t$$

That is,

$$\frac{p(t + \Delta t) - p(t)}{\Delta t} \doteq \lambda - \alpha \psi(t) q(t)$$

Thus as $\Delta t \to 0^+$, we can take the kinematic equation to be

$$\frac{dp}{dt} = \lambda - \alpha \psi(t) q(t)$$

Similarly, for nation B it is

$$\frac{dq}{dt} = \mu - \beta \varphi(t) p(t)$$

Let us assume that the excess force of one over the other used for attack measures the superiority of a side at time t and that a suitable payoff for nation B is

$$\int_0^N [(1 - \psi) q - (1 - \varphi) p] \, dt$$

Thus we have the following problem.

How optimally should nation B choose the attrition proportion $\psi = \psi(t)$ at each time t so that, independent of what nation A does, it could attain a certain maximum value for the integral payoff above, when the two sides are constrained by such forward differential equations as given above? Since the time t has to satisfy $0 \leq t \leq N$, we can incorporate this constraint also into our system by defining $dT/dt = -1$ so that $T = \text{const} - t$. If $T(0) = N$, then $T = N - t$ expresses the amount of time left over for the end of the battle. Counting time from back (namely, counting the leftover time) has an advantage that will be apparent in our analysis.

We shall make the following assumptions.

1. The physical restrictions of the combat permit each side to change the values of φ and ψ at most a finite number of times. That is, the admissible pure strategies for the two sides are the piecewise constant right continuous functions φ and ψ on $[0, N]$ with $0 \leq \varphi, \psi \leq 1$.

2. For any admissible pure strategies (or admissible controls) φ and ψ, the available force at any time t in $0 \leq t \leq N$ for any side is never zero; that is, $p(t) > 0$, $q(t) > 0$, for any φ, ψ chosen from the piecewise constant controls.

3. The loss rate for nation A is more than the loss rate for nation B.

4. There exists optimal controls φ^*, ψ^* (or optimal pure strategies) for the two nations satisfying condition 1 and the minimax condition.

$$\min_{\varphi} \max_{\psi} \int_0^N [(1 - \psi)\, q - (1 - \varphi)\, p]\, dt$$

$$= \max_{\psi} \min_{\varphi} \int_0^N [(1 - \psi)\, q - (1 - \varphi)\, p]\, dt$$

$$= \int_0^N [(1 - \psi^*)\, q - (1 - \varphi^*)\, p]\, dt$$

where φ and ψ range over piecewise constant functions on $[0, N]$ with $0 \leq \varphi$, $\psi \leq 1$.

5. There exists a unique function $V(p, q, T)$ satisfying the functional equation

$$\frac{\partial V}{\partial T} = \operatorname*{value}_{\varphi,\,\psi} \left[(1 - \psi)\, q - (1 - \varphi)\, p + \frac{\partial V}{\partial p} (\lambda - \alpha \psi q) + \frac{\partial V}{\partial q} (\mu - \beta \varphi p) \right]$$

for all $0 \leq T \leq N$, with $V(p, q, 0) = 0$. V has continuous partial derivatives in p, q, T on $\mathcal{H}$, the permissible (p, q, T) space. Before trying to solve the problem we shall examine our assumptions.

Assumptions (1) and (3) are conditions to be fulfilled, and the restriction of φ and ψ to piecewise constant controls is not unreasonable. In fact, we are treating just a discrete problem with a continuous argument. Assumption (2) is also not unreasonable if we agree that a nation cannot continue to fight even for a minute more if its total weapon force is annihilated at any time.

Assumption (5) is the crux assumption. We shall solve for $V(p, q, T)$, and this functional equation is just the continuous analog of the functional equation of theorem 8.4.1. Thus this functional equation is the one stated in remark 8.4.1 for our problem. In our equation, the function that appears inside the brackets is bilinear in φ, ψ, and thus we have optimal pure strategies for the bilinear payoff. As in the discrete case we naturally feel that $V(p, q, T)$ should be the value of the game that has initial state p, q, and duration T.

In fact, we shall check that V has continuous partial derivatives in our (p, q, T) space under consideration and that by a theorem of Fleming (theorem 2, [13]) V is actually the value of the game with duration T and payoff

$$\int_{N-T}^{N} [(1 - \psi) q - (1 - \varphi) p] \, dt$$

(Notice that in our theorems, such as theorem 8.5.1, we have proved the value of the game to be the limit of the time-discrete game without any assumption on the value function. We have not demanded the existence of the solution to the partial differential equation. We have imposed conditions on the payoff which with all other conditions imposed there are fulfilled in our problem.)

In practice, for many games we encounter we would have optimal pure strategies for the payoff in the bracketed expression. This happens in our general case too, under the assumptions we imposed in Section 8.4. These optimal pure strategies are not to be confused with the optimal pure strategies of our game with integral payoff. They depend on p, q and $\partial V/\partial p$, $\partial V/\partial q$, which we do not know. But we see that, once V is known with all its partial derivatives inserted, these functions turn out to be the optimal pure strategies. Thus we would not use a different symbol, and we would assume that they furnish the required strategies. (For a formal proof of this statement, see theorem 4.4.1 [19]).

Let us take the partial differential equation:

$$\frac{\partial}{\partial T} V(p, q, T)$$

$$= \operatorname*{value}_{\varphi, \psi} \left[(1 - \psi) q - (1 - \varphi) p + \frac{\partial V}{\partial p} (\lambda - \alpha \psi q) + \frac{\partial V}{\partial q} (\mu - \beta \varphi p) \right]$$

for fixed p, q, with T giving the duration of the game. Our space of action, $\mathscr{H}$, is given by

$$p > 0, \qquad q > 0, \qquad T \geq 0$$

We stop playing (the nations stop fighting) as soon as $T = 0$ (that is, at $t = N$). We demand as part of our assumption that

$$p = s_1 > 0, \qquad q = s_2 > 0$$

This terminator is given by the surface $\mathcal{B}$ in $\mathcal{H}$, where

$$\mathcal{B} = \{(p, q, T): p = s_1 > 0, q = s_2 > 0, T = 0\}$$

For fixed p, q, the bilinearity of the expression inside the brackets of our main partial differential equation yields

$$\underset{0 \leq \varphi, \psi \leq 1}{\text{Value}} \left[(1 - \psi) q - (1 - \varphi) p + \frac{\partial V}{\partial p} (\lambda - \alpha \psi q) + \frac{\partial V}{\partial q} (\mu - \beta \varphi p) \right]$$

$$= \frac{\partial V}{\partial T} = \left[(1 - \psi^*) q - (1 - \varphi^*) p + \frac{\partial V}{\partial p} (\lambda - \alpha \psi^* q) + \right.$$

$$\left. + \frac{\partial V}{\partial q} (\mu - \beta \varphi^* p) \right] \ldots \text{(ME)}$$

Since p, q are always positive in $\mathcal{H}$, φ^* being the minimizing and ψ^* being the maximizing strategy, we have

$$\varphi^* = 0 \text{ for } 1 - \beta \frac{\partial V}{\partial q} = A > 0; \qquad \psi^* = 0 \text{ for } \left[-1 - \alpha \frac{\partial V}{\partial p} \right] = B < 0$$

$$= 1 \qquad \text{for} \qquad A < 0; \qquad\qquad = 1 \qquad \text{for} \qquad B > 0$$

If a game starts at t, the duration is $\tau = N - t$, and therefore it is advantageous to consider the kinematic equations with respect to τ. This yields

$$\frac{dp}{d\tau} = \alpha \psi q - \lambda, \qquad \frac{dq}{d\tau} = \beta \varphi p - \mu, \qquad \frac{dT}{d\tau} = 1 \ldots \ldots \text{(KE)}$$

Since we do not know $\partial V/\partial p$ and $\partial V/\partial q$, we get them also by the following argument in terms of differential equations in τ. For that, consider our main equation ME above.

Now, partially differentiate, with respect to p, both sides of the equation ME. Surprisingly, one can prove, because of assumption (4) (see p. 80 [20]), that $\frac{d}{dt}\left(\frac{\partial V}{\partial p}\right)$ and $\frac{d}{dt}\left(\frac{\partial V}{\partial q}\right)$ are the negatives of the formal derivatives of the right-hand side of the main partial differential equation ME with respect to p and q, respectively, that explicitly appear, and that

$$\frac{d}{d\tau}\left(\frac{\partial V}{\partial p}\right) = A\varphi^* - 1, \qquad \frac{d}{d\tau}\left(\frac{\partial V}{\partial q}\right) = B\psi^* + 1$$

When the duration of the game is zero, then we have $V(p, q, 0) = 0$ for any $p, q \geq 0$, and $V(s_1, s_2, 0) = 0$ for any possible s_1, s_2 at which one could arrive by some φ, ψ. Thus

$$\left. \frac{\partial V}{\partial p} \right|_{p = s_1} = \left. \frac{\partial V}{\partial q} \right|_{q = s_2} = 0$$

Hence $A = 1$, $B = -1$, and thus $\varphi^* = \psi^* = 0$ when the duration is zero. By assumption, they are piecewise constant right continuous functions (in t), and just before the end of the war the two sides launch a full-scale attack. The kinematic equations are

$$p = s_1 - \lambda\tau, \qquad q = s_2 - \mu\tau, \qquad \frac{\partial V}{\partial p} = -\tau, \qquad \frac{\partial V}{\partial q} = \tau$$

$$B = -1 + \alpha\tau, \qquad A = 1 - \beta\tau \text{ in a neighborhood of } \tau = 0$$

(Since $\tau \geq 0$, the neighborhood is only to the right of $\tau = 0$.) But B ceases to be negative for $\tau \geq 1/\alpha$, and, since $\beta > \alpha$, A remains negative even after $\tau \geq 1/\alpha$ for some more τ's. Intuitively we expect a change at $\tau = 1/\alpha$ for the control ψ^*, but not for φ^*. We feel that, since $B > 0$ for $\tau > 1/\alpha$ in some interval $(1/\alpha, \tau_0)$ for τ, $\psi^* = 1$, but $\varphi^* = 0$. When the duration $T = \tau \leq 1/\alpha$, the optimal strategy $\varphi^* = \psi^* = 0$ yields the value

$$\int_{N-T}^{N} [(1 - 0)\, q - (1 - 0)\, p]\, dt$$

That is,

$$\int_0^T (q - p)\, d\tau = \int_0^T [(s_2 - \mu\tau) - (s_1 - \lambda\tau)]\, d\tau$$

$$= (s_2 - s_1)\, T - \tfrac{1}{2}(\mu - \lambda)\, T^2$$

$$= (q - p)\, T + \tfrac{1}{2}(\mu - \lambda)\, T^2 \quad \text{(after eliminating } s_1, s_2\text{)}$$

(Here we tacitly exploit the standard principle of optimality in dynamic programming. If φ^*, ψ^* are optimal throughout and if s_1, s_2 are the positions short of T time units to the end of the game, the game with duration T and initial condition s_1, s_2 also yields φ^*, ψ^* as an optimal strategy with payoff defined as above.)

Let the state of the system be $p = s_1{}^*$, $q = s_2{}^*$, at time $T = 1/\alpha$. Here we start afresh with the above as the terminal surface to get $p(\tau)$, $q(\tau)$, etc. By the continuity assumption of $\partial V/\partial p$, $\partial V/\partial q$, we have

$$\frac{\partial V}{\partial p}\bigg|_{T=1/\alpha} = -\frac{1}{\alpha} \frac{\partial V}{\partial q}\bigg|_{T=1/\alpha} = \frac{1}{\alpha} \text{ from above}$$

For $\tau > 1/\alpha$ let us use our heuristic $\varphi^* = 0$, $\psi^* = 1$. Since

$$dq/d\tau = -\mu + \beta\varphi^* p$$

then for τ in the right neighborhood of $1/\alpha$,

$$q = -\mu\tau + \text{constant}$$

$$T = \text{constant} + \tau = \frac{1}{\alpha} + \tau$$

Since $q = s_2^*$ at $T = 1/\alpha$, we have $q = s_2^* - \mu\tau$. Similarly we get

$$p = s_1^* + (\alpha s_2^* - \lambda)\,\tau - \tfrac{1}{2}\alpha\mu\tau^2$$

These are all obtained from our kinematic equations KE. Further

$$\frac{d}{d\tau}\left(\frac{\partial V}{\partial p}\right) = A\varphi^* - 1 \quad \text{gives} \quad \frac{\partial V}{\partial p} = -\frac{1}{\alpha} - \tau$$

Similarly

$$\frac{\partial V}{\partial q} = \frac{1}{\alpha} + \tau + \frac{1}{2}\alpha\tau^2$$

$$A = 1 - \frac{\beta}{\alpha} - \beta\tau - \frac{1}{2}\alpha\beta\tau^2$$

$$B = \alpha\tau$$

Since for duration $T > 1/\alpha$, $\tau > 0$, and $B > 0$, A also remains positive for $T > 1/\alpha$ for some more time T, and A changes from positive to negative when $\tau = \tau^0$, where

$$1 - \frac{\beta}{\alpha} - \beta\tau - \frac{1}{2}\alpha\beta\tau^2 = 0, \qquad \text{at } \tau = \tau_0 > 0$$

Here

$$\tau_0 = \frac{1}{\alpha}\left(\sqrt{\frac{2\alpha - \beta}{\beta}} - 1\right)$$

Let us assume that the strategy φ^* changes from 0 to 1, and let us assume that there are no more changes; then we have determined φ^*, ψ^* completely. Now to evaluate the value of the game whose duration is $1/\alpha < T = 1/\alpha + \tau$, we use the actual main partial differential equation, with $\tau = T - 1/\alpha$ used to plug in $\partial V/\partial p$ and $\partial V/\partial q$. Thus

$$\frac{\partial V}{\partial p} = -\frac{1}{\alpha} - \tau = -T, \qquad \frac{\partial V}{\partial q} = T + \frac{\alpha}{2}\left(T - \frac{1}{\alpha}\right)^2$$

This with the main equation at $\varphi^* = 0$, $\psi^* = 1$, yields

$$\frac{\partial V}{\partial T} = q(-1 + \alpha T) - \lambda T + \mu\left(T + \frac{\alpha}{2}\left(T - \frac{1}{\alpha}\right)^2\right) + (q - p)$$

Thus

$$V(p, q, T) = -qT + \alpha q\frac{T^2}{2} - \lambda\frac{T^2}{2} + \mu\frac{T^2}{2} + \frac{\alpha\mu}{6}\left(T - \frac{1}{\alpha}\right)^3 +$$

$$+ (q - p)\,T + \text{terms independent of } T$$

This simplifies to

$$-pT + \alpha q \frac{T^2}{2} + (\mu - \lambda) \frac{T^2}{2} + \frac{\alpha\mu}{6}\left(T - \frac{1}{\alpha}\right)^3 + \text{terms independent of } T$$

To determine those terms, we take $T = 1/\alpha$ and equate it with the value we obtained earlier for $T < 1/\alpha$. We are approaching $T = 1/\alpha$ from right and left, which yields

$$-p \cdot \frac{1}{\alpha} + \alpha q \cdot \frac{1}{2\alpha^2} + (\mu - \lambda)\frac{1}{2\alpha^2} + \text{constant} = (q - p)\frac{1}{\alpha} + \frac{1}{2}(\mu - \lambda)\frac{1}{\alpha^2}$$

That is,

$$\text{Constant} = \frac{q}{2\alpha}$$

Substituting this yields, after simplification,

$$V(p, q, T) = \left(\frac{q}{2\alpha} - \frac{\mu}{6\alpha^2}\right) + \left(\frac{\mu}{2\alpha} - p\right)T + \left(\frac{\alpha q - \lambda}{2}\right) \cdot T^2 + \alpha\mu \frac{T^3}{6}$$

for $1/\alpha \le T \le \tau_0$. But, since the intuitive choice beyond τ_0 is $\varphi^* = \psi^* = 1$, which in fact contributes nothing to the payoff, we have effectively determined the value function and φ^*, ψ^* when $p, q, > 0$ for all possible φ, ψ and for all t. The following is the verbal instruction for the optimal strategy.

Nations A and B should start concentrating their entire force for attrition purposes. When the play reaches that time when the leftover duration is only $\frac{1}{\alpha}\sqrt{\frac{2\alpha - \beta}{\beta}}$, or equivalently at time $N - \frac{1}{\alpha}\sqrt{\left(\frac{2\alpha - \beta}{\beta}\right)}$, nation A should first switch over to concentrating entirely on attack. Nation B should continue with its whole force concentrated on attrition for some more time. But when the duration remaining is $1/\alpha$, suddenly nation B should also switch over to attack with the entire force.

REFERENCES

[1] Bellman, R. (1957). "Dynamic Programming," Princeton University Press, Princeton, New Jersey.
[2] Berkovitz, L. D. (1961). Variational methods in problems of control and programming. *J. Math. Anal. Appl.* **13**, 145–169.
[3] Berkovitz, L. D. (1964). A differential game with no pure strategy solutions. "Advances in Game Theory," pp. 175–194 (*Ann. Math. Studies No. 52*), edited by M. Dresher, L. S. Shapley, and A. W. Tucker, Princeton University Press, Princeton, New Jersey.
[4] Berkovitz, L. D. (1964). A variational approach to differential games. "Advances in Game Theory," pp. 127–194 (*Ann. Math. Studies No. 52*), edited by M. Dresher, L. S. Shapley, and A. W. Tucker, Princeton University Press, Princeton, New Jersey.

[5] Berkovitz, L. D. (1967). Necessary conditions for optimal strategies in a class of differential games and control problems. *J. SIAM Control* **5**, 1–24.

[6] Berkovitz, L. D., and Dreyfus, S. E. (1966). A dynamic programming approach to the non-parametric problem in the calculus of variations. *J. Math. Mech.* **15**, 83–100.

[7] Berkovitz, L. D., and Fleming, W. H. (1957). On differential games with integral payoff. In "Contributions to the Theory of Games," Vol. 3, pp. 413–435 (*Ann. Math. Studies No. 39*), edited by M. Dresher, A. W. Tucker, and P. Wolfe, Princeton University Press, Princeton, New Jersey.

[8] Bliss, G. A. (1946). "Calculus of Variations," University of Chicago Press, Chicago, Illinois.

[9] Dresher, M. (1961). "Games of Strategy," Prentice-Hall, Englewood Cliffs, New Jersey.

[10] Dresher, M., and Berkovitz, L. D. (1959). A game theory analysis of tactical air war. *Operations Res.* **7**, 599–620.

[11] Dresher, M., and Berkovitz, L. D. (1960). A multimove infinite game with linear payoff. *Pacific J. Math.* **10**, 743–765.

[12] Dubins, L. E. (1957). A discrete evasion game. In "Contributions to the Theory of Games," Vol. 3, pp. 231–255 (*Ann. Math Studies No. 39*), edited by M. Dresher, A. W. Tucker, and P. Wolfe, Princeton University Press, Princeton, New Jersey.

[13] Fleming, W. H. (1957). A note on differential games of prescribed duration. In "Contributions to the Theory of Games," Vol. 3, pp. 407–416 (*Ann. Math. Studies No. 39*), edited by M. Dresher, A. W. Tucker, and P. Wolfe, Princeton University Press, Princeton, New Jersey.

[14] Fleming, W. H. (1961). The convergence problem for differential games. *J. Math. Anal. Appl.* **3**, 103–116.

[15] Fleming, W. H. (1963). Some Markovian optimization problems. *J. Math. Mech.* **12**, 131–140.

[16] Fleming, W. H. (1964). The convergence problem for differential games II. "Advances in Game Theory," pp. 195–210 (*Ann. Math. Studies No. 52*), edited by M. Dresher, L. S. Shapley, and A. W. Tucker, Princeton University Press, Princeton, New Jersey.

[17] Friedman, A. (1958). On quasi-linear parabolic equations of the second order. *J. Math. Mech* Part I, pp. 793–809.

[18] Grenander, U. (March 1963). A Tactical Study of Evasive Maneuvers, FOAP Report 126, Research Institute of National Defense, Stockholm 80, Sweden.

[19] Isaacs, R. (1954). "Differential Games," I, II, III, IV, The Rand Corporation, Research Memoranda RM–1391, RM–1399, RM–1411, RM–1486.

[20] Isaacs, R. (1965). "Differential Games," John Wiley & Sons, New York.

[21] Karlin, S. (1957). An infinite move game with a lag. In "Contributions to the Theory of Games," Vol. 3, pp. 257–272 (*Ann. Math. Studies No. 39*), edited by M. Dresher, A. W. Tucker, and P. Wolfe, Princeton University Press, Princeton, New Jersey.

[22] Kolmogorov, A. N., and Fomin, F. V. (1957). *Functional Analysis,* Vol. 1. Grelock, New York.

[23] Scarf, H. (1957). On differential games with survival payoff. In "Contributions to the Theory of Games," Vol. 3, pp. 393–405 (*Ann. Math. Studies No 39*). edited by M. Dresher, A. W. Tucker, and P. Wolfe, Princeton University Press, Princeton, New Jersey.

[24] Pontryagin, L. S., Boltyanskii, V. G., Gamkrelidze, R. V., and Mishchenko, E. F. (1964). "The Mathematical Theory of Optimal Processes," Pergamon Press, New York.

Chapter Nine

N-PERSON GAMES AND PRODUCT SOLUTIONS
FOR SIMPLE GAMES

The main purpose of this chapter is to give an introduction to a theory of solutions for n-person games in characteristic function form given by von Neumann and Morgenstern. The main mathematical question concerning their model is whether every game has at least one solution. We shall present an example of a ten-person game due to Lucas, which has no solution. We shall also investigate the solutions of the games that are formed by combining two or more simple games played by separate groups of individuals.

9.1 n-PERSON GAMES

In the case of two-person games we have been able to give intuitively acceptable definitions of the value of the game to each player and of optimal strategies. We now turn our attention to finite games with more than two players. For this wider class of games there is not available, unfortunately, any theory that is intuitively as acceptable as is the theory for two-person games. We shall present in this section the von Neumann–Morgenstern approach to games with more than two players. This approach has its own pitfalls, but there are also some elements of soundness. The example of a ten-person game due to Lucas, which has no solution, certainly shattered the hopes of this theory to a great extent.

We shall now describe a rectangular zero-sum n-person game—that is, a game in which each player makes just one choice from a finite set and in which each player chooses in ignorance of the choices of the other players. Such a game has just n moves; in the ith move, player i $(i = 1, 2, \cdots, n)$, not being informed about the outcome of any of the previous moves, chooses a number x_i from the finite set C_i. After the n moves are completed, player i $(i = 1, 2, \cdots, n)$ gets the amount $M_i(x_1, x_2, \cdots, x_n)$. Since the game is zero-sum;

$$\sum_{i=1}^{n} M_i(x_1, x_2, \cdots, x_n) = 0 \quad \text{identically in} \quad (x_1, x_2, \cdots, x_n)$$

The ideas that center around the theory of n-person games include the following: what combinations of players ("*coalitions*") will be formed, and

what payments the players can be expected to make to each other as induce-
ments to join the various coalitions.

Suppose that the n players split themselves into two groups T_1 and T_2 such
that $T_1 \cup T_2 = (1, 2, \cdots, n)$, $T_1 \cap T_2 = \emptyset$. Suppose that $T_1 = (i_1, i_2, \cdots, i_r)$
and $T_2 = (j_1, j_2, \cdots, j_s)$. We are going to regard T_1 and T_2 as two players;
T_1 chooses an element from $C_{i_1} \times C_{i_2} \times C_{i_r}$, and T_2 chooses from
$C_{j_1} \times C_{j_2} \times \cdots \times C_{j_s}$. If T_1 uses strategy $t_1 = (x_{i_1}, x_{i_2}, \cdots, x_{i_r})$ and
T_2 uses strategy $t_2 = (x_{j1}, x_{j2}, \cdots, x_{js})$, then the total payoff to T_1 is simply
$\sum_{\epsilon T_1} M_i(x_1, x_2, \cdots, x_n)$.

Let us write $M_i(x_1, x_2, \cdots, x_n) = M_i(t_1, t_2)$. We shall denote the payoff
to T_1 by

$$M_{T_1}(t_1, t_2) = \sum_{i \epsilon T_1} M_i(t_1, t_2)$$

From the fact that the original game was zero-sum, it follows immediately
that

$$M_{T_2}(t_1, t_2) = -M_{T_1}(t_1, t_2)$$

where M_{T_2} has a definition similar to that of M_{T_1}. Since each C_i is finite,
the strategy spaces of T_1 and T_2 are finite. Let $\{a_1, a_2, \cdots, a_k\}$ and
$\{b_1, b_2, \cdots, b_l\}$ be the strategy spaces of T_1 and T_2, respectively. If T_1 uses
the mixed strategy $(\xi_1, \xi_2, \cdots, \xi_k)$ and T_2 uses the mixed strategy
$(\eta_1, \eta_2, \cdots, \eta_l)$, then the expected payoff to T_1 is

$$\sum_{i=1}^{k} \sum_{j=1}^{l} \xi_i \eta_j M_{T_1}(a_i, b_j)$$

Let us denote this quantity by $M_{T_1}(\xi, \eta)$. From von Neumann's theorem it
follows that

$$\max_{\xi} \min_{\eta} M_{T_1}(\xi, \eta) = \min_{\eta} \max_{\xi} M_{T_1}(\xi, \eta)$$

We set $v(T_1) = \max_{\xi} \min_{\eta} M_{T_1}(\xi, \eta)$.

From the discussion above we see that for every subset $T_1 \subseteq \{1, 2, \cdots, n\}$
we can define a real valued function, which in a sense represents the total
amount the members of T_1 can expect to get if they make a coalition. The
function v is called the *characteristic function* of the game. We shall now
derive some of the mathematical properties of characteristic functions.

Theorem 9.1.1

Let v be the characteristic function of an n-person zero-sum game, where N stands for the set of players $\{1, 2, \cdots, n\}$. Then

(1) $v(N) = 0$.

(2) $v(T) = -v(N - T)$, where $T \subseteq N$.

(3) $v(R \cup T) \geq v(R) + v(T)$, whenever $R \cap T = \varnothing$.

PROOF OF THEOREM: $v(N) = 0$ follows from the fact that the n-person game is zero-sum. To prove (2) we have

$$v(N - T) = \max_{\xi} \min_{\eta} M_{N-T}(\xi, \eta)$$

$$= \max_{\xi} \min_{\eta} (-M_T(\xi, \eta)) \; [\because M_T = -M_{N-T}]$$

$$= -\min_{\xi} \max_{\eta} M_T(\xi, \eta)$$

$$= -\max_{\eta} \min_{\xi} M_T(\xi, \eta) = -v(T)$$

We leave the proof of (3) to the reader. See [3].

Remark 9.1.1: It is not difficult to observe the following facts: (1) $v(\varphi) = 0$. (2) If $T_1, T_2, \cdots, T_r$ are mutually disjoint subsets of N, then

$$v(T_1 \cup T_2 \cup \cdots \cup T_r) \geq v(T_1) + v(T_2) + \cdots + v(T_r)$$

And (3) if $(T_1, T_2, \cdots, T_r)$ is a partition of N—that is, they are mutually disjoint and their union is N—then $v(T_1) + v(T_2) + \cdots + v(T_r) \leq 0$.

We shall now derive the converse of theorem 9.1.1—that is, every function v that satisfies the three conditions of theorem 9.1.1 is in fact the characteristic function of some zero-sum n-person game.

Theorem 9.1.2

Let N be a finite set containing n persons, and let v be a real valued function which is defined for every subset T of N and which satisfies the three conditions of theorem 9.1.1. Then there exists an n-person zero-sum game of which v is the characteristic function.

PROOF OF THE THEOREM: We reproduce the proof as given in McKinsey [3]. We define a game whose players are the members of N as follows. Each member, x, of N, makes just one move, which consists in choosing a subset T_x of N such that $x \in T_x$; each of these moves is made in ignorance of the moves of the other players.

To define the payoff functions, we first introduce the auxiliary notion of a distinguished subset of N. A subset, T, of N is called distinguished (with respect to a given play of the game) if either (1) for every $x \in T$, we have $T_x = T$, or (2) T is a set containing just one element x, and x belongs to no set satisfying condition 1.

It is easily seen that, for a given play of the game, the distinguished subsets of N are mutually disjoint and their union is N. Now suppose that, for a given play of the game, the distinguished subsets of N are $T_1, T_2, \cdots, T_p$, and that T_j (for $j = 1, 2, \cdots, p$) contains n_j elements. Then if player i ($i = 1, 2, \cdots, n$) belongs to T_j, the payoff to player i is defined as

$$\frac{1}{n_j} v(T_j) - \frac{1}{n} \sum_{r=1}^{p} v(T_r)$$

where v is the given function.

We shall show first that the game defined in this way is zero-sum. Suppose, as before, that, for a given play, the distinguished subsets are $T_1, T_2, \cdots, T_p$, where T_j contains n_j elements. We notice that the payoff to any two members of the same set T_j is the same. Hence, for each j, the sum of the payments to the members of T_j is simply the product of n_j by the payoff to each member of T_j; that is,

$$n_j \left\{ \frac{1}{n_j} v(T_j) - \frac{1}{n} \sum_{r=1}^{p} v(T_r) \right\}$$

The sum of the payments to all the members of N is therefore

$$\sum_{j=1}^{p} n_j \left[\frac{1}{n_j} v(T_j) - \frac{1}{n} \sum_{r=1}^{p} v(T_r) \right]$$

$$= \sum_{j=1}^{p} v(T_j) - \frac{1}{n} \sum_{j=1}^{p} n_j \sum_{r=1}^{p} v(T_r)$$

$$= \sum_{1}^{p} v(T_j) - \sum_{1}^{p} v(T_j) = 0$$

as was to be shown.

To complete the proof of our theorem, it now remains to show that v is the characteristic function of the game defined above. Let $\bar{v}$ be the characteristic function of the game defined; we are to prove that, for every subset T of N, $\bar{v}(T) = v(T)$.

By hypothesis, v satisfies the three conditions of theorem 9.1.1, and $\bar{v}$ satisfies these conditions because it is the characteristic function of a zero-sum game. It follows that v and $\bar{v}$ also satisfy the three conditions of the remarks.

We shall first show that, for every subset T of N,

$$\bar{v}(T) \geq v(T)$$

For $T = \emptyset$, this follows from condition (1) given in the remarks. If $T \neq \emptyset$, the players in T can form a coalition and can agree that each member x of T will choose $T_x = T$. This will make T a distinguished set with respect to the play in question. Suppose, now, that the distinguished sets of such a play are $T_1, T_2, \cdots, T_p$, with $T_1 = T$. Then each player in T gets exactly

$$\frac{1}{n_1} v(T) - \frac{1}{n} \sum_{r=1}^{p} v(T_r)$$

and hence the total payment to the members of T is equal to

$$v(T) - \frac{n_1}{n} \sum_{r=1}^{p} v(T_r)$$

Since the members of T can thus ensure that they will get at least

$$v(T) - \frac{n_1}{n} \sum_{r=1}^{p} v(T_r)$$

we conclude that

$$\bar{v}(T) \geq v(T) - \frac{n_1}{n} \sum_{r=1}^{p} v(T_r)$$

Since $\sum_{1}^{p} v(T_r) \leq 0$, it follows that $\bar{v}(T) \geq v(T)$. Now we shall establish the reverse inequality.

Since $v(T) \geq v(T)$ for every $T \subseteq N$, it follows that $\bar{v}(N - T) \geq v(N - T)$. But

$$\bar{v}(N - T) = -\bar{v}(T)$$

$$v(N - T) = -v(T)$$

Hence

$$-v(T) \geq -v(T)$$

That is

$$\bar{v}(T) \leq v(T)$$

which completes the proof of our theorem.

Remark 9.1.2: Because of theorems 9.1.1 and 9.1.2, it is enough to speak of characteristic functions, and we do not have to refer to any game that generates them.

Definition 9.1.1: Two *n*-person games are said to be *strategically equivalent* if there is a positive constant K and constants $a_1, a_2 \cdots, a_n$, with $\Sigma a_i = 0$ such that

$$v(T) = Kv'(T) + \sum_{i \in T} a_i$$

where v and v' are the characteristic functions of the two games.

Definition 9.1.2: An *n*-person game with a characteristic function v is said to be in *reduced form* if

$$v(1) = v(2) = \cdots = v(n) = r$$

where $r = 0$ or $r = -1$. When these equations hold, we also say that the characteristic function v is in reduced form with modulus r. Now we can establish the following theorem.

Theorem 9.1.3

Every characteristic function is strategically equivalent to one, and only one, characteristic function in reduced form.

PROOF: Let v be a characteristic function. If $\sum_{i=1}^{n} v(i) = 0$, we put $K = 1$ and $a_i = -v(i)$ for $i = 1, 2, \cdots, n$.

Define $v'(T) = v(T) + \sum_{i \in T} a_i$. Then it follows that $v'(i) = 0$ for every i.

Further it is clear that $\sum_{1}^{n} a_i = -\sum_{1}^{n} v(i) = 0$ by assumption. If $\sum_{1}^{n} v(i) \neq 0$,

put

$$K = -\frac{n}{\displaystyle\sum_{1}^{n} v(i)} \quad \text{and} \quad a_i = -1 + \frac{nv(i)}{\displaystyle\sum_{i=1}^{n} v(i)}$$

Then

$$\sum_{i=1}^{n} a_i = -n + \frac{n \displaystyle\sum_{1}^{n} v(i)}{\displaystyle\sum_{1}^{n} v(i)} = 0 \quad \text{and} \quad v'(i) = Kv(i) + a_i = -1$$

$$\text{for every} \quad i = 1, 2, \cdots, n$$

It is clear, for both $\sum_{1}^{n} v(i) = 0$ and $\sum_{i=1}^{n} v(i) \neq 0$, that v' and v are strategically equivalent, and v' is in reduced form.

Remark 9.1.3: In essence, theorem 9.1.3 says that it is enough to consider *n*-person zero-sum games in reduced form. If $r = 0$, then it is not hard to check $v(T) = 0$ for every $T \subseteq N$. Such games are called *inessential*. Thus there is no point in forming coalitions in such a game, and hence no theory is required for such games. We call a game *essential* if $r = -1$; from now on we shall restrict our attention to essential games in reduced form.

It is interesting to note that, when $n = 3$, there is exactly one essential *n*-person game in reduced form. For if *v* is the characteristic function of such a game, then we have

$$v(1) = v(2) = v(3) = -1$$

and hence it follows that

$$v(12) = v(23) = v(13) = +1$$

$$v(123) = 0 = v(\emptyset)$$

In other words, *v* is completely specified, and hence we can talk about the essential three-person game in reduced form.

But for $n > 3$, there are infinitely many essential *n*-person games in reduced form. We shall consider $n = 4$, $v(1) = v(2) = v(3) = v(4) = -1$. Also

$$v(234) = v(134) = v(124) = v(123) = +1$$

$$v(1234) = 0 = v(\emptyset)$$

and

$$v(12) = -v(34)$$

$$v(13) = -v(24)$$

$$v(23) = -v(14)$$

Thus *v* will be completely known once we specify values for $v(12)$, $v(13)$, and $v(2, 3)$.

Since $v(12) \geq v(1) + v(2) = -2$, and $-v(12) = v(34) \geq -2$, hence $-2 \leq v(12) \leq 2$. In other words, $v(12)$, $v(13)$, and $v(23)$ will lie between -2 and $+2$. Let $v(12) = 2x_1$, $v(13) = 2x_2$, and $v(23) = 2x_3$. Then $-1 \leq x_i \leq +1$ for $i = 1, 2, 3$.

It is easily checked that, if x_1, x_2, x_3 are any numbers with $-1 \leq x_i \leq +1$, and if we define $v(T)$ as above for every $T \subseteq (1, 2, 3, 4)$, then *v* is the characteristic function of an essential four-person game in reduced form. Thus we see that the totality of essential four-person games in reduced form can be put in a one-to-one correspondence to the points (x_1, x_2, x_3) of a certain cube in three-dimensional Euclidean space.

Definition 9.1.3: An *n*-person game is called *symmetric* if, whenever T_1 and T_2 are two subsets of *N* with the same number of elements, $v(T_1) = v(T_2)$.

Remark 9.1.4: There is only one symmetric essential four-person game in reduced form, whereas there are infinitely many symmetric essential five-person games in reduced form.

9.2 SOLUTIONS OF *n*-PERSON GAMES

As has been mentioned, we are interested, in the case of *n*-person games, in the questions of what coalitions will tend to form and what each player will be paid (after all side-payments are made) in the event a given coalition forms. The payments to the various players, for given coalitions and side payments, can be represented as a vector of real numbers $(x_1, x_2, \cdots, x_n)$, where x_i is the amount the *i*th player receives. Since the game is zero-sum, $\sum_1^n x_i = 0$.

Also, it is clear that $x_i \geq v(i)$, for player *i* can see to it that he gets at least $v(i)$, and he would certainly reject any system of distribution that would give him less than $v(i)$. Call a vector $(x_1, x_2, \cdots, x_n)$ an *imputation* if it satisfies the following two conditions: (1) $\sum_1^n x_i = 0$ and (2) $x_i \geq v(i)$.

Remark 9.2.1: One might think that the second condition can be strengthened so as to say that, if T is any subset of N, then $\sum_{i \in T} x_i \geq v(T)$.

But such a condition in general would make the class of imputations empty. For instance, if this condition were satisfied by an imputation (x_1, x_2, x_3) for the essential three-person game in reduced form, then we should have

$$x_1 + x_2 \geq v(12) = 1$$

Since $x_1 + x_2 + x_3 = 0$, $x_3 \leq -1$. Also, $x_3 \geq v(1) = -1$. Hence $x_3 = -1$. Similarly, $x_1 = x_2 = -1$. This contradicts the fact that the game is zero-sum. Thus there would exist no imputation for this game.

It can be quickly checked that the set of all imputations for an *n*-person game is a convex subset of *n*-dimensional Euclidean space. Observe that an inessential game has just one imputation $(v(1), v(2), \cdots, v(n))$, whereas an essential game has infinitely many imputations.

An imputation $(y_1, y_2, \cdots, y_n)$ will clearly be preferred by player i_0 to the imputation $(x_1, x_2, \cdots, x_n)$ if $y_{i_0} > x_{i_0}$. Similarly, $(y_1, y_2, \cdots, y_n)$ will be preferred by a subset T of N if $y_i > x_i$ for all $i \in T$.

Definition 9.2.1: Let $(y_1, y_2, \cdots, y_n)$ and $(x_1, x_2, \cdots, x_n)$ be imputations for a game whose characteristic function is v, and let T be a subset of the players. Then we say that $(y_1, y_2, \cdots, y_n)$ *dominates* $(x_1, x_2 \cdots, x_n)$ with respect to T if the following conditions are satisfied.

(1) $T \neq \emptyset$

(2) $v(T) \geq \displaystyle\sum_{i \in T} y_i.$

(3) $y_i > x_i$ for all $i \in T$.

We say that imputation $y = (y_1, y_2, \cdots, y_n)$ dominates imputation $x = (x_1, x_2, \cdots, x_n)$ (in notation $y > x$) if there exists a nonempty set $T \subseteq N$ such that $y_i > x_i$ for all $i \in T$ and $\Sigma y_i \leq v(T)$. It is clear from the definition that, if $y > x$ via T, then T must contain at least 2 and, at most, $(n - 1)$ members. It is also seen that no imputation can dominate itself. Observe that the relation of dominance is not in general transitive.

Definition of a Solution: Now we shall consider the question of what imputations are likely to arise from actual plays of a game. The first answer that anyone is tempted to give is the following: An imputation $\alpha = (\alpha_1, \alpha_2, \cdots, \alpha_n)$ can be realized in an actual play if there exists no imputation β which dominates α. Unfortunately, however, such an imputation does not, in general, exist. Indeed one can show, for essential games, that every imputation is dominated by some other imputation!

Let us for the present look at the essential three-person game in reduced form. There appear to be just three possibilities here: the three ways in which a coalition of two can be formed. If players 1 and 2 agree to form a coalition, then they will take as much as they possibly can (namely $+1$) and player 3 will accordingly be given -1. Moreover, from the fact that the game is completely symmetrical (so that neither player 1 nor player 2 is in a more advantageous position), we would intuitively expect them to divide their winnings equally; hence if players 1 and 2 agree to cooperate, we arrive at the imputation $(^1/_2, \, ^1/_2, \, -1)$. Similarly, if players 1 and 3 agree or if players 2 and 3 agree, we arrive at $(^1/_2, \, -1, \, ^1/_2)$ or $(-1, \, ^1/_2, \, ^1/_2)$. Thus we obtain the following set of three imputations:

$$X = [(\tfrac{1}{2}, \tfrac{1}{2}, -1), (\tfrac{1}{2}, -1, \tfrac{1}{2}), (-1, \tfrac{1}{2}, \tfrac{1}{2})]$$

It is clear that no one of them dominates another. It is not hard to check that every other imputation not in X is dominated by at least one member of X. The two properties of the set X occupies the central role in von Neumann and Morgenstern's development of the theory of *n*-person games. Now we have the following definition.

Definition 9.2.2: A set X of imputations for a given *n*-person game is called a solution of the game if (1) no member of X dominates another member of X and (2) every imputation not in X is dominated by some member of X.

Remark 9.2.2: The theory consists essentially in the search for solutions and a discussion of their properties. The intuitive justification of the use of the word "solution" is, of course, now very different from what it was in the case of two-person games. In two-person games a solution meant a set of probabilities with which the player should play his various pure strategies in order to maximize his expectation of gain. But in n-person games (for $n > 2$), a solution merely gives a set of possible ways in which winnings can be divided at the end of a play. Some people have felt dissatisfied with the intuitive basis of this notion, however; and the question has been raised as to whether knowing a solution of a given n-person game would enable a person to play it with greater expectation of profit than if he were quite ignorant of this theory. Lucas' example of a ten-person game, which has no solution in the sense of von Neumann–Morgenstern, justifies that the notion of a solution does not constitute an adequate foundation for the theory of n-person games. We now introduce the following definition.

Definition 9.2.3: Two n-person games, v and v', are called *isomorphic* if there exists a one-to-one correspondence $\leftrightarrow$ between the imputations of v and the imputations of v' such that, for every subset T of the players, if α and β are imputations for v, and α' and β' are imputations for v', and $\alpha \leftrightarrow \alpha'$, $\beta \leftrightarrow \beta'$, then $\alpha \underset{T}{>} \beta$ in game v if and only if $\alpha' \underset{T}{>} \beta'$ in game v'. We should like to make the following observations:

(1) The relation of isomorphism of games is reflexive, symmetric, and transitive.

(2) If v and v' are isomorphic under the relation $\leftrightarrow$ and if A is a solution of v, then A' is a solution of v' where $A' =$ the set of all imputations α' such that, for some $\alpha \in A$, $\alpha \leftrightarrow \alpha'$.

(3) Let v and v' be two games that are strategically equivalent with the constants $K > 0$, and $a_1, a_2, \cdots, a_n$.

If $(x_1, x_2, \cdots, x_n) \rightarrow (Kx_1 + a_1, \cdots, Kx_n + a_n)$, then the relation $\rightarrow$ establishes an isomorphism between v and v'. We shall prove observation (3), as (1) and (2) are easy to check.

PROOF: If $(x_1, x_2, \cdots, x_n)$ is any imputation of v, then

$$(Kx_1 + a_1, Kx_2 + a_2, \cdots, Kx_n + a_n)$$

is an imputation for v'. Since $\Sigma a_i = 0$, $\Sigma x_i = 0$,

$$\sum_1^n (Kx_i + a_i) = K \sum x_i + \sum a_i = 0$$

Also, for each i, from the fact that v and v' are strategically equivalent,

$$v'(i) = Kv(i) + a_i$$

and, since $x_i \geq v(i)$ ($\because (x_1, \cdots, x_n)$ is an imputation), we have

$$Kx_i + a_i \geq Kv(i) + a_i = v'(i)$$

as was to be shown. To complete our proof, it is sufficient to show that, if $x \underset{T}{>} y$ in v, then

$$Kx + a > Ky + a \quad \text{in} \quad v'$$

where $x = (x_1, \cdots, x_n)$, $a = (a_1, a_2, \cdots, a_n)$, K is a scalar,

$$Kx + a = (Kx_1 + a_1, Kx_2 + a_2, \cdots) \text{ etc.}$$

Since $x_i > y_i$ for all $i \in T$ and $K > 0$,

$$Kx_i + a_i > Ky_i + a_i \quad \text{for all} \quad i \in T$$

Further

$$\sum_{i \in T} x_i \leq v(T)$$

Hence

$$\sum_{i \in T} (Kx_i + a_i) \leq Kv(T) + \sum_{i \in T} a_i = v'(T)$$

which completes the proof of observation (3).

Combining observations (2) and (3), we can conclude that if every game in reduced form has a solution then every game has a solution. To find all the solutions of the essential three-person games in reduced form, it is convenient to introduce a new coordinate system for the Euclidean plane as shown in Fig. 1. We take as axes three concurrent lines which make angles of 60° with each other; and by the coordinates of an arbitrary point we mean the perpendicular distances from it to these three lines, the distances being called positive or negative, as indicated in Fig. 1.

It is well known that the points of the Euclidean plane can be represented by using only two coordinates (as in the usual Cartesian system, for example). We should expect these three coordinates not to be mutually independent. And indeed, it is not hard to check that for every point (x_1, x_2, x_3) we have $x_1 + x_2 + x_3 = 0$. Thus the shaded area, which we shall call the fundamental triangle, represents all imputations for this game.

If (x_1, x_2, x_3) is an imputation, then $x_1 + x_2 \leq v(12)$. For if we had $x_1 + x_2 > v(12)$, it follows that $-x_3 > -v(3)$ or $x_3 < v(3)$, contradicting

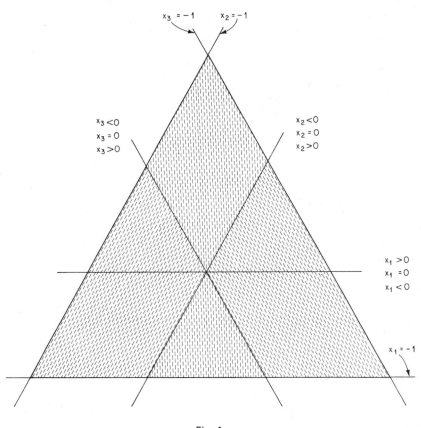

Fig. 1.

that (x_1, x_2, x_3) is an imputation. Similarly, $x_2 + x_3 \leq v(23)$ and $x_1 + x_3 \leq v(13)$.

We have already seen that domination cannot occur with respect to an empty set or a 1-element set or an $(n - 1)$-element set. Hence, in three-person games domination can occur only with respect to 2-element sets. Imputation $x = (x_1, x_2, x_3)$ dominates $y = (y_1, y_2, y_3)$ if and only if either

$$x_1 > y_1 \quad \text{and} \quad x_2 > y_2$$

or

$$x_2 > y_2 \quad \text{and} \quad x_3 > y_3$$

or

$$x_1 > y_1 \quad \text{and} \quad x_3 > y_3$$

From this we conclude that an imputation (x_1, x_2, x_3) dominates just those imputations that lie in the shaded areas of Fig. 2 [the shaded areas exclusive of the three boundary lines passing through (x_1, x_2, x_3)]. We see, moreover, that every point in the unshaded areas represents an imputation that dominates (x_1, x_2, x_3). Thus, if (x_1, x_2, x_3) and (y_1, y_2, y_3) are two imputations, neither one of which dominates the other, then the corresponding points lie on a line parallel to one of the coordinate axes.

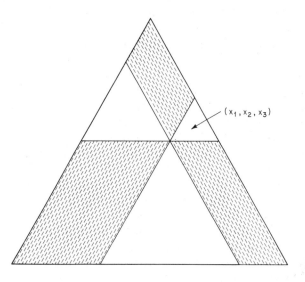

(x_1, x_2, x_3)

Fig. 2.

Let us now turn our attention to the problem of determining all solutions to this game. Since the game is essential, every solution X must contain at least two imputations. Moreover, we have seen that every two members of X must lie on a line parallel to one of the coordinate axes.

We now consider two cases according to whether all points of X lie on the same line or not. In the second case we are led to the following solution:

$$X = [(\tfrac{1}{2}, \tfrac{1}{2}, -1), (\tfrac{1}{2}, -1, \tfrac{1}{2}), (-1, \tfrac{1}{2}, \tfrac{1}{2})]$$

In the first case, we conclude that A must contain all points within the fundamental triangle that lie on the given line, and also one of the following three conditions should be satisfied. Either (1) X consists of all imputations (c, x_2, x_3) where c is fixed and satisfies $-1 \le c < {}^1/_2$; or (2) X consists of all imputations (x_1, c, x_3) where c is fixed and $-1 \le c < {}^1/_2$; or (3) X consists of all imputations (x_1, x_2, c) where c is fixed and $-1 \le c < {}^1/_2$.

Remark 9.2.3: In an essential three-person game we have an embarrassing richness of solutions. Von Neumann accounts for this situation by regarding the various solutions as representing various standards of social behavior. In the three-person game in question, he calls the unique finite solution the "nondiscriminatory" solution, and the others "discriminatory." Thus, for instance, a solution of the form (3) above represents a social arrangement where players 1 and 2 have decided to exclude player 3 from their negotiations but to allow him a fixed amount c (the smaller the c, of course, the worse for player 3). How players 1 and 2 will divide the $-c$ among them is not decided by this theory.

We have previously seen that the totality of essential four-person games in reduced form corresponds in a one-to-one way to the points (x_1, x_2, x_3) of a certain cube in three-dimensional space. Ever since von Neumann and Morgenstern introduced their theory of the n-person game, there has been a tendency to place special emphasis on finite solutions. In part this is because they do not exhibit the bewildering variety that has been observed for arbitrary solutions, in part because they are frequently easier to interpret, and in part because of the inspiration of the three-person game. Unfortunately, even for the four-person game, our knowledge of finite solutions is limited. It is known, however, that those four-person zero-sum games that correspond to the vertices of the von Neumann cube have unique finite solutions of a simple nature [7], whereas those games that correspond to the interior points of the edges of the cube have no finite solutions whatsoever. Each face of the cube has two opposite vertices that correspond to the three-person game with a dummy. Let D be the diagonal joining these two vertices. Then each interior point of D corresponds to a game with a unique finite solution, and this finite solution consists of seven points. Except for the points of D and the vertices, there are no games whatsoever on the face of the cube that have finite solutions.

These interesting results are due to Mills [4]. The proofs are lengthy, and the interested reader can refer to [4]. It is still an open problem whether all five-person games have solutions!

9.3 A GAME WITH NO SOLUTION

This section describes a ten-person game which has no solution. Call an n-person game a constant-sum game if the characteristic function, v, satisfies the following conditions:

(1) $v(N) = C$.

(2) $v(T) = C - v(N - T)$.

(3) $v(S \cup T) \geq v(S) + v(T)$, if $S \cap T = \emptyset$.

The concept of solution, etc., to constant-sum games can be defined in exactly the same way as was defined for the zero-sum games.

Consider the game (N, v), where $N = \{1, 2, 3, 4, 5, 6, 7, 8, 9, 10\}$ and v is given by

$$v(N) = 5, \qquad v(13579) = 4$$

$$v(12) \;\; = v(34) \;\; = v(56) = v(78) = v(9, 10) = 1$$

$$v(3579) = v(1579) = v(1379) = 3$$

$$v(357) \;\; = v(157) \;\; = v(137) \;\; = 2$$

$$v(359) \;\; = v(159) \;\; = v(139) \;\; = 2$$

$$v(1479) = v(3679) = v(5279) = 2$$

$$v(S) = 0 \quad \text{for all other} \quad S \subset N$$

Remark 9.3.1: It is clear that the function v defined above is not super-additive. However, it is equivalent solution-wise to a game with a super-additive characteristic function (see Gillies [1, p. 68]).

Definition 9.3.1: The core of an n-person game is the set of imputations, C, defined as below.

$$C = \left\{ x \in A : \sum_{i \in S} x_i \geq v(S) \quad \text{for all} \quad S \subset N \right\}$$

It is clear that C will be contained in every solution of the game. It is quite possible that the core of the game could be empty.

Let A stand for the set of all imputations. That is,

$$A = \left\{ x \,\middle|\, \sum_{i \in N} x_i = v(N) \quad \text{and} \quad x_i \geq v(i) \quad \text{for} \quad i \in N \right\}$$

For any $X \subset A$ and nonempty $S \subset N$, define $\text{dom}_S X$ to be the set of all vectors $x \in A$ such that there exists a $y \in X$ with $y_i > x_i$ for all $i \in S$ and $\sum_{i \in S} y_i \leq v(S)$. Let $\text{dom } X = \bigcup_{S \subset N} \text{dom}_S X$. Also let $\text{dom}^{-1} X$ be the set of all $y \in A$ such that there exists $x \in X$ with $x \in \text{dom } y$. If $X \subset A$ and $K' \subset X$, then K' is a *solution for X* if $K' \cap \text{dom } K' = \emptyset$ and $K' \cup \text{dom } K' \supset X$. For any solution K of A, $K \cap \text{dom } C = \emptyset$.

For the ten-person game given above, one can show that the core, C, of the game is the convex hull of the six imputations

$$(1, 0, 1, 0, 1, 0, 1, 0, 1, 0), \; (0, 1, 1, 0, 1, 0, 1, 0, 1, 0)$$

$$(1, 0, 0, 1, 1, 0, 1, 0, 1, 0), \; (1, 0, 1, 0, 0, 1, 1, 0, 1, 0)$$

$$(1, 0, 1, 0, 1, 0, 0, 1, 1, 0), \; (1, 0, 1, 0, 1, 0, 1, 0, 0, 1)$$

We shall reproduce the outline of the proof given by Lucas [2].

OUTLINE OF PROOF: Consider the following subsets of A.

$$B = \{x \in A: x_1 + x_2 = x_3 + x_4 = x_5 + x_6 = x_7 + x_8 = x_9 + x_{10} = 1\}$$

$$E_i = \{x \in B: x_j = x_k = 1, x_i < 1, x_7 + x_9 < 1\}$$

$$E = \bigcup_i E_i, \qquad i = 1, 3, 5$$

$$F = \left[\bigcup_{j, K} \{x \in B: x_j = x_k = 1, x_7 + x_9 \geq 1\} \right.$$

$$\bigcup \bigcup_{(p, q)} \{x \in B: x_p = 1, x_q < 1, x_3 + x_5 + x_q \geq 2,$$

$$x_1 + x_5 + x_q \geq 2, x_1 + x_3 + x_q \geq 2\}$$

$$\bigcup \{x \in B: x_7 = x_9 = 1\}$$

$$\left. \bigcup \{x \in B: x_1 = x_3 = x_5 = 1\} \right] - C$$

where $(i, j, k) = (1, 3, 5)$, $(3, 5, 1)$, and $(5, 1, 3)$, and $(p, q) = (7, 9)$ and $(9, 7)$. One can check that $A - B$, $B - (C \cup E \cup F)$, C, E, and F form a partition of A.

To prove that this game has no solution, it is sufficient to prove that

(1) dom $C \supset [A - B] \cup [B - (C \cup E \cup F)]$.

(2) $E \cap$ dom $(C \cup F) = \emptyset$.

(3) There is no solution for E.

One can prove (1) and (2) by checking various subsets S of N. In fact, one can prove in addition that

$$\text{dom } C = A - (C \cup E \cup F)$$

and

$$F \cap \text{dom } (C \cup E \cup F) = \emptyset$$

and thus $C \cup F$ is contained in every solution. Now consider the region E. One can check that $E_i \cap \text{dom}_S E = \emptyset$ for all S except $(i, r, 7, 9)$, and $E_i \cap \text{dom}_{(i,r,7,9)} (E_i \cup E_k) = \emptyset$, where $(i, r, k) = (1, 4, 5)$, $(3, 6, 1)$, and $(5, 2, 3)$. Thus the "dom" pattern in E is cyclic, as illustrated by the diagram:

$$E \xrightarrow{\ \ \ } E \xrightarrow{\ \ \ } E \xrightarrow{\ \ \ } E_5$$
$$_{5\{3,6,7,9\}} _{3\{1,4,7,9\}} _{1\{5,2,7,9\}}$$

To prove (3), assume that $K'(\neq \emptyset)$ is a solution for E, and pick any $y \in K'$. Using the symmetry in E, one can assume that $y \in E_3$.
    Define:

$$G_i(y) = \{x \in E_i: x_7 > y_7, x_9 > y_9, x_k + x_r + x_7 + x_9 \leq 2\}$$

where $(i, k, r) = (1, 5, 2), (3, 1, 4)$, and $(5, 3, 6)$. Then one can verify that $E \cap \mathrm{dom}^{-1}\{y\} = G_5(y)$, and so $K' \cap G_5(y) = \emptyset$. However,

$$E \cap \mathrm{dom}^{-1} G_5(y) = G_1(y)$$

and so $K' \cap G_1(y) \neq \emptyset$.

On the other hand, $G_3(y) \cap \mathrm{dom}\,(E_5 - G_5(y)) = \emptyset$ and so $G_3(y) \subset K'$. However, $G_1(y) \subset \mathrm{dom}\, G_3(y)$, and so $K' \cap G_1(y) = \emptyset$, which gives a contradiction. Therefore, there is no solution K' for E.

Remark 9.3.2: The ten-person game described in this section does not have a superadditive v. However, one can use a method of Gillies [1, pp. 68–69] to extend this game to a game with a superadditive v'. For every $S \subset N$, define

$$v'(S) = \max \sum_{h=1}^{m} v(S_h)$$

where the maximum is taken over all partitions $\{S_1, S_2, \cdots, S_m\}$ of S. The resulting v' is superadditive. One can also check that for this particular game $v'(S) = v(S)$ for all $S \subset N$ which have $v(S) > 0$. In particular, $v'(N) = v(N)$, and thus the set A remains the same. One can also show that, if $y \mathrm{\,dom\,} x$ in the game (N, v'), then $y \mathrm{\,dom\,} x$ in the game (N, v). Thus the game (N, v') has the same core and solutions as (N, v). Therefore, the game (N, v') has a superadditive v', and it has no solution.

9.4 SOLUTIONS OF PRODUCT SIMPLE GAMES

A game is called simple if every coalition of players either can win outright or is completely helpless; that is, $v(s) = 0$ or 1 for any S. [We are going to assume that $v(N) = 1$.] A simple game can therefore be described by specifying its players and its winning coalitions. We shall denote a simple game by the symbol $\Gamma(P, W)$, where P is a finite set (the players) and W is a collection of subsets of P (the winning coalitions). We shall require that W includes P and excludes $\emptyset$ and that it be monotonic in the following sense:

$$S \in W, T \supseteq S \Rightarrow T \in W$$

In view of this condition the collection of minimal winning coalitions, denoted by W^m, is sufficient to specify the game. If W has the further property that

$$S \in W \Rightarrow P - S \notin W$$

then the game is said to be *proper*.

If W enjoys the converse property, $S \notin W \Rightarrow P - S \in W$, then the game is said to be strong. The proper strong games are constant-sum in the classical theory; they constitute the class of simple games that was originally defined and investigated by von Neumann and Morgenstern.

Sums and Products: A pair of simple games with different players can be combined into a larger game in two natural ways, related to the Boolean operations of addition and multiplication.

Let $P_1 \cup P_2 = P$ and $P \cap P_2 = \emptyset$. Then the *sum* $\Gamma(P, W) \oplus \Gamma(P_2 \ W_2)$ is defined as the game $\Gamma(P, W)$, where W consists of all $S \subseteq P$ such that either $S \cap P_1 \in W_1$ or $S \cap P_2 \in W_2$.

Similarly, the *product* $\Gamma(P_1, W_1) \otimes \Gamma(P_2, W_2)$ is defined as the game $\Gamma(P, W)$, where W consists of all $S \subseteq P$ such that $S \cap P_1 \in W_1$ and $S \cap P_2 \in W_2$. These concepts are due to Shapley.

Products are always weaker than the corresponding sums. In general, sums are improper and products are nonstrong. This may explain why these operations do *not* appear in the simple-game theory of von Neumann and Morgenstern [7], which is limited to the constant-sum (that is, strong and proper) case.

Imputations: Let A_P denote the simplex of nonnegative vectors x such that $\Sigma x_i = 1$. These vectors are traditionally called imputations, where P is the set of players in a simple game. Let us write $x(S)$ for $\displaystyle\sum_{i \in S} x_i$, where $S \subseteq P$. Let $R_S x$ be the restriction of x to S. Thus

$$R_S x = x_i \quad \text{if} \quad i \in S$$
$$= 0 \quad \text{if} \quad i \notin S$$

The barycentric projection of x on A_S is given by

$$B_S x = \frac{1}{x(S)} \cdot R_S x$$

This is well defined provided that $x(S) > 0$. Let P_1 and P_2 be fixed disjoint sets. Let $P = P_1 \cup P_2$ and $A_{P_i} = \{x \mid x \in A_P \text{ and } x(P_i) = 1\}$. If X and Y are subsets of A_{P_1} and A_{P_2}, respectively, we define our operation $\underset{\alpha}{\times}$ ($0 \leq \alpha \leq 1$):

$$X \underset{\alpha}{\times} Y = \{z \mid z = \alpha x + (1 - \alpha) y \quad \text{for some} \quad x \in X, y \in Y\}$$

We recall that a solution of the game $\Gamma(P, W)$ is a set X of imputations such that $X = A_P - \text{dom } X$, where dom X denotes the set of all $y \in A_P$ such that, for some $x \in X$, the set $[i \mid x_i > y_i]$ is an element of W. It is not hard to see that every simple game has at least one solution. The notation dom_1 and dom_2 will be used for domination with respect to special classes W_1 and W_2.

One can prove the following theorem for sums [6].

Theorem 9.4.1

If X_i is a solution of $\Gamma(P_i, W_i)$, $i = 1, 2$, and if $0 \leq \alpha \leq 1$, then $X_1 \underset{\alpha}{\times} X_2$ is a solution of $\Gamma(P, W) = \Gamma(P_1, W_1) \oplus \Gamma(P_2, W_2)$. Conversely, if X is a solution of $\Gamma(P, W) = \Gamma(P_1, W_1) \oplus \Gamma(P_2, W_2)$, then there exist solutions X_1 and X_2 of $\Gamma(P_1, W_1)$ and $\Gamma(P_2, W_2)$, respectively, and a real number α, $0 \leq \alpha \leq 1$, such that $X = X_1 \underset{\alpha}{\times} X_2$. For a proof, see Shapley [6].

Remark 9.4.1: This theorem gives a complete characterization of the solutions to sums of simple games. Such a complete characterization to product simple games is not known. However, one can prove certain partial results. We shall start with some definitions.

Definition 9.4.1: A parametrized family of sets of imputations

$$[Y(\alpha): 0 \leq \alpha \leq 1]$$

will be called semimonotonic if for every α, β, x such that $0 \leq \alpha \leq \beta \leq 1$ and $x \in Y(\beta)$, there exists $y \in Y(\alpha)$ such that

$$\alpha y \leq \beta x$$

Definition 9.4.2: A semimonotonic family $\{y(\alpha)\}$ is called fully monotonic if for every α, β, y such that $0 \leq \alpha \leq \beta \leq 1$ and $y \in Y(\alpha)$ there exists $x \in Y(\beta)$ such that

$$\alpha y \leq \beta x$$

We shall now generalize the concept of full monotonicity.

Definition 9.4.3: A semimonotonic family $Y(\alpha)$ is called δ-monotonic ($0 \leq \delta \leq 1$) if, for every α, β, y such that $\delta \leq \alpha \leq \beta \leq 1$ and $y \in Y(\alpha)$, there exists $x \in Y(\beta)$ such that $\alpha y \leq \beta x$.

Definition 9.4.4: Let $\Gamma(P, W)$ be a simple game. Let $X \subseteq A_P$. Call X externally stable if $X \cup \text{dom } X = A_P$. Call X internally stable if $X \cap \text{dom } X = \emptyset$. Call X a solution if X is both internally and externally stable.

Remark 9.4.2: Any δ-monotone family is necessarily δ_1-monotonic if $\delta_1 \geq \delta$.

Shapley has proved the following two theorems [6] for product simple games.

Theorem 9.4.2

For $i = 1, 2$, let $\{X_i(\alpha): 0 \le \alpha \le 1\}$ be a semimonotonic family of solutions of $\Gamma(P_i, W_i)$. Then

$$X = \bigcup_{0 \le \alpha \le 1} \; X_1(\alpha) \times X_2(1 - \alpha)$$

is a solution of $\Gamma(P, W) = \Gamma(P_1, W_1) \otimes \Gamma(P_2, W_2)$.

Theorem 9.4.3

For $i = 1, 2$, let $\{Y_i(\alpha): 0 < \alpha < 1\}$ be a fully monotonic family of solutions to $\Gamma(P_i, W_i)$ except that $Y_i(1)$ need not be externally stable, and let $X_i(\alpha) = A_{P_i} - \text{dom}_i \; Y_i(\alpha)$. Then

$$X = \bigcup_{0 \le \alpha \le 1} \; X_1(\alpha) \times X_2(1 - \alpha)$$

is a solution of $\Gamma(P, W) = \Gamma(P_1, W_1) \otimes \Gamma(P_2, W_2)$.

The purpose of this section is to prove a theorem that includes both of these theorems.

Let $Y_i(\alpha) \subseteq A_{P_i}$ for $i = 1, 2$. Further, suppose the families $\{Y_i(\alpha)\}$ to be semimonotonic. Let $X_i(\alpha) = A_{P_i} - \text{dom}_i \; Y_i(\alpha)$. For every $S \in W_1$, consider the following set.

$$\Lambda_1(S) = \{\alpha: 1 > \alpha > 0; \text{ there exists } x_1 \in Y_1(\alpha) \text{ and } y_1 \in X_1(1) - Y_1(1)$$

$$\text{such that } \alpha x_1 > y_1 \text{ on } S\}$$

Now choose one α_S from those $\Lambda_1(S)$ that are not empty. Let α_0 stand for the minimum of the α_S chosen. We fix these α_S. It is clear that $\alpha_0 > 0$. We wish to remark that there is a certain amount of arbitrariness in choosing the α_S. Now we shall prove the following simple but useful lemma [5].

LEMMA: Suppose that $Y_1(\alpha)$ is α_0-monotonic. For every S for which $\Lambda_1(S) \ne \emptyset$, there exist a vector $z_S \in Y_1(1)$ and a δ_1 with $0 < \delta_1 \le \alpha_0$ such that $z_S \ge \delta_1$ on S. Further, if we take any $\alpha \in \Lambda_1(S)$, $\alpha \ge \delta_1$.

PROOF OF THE LEMMA: Since $\alpha_S \in \Lambda_1(S)$, there exists $x_S \in Y_1(\alpha_S)$ such that $\alpha_S x_S > 0$ on S. Since $\alpha_0 \le \alpha_S$ and the family is α_0-monotonic, there exists $z_S \in Y_1(1)$ such that $z_S \ge \alpha_S x_S$.

Let e_S denote the least component among the S-components of x_S. Define $\delta_1 = \min \alpha_S \, e_S$; then we have $z_S \ge \delta_1$ on S.

Also it is easy to check that $0 < \delta_1 \le \alpha_0$. Hence the first part of the proof of the lemma is complete. We shall now prove that any $\alpha \in \Lambda_1(S)$ will satisfy the inequality $\alpha \ge \delta_1$.

Take any $\alpha \in \Lambda_1(S)$. This means that there exists an $x_1 \in Y_1(\alpha)$ and a $y_1 \in X_1(1) - Y_1(1)$ such that $\alpha x_1 > y_1$ on S. We claim that at least one of the S_j components of y_1 must be greater than or equal to δ_1; otherwise $z_S > y_1$ on S and therefore $y_1 \in \text{dom}_1 z_1 \subseteq \text{dom}_1 Y_1(1)$, which contradicts the assumption that $y_1 \in X_1(1) = A_{P_1} - \text{dom}_1 Y_1(1)$. Therefore we have $\alpha > y_1(S) \geq \delta_1$. Hence the proof of the lemma is complete.

Similarly, we can get hold of a δ_2 by constructing $\Lambda_2(S)$ for $S \in W_2$. If all the sets Λ_1 and Λ_2 are empty, we define δ_0 to be any positive number in the interval $(0, 1]$; in fact, we can take $\delta_0 = 1$. If $\Lambda_1(S) \neq \emptyset$ for some $S \in W_1$ and all Λ_2 are empty, define $\delta_0 = \delta_1$; and if all Λ_1 are empty and $\Lambda_2(S) \neq \emptyset$ for some $S \in W_2$, define $\delta_0 = \delta_2$. If Λ_1 and Λ_2 both are nonempty for certain coalitions, define $\delta_0 = \min [\delta_1, \delta_2]$. In all these cases we see that $\delta_0 > 0$. Now we are in a position to state and prove the following theorem [5].

Theorem 9.4.4

Let $Y_1(\alpha)$ and $Y_2(\alpha)$ be a semimonotonic family of solutions to $\Gamma(P_1, W_1)$ and $\Gamma(P_2, W_2)$, respectively, except that $Y_1(1)$ and $Y_2(1)$ need not be externally stable. Further, suppose the family to be δ_0-monotonic. Then

$$X = \bigcup_{0 \leq \alpha \leq 1} X_1(\alpha) \times_\alpha X_2(1 - \alpha)$$

is a solution for the product simple game $\Gamma(P, W) = \Gamma(P_1, W_1) \otimes \Gamma(P_2, W_2)$.

Remark 9.4.3: Since $\{Y_i(\alpha)\}$ is semimonotonic, it follows that $Y_i(1)$ is internally stable. If $Y_i(1)$ for $i = 1, 2$ is also externally stable—that is, if $Y_i(1)$ is a solution—then it is not hard to check that Λ-sets are all empty, and hence we can take $\delta_0 = 1$. In other words, theorem 9.4.2 is included in this theorem. It is also quickly seen that Λ-sets are empty if $Y_i(\alpha)$ are fully monotonic, and hence theorem 9.4.4 includes theorem 9.4.3 also. We shall show by presenting an example that theorem 9.4.4 is actually a generalization of theorems 9.4.2 and 9.4.3.

PROOF OF THEOREM 9.4.4: Our proof follows along the same line as the one given for theorem 9.4.2 [6]. Note that $X_i(\alpha) = Y_i(\alpha)$ for all α except $\alpha = 1$.

External Stability of X: Take any $y \in A_P$; define $\beta_i = y(P_i)$, and let y_i be the barycentric projection of y on A_{P_i}. We shall now show that y belongs either to X or to dom X.

Case 1: $\qquad\qquad\qquad\qquad 0 < \beta_1 < 1$

Case 1a: $\qquad\qquad\qquad\qquad y_1 \in X_1(\beta_1), y_2 \in X_2(\beta_2)$

Then $y = \beta_1 y_1 + \beta_2 y_2 \in X$.

Case 1b: $\qquad\qquad\qquad\qquad y_1 \notin X_1(\beta_1), y_2 \notin X_2(\beta_2)$

such that $y_i \in \text{dom}_i\, x_i$ for $i = 1, 2$. Then the imputation defined by

$$x = \beta_1 x_1 + \beta_2 x_2$$

which is in X, clearly dominates y, or $y \in \text{dom}\, X$.

Case 1c: $\qquad\qquad\qquad y_1 \in X_1(\beta_1),\ y_2 \notin X_2(\beta_2)$

Find $x_2 \in Y_2(\beta_2)$ such that $y_2 \in \text{dom}_2\, x_2$. Let $x_2 > y_2$ on $S_2 \in W_2$. Choose $e > 0$ so that $\beta_2 x_2 - \beta_2 y_2 > e$ on S_2. Using the semimonotonic property of $Y_2(\alpha)$, find $x_2^1 \in Y_2(\beta_2 - e)$ such that $\beta_2 x_2 - (\beta_2 - e)\, x_2^1 \geq 0$. This vector must be $\leq e$ in all components, since no component of a nonnegative vector can exceed the sum of all components. It follows that we have

$$(\beta_2 - e)\, x_2^1 > \beta_2 x_2 \quad \text{on} \quad S_2$$

Let u_1 be any interior point of A_{P_1}; then the imputation x defined by

$$x = \beta_1 y_1 + e u_1 + (\beta_2 - e)\, x_2^1$$

dominates y on $P_1 \cup S_2$. If $x \in X$, then $y \in \text{dom}\, X$, and we are through. Suppose that $x \notin X$; then it means that the vector x_0 defined below

$$x_0 = \frac{\beta_1 y_1 + e u_1}{\beta_1 + e} \notin X_1(\beta_1 + e) = Y_1(\beta_1 + e)$$

Hence we can find $x_1 \in Y_1(\beta_1 + e)$ such that $x_0 \in \text{dom}_1\, x_1$. Then it is clear that the imputation z defined by $z = (\beta_1 + e)\, x_1 + (\beta_2 - e)\, x_2^1$ dominates y. Since $z \in X$, it follows that $y \in \text{dom}\, X$.

Case 1d: $\qquad\qquad y_1 \notin X_1(\beta_1),\ y_2 \in X_2(\beta_2)$ (like case 1c)

Case 2: $\qquad\qquad\qquad\qquad\qquad \beta_1 = 0$

Case 2a: $\qquad\qquad\qquad\qquad\qquad y_2 \in X_2(1)$

Then $y \in X$.

Case 2b: $\qquad\qquad\qquad\qquad\qquad y_2 \notin X_2(1)$

This means that $y_2 \in \text{dom}_2\, Y_2(1)$, and hence the argument of case 1c can be repeated with the understanding that $\beta_1 y_1 = 0$.

Case 3: $\qquad\qquad\qquad\qquad \beta_1 = 1$ (like case 2)

This completes the proof of external stability.

Internal Stability of X: We shall now show that $X \cap \text{dom}\, X = \emptyset$. Suppose that there exists $x, y \in X$, such that $x > y$ on $S \in W$. Let $S_1 = S \cap P_1$ and

$S_2 = S \cap P_2$; $x = \alpha x_1 + (1 - \alpha) x_2$ and $y = \beta y_1 + (1 - \beta) y_2$. Since $x > y$ on $S_1 \cup S_2$, it follows that $0 < \alpha < 1$.

Case 1a: $1 > \beta \geq \alpha$

Since $\alpha > 0$, β is also positive. Also, $\alpha x_1 > \beta y_1$ on S_1. Since $y_1(\alpha)$ is semi-monotonic, there exists $x_1' \in Y_1(\alpha)$ such that $\beta y_1 \geq \alpha x_1'$. Hence $\alpha x_1 > \alpha x_1'$ on S_1 or $x_1 > x_1'$ on S_1, contradicting the internal stability of $Y_1(\alpha)$.

Case 1b: $\alpha \geq \beta > 0$

This means $1 > 1 - \beta \geq 1 - \alpha$. Since $\alpha < 1$, this case is similar to the previous case except that we have to utilize the semimonotonic property of $Y_2(\alpha)$.

Case 2a: $\beta_1 = 1$

That is, $\alpha x_1 > y = y_1$ on S_1. If $y_1 \in Y_1(1)$, then, using the semimonotonic property of $Y_1(\alpha)$, we arrive at a contradiction. If $y_1 \in X_1(1) - Y_1(1)$, then $\alpha \in \Lambda_1(S_1)$ and hence, by lemma, $\alpha \geq \delta_1 \geq \delta_0$. Since $Y_1(\alpha)$ is δ_0-monotonic, one can find $y_1' \in Y_1(1)$ such that $y_1' \geq \alpha x_1 > y_1$. This implies $y_1 \in \mathrm{dom}_1 Y_1(1)$ or $y_1 \notin X_1(1)$, which contradicts our assumption regarding y_1.

Case 2b: $\beta = 0$ or $1 - \beta = 1$

and the argument can be carried over as in case 2a. Hence the proof of the theorem is complete.

We shall now present an example of a solution to product simple games that satisfies the conditions of theorem 9.4.4 but not those of theorem 9.4.2 or 9.4.3. In other words, every solution that satisfies the conditions of theorem 9.4.4 need not have the property of full monotonicity.

Example: The four-person game, *J*, is defined by

$$\Gamma(1234, \{124, 134, 234\})$$

where members in the braces denote the winning coalitions of the game *J*.

$$J = M_3 \otimes B_4$$

Let M_3 denote the three-person simple majority game and B_4 the one-person pure bargaining game. Define, for $0 \leq \alpha \leq {}^3/_4$,

$$X_1(\alpha) = \bigcup_{0 \leq \beta \leq 1} Y(\beta)$$

where

$$Y(\beta) = \left\{ \left(\frac{\beta^2}{2}, \beta t, \beta(1 - t) - \frac{\beta^2}{2}, 1 - \beta \right) : 0 \leq t \leq 1 - \frac{\beta}{2} \right\}$$

for $0 \leq \beta < 1$, and

$$Y(1) = \{\tfrac{1}{2}, t, \tfrac{1}{2} - t, 0\colon 0 \leq t \leq \tfrac{1}{2}\} \cup (0, \tfrac{1}{2}, \tfrac{1}{2}, 0)$$

Define for $^3/_4 < \alpha \leq \alpha_0$, where α_0 is so chosen that

$$\alpha_0(1 - \tfrac{7}{8}(1 + \alpha_0)) = \tfrac{1}{2}$$

$$X_1(\alpha) = \bigcup_{0 \leq \beta \leq 1} Y(\beta)$$

where

$$Y(\beta) = \left\{\left(\frac{7\beta^2}{8(1 + \alpha)}, \beta t, \beta(1 - t) - \frac{7\beta^2}{8(1 + \alpha)}, 1 - \beta\right)\right.$$

$$\left. \text{where} \quad 0 \leq t \leq 1 - \frac{7\beta}{8(1 + \alpha)}\right\} \quad \text{for} \quad 0 \leq \beta \leq 1$$

For $\alpha_0 < \alpha < 1$, define

$$X_1(\alpha) = \bigcup_{0 \leq \beta \leq} Y(\beta)$$

where

$$Y(\beta) = \left\{\left(\left(1 - \frac{1}{2\alpha}\right)\beta^2, \beta t, \beta(1 - t) - \left(1 - \frac{1}{2\alpha}\right)\beta^2, 1 - \beta\right),\right.$$

$$\left. \text{where} \quad 0 \leq t \leq 1 - \left(1 - \frac{1}{2\alpha}\right)\beta\right\} \quad \text{for} \quad 0 \leq \beta \leq 1$$

For $\alpha = 1$, define $X_1(1) = \bigcup_{0 \leq \beta \leq 1} Y(\beta)$, where

$$Y(\beta) = \left\{\left(\frac{\beta^2}{2}, \beta t, \beta(1 - t) - \frac{\beta^2}{2}, 1 - \beta\right)\colon 0 \leq t \leq 1 - \frac{\beta}{2}\right\}$$

Now it is not hard to check that the family $X_1(\alpha)$ is semimonotonic and that each $X_1(\alpha)$ is a solution to the game J except $X_1(1)$ because the element $(0, {}^1/_2, {}^1/_2, 0) \notin X_1(1) \cup \mathrm{dom}_1 X_1(1)$. If $X_1'(\alpha) = X_1(1) \cup (0, {}^1/_2, {}^1/_2, 0)$ then $X_1'(1)$ is a solution to J. But $\{X_1(\alpha)\colon 0 \leq \alpha < 1\}$ together with $X_1'(1)$ is not semimonotonic, for corresponding to $(0, {}^1/_2, {}^1/_2, 0)$ there exists no element $x \in X_1(\alpha)$ for any $\alpha > {}^3/_4$ with $\alpha x \leq (0, {}^1/_2, {}^1/_2, 0)$.

Further, the family $(X_1(\alpha)\colon 0 \leq \alpha \leq 1)$ is not fully monotonic, for corresponding to $(0, {}^1/_2, {}^1/_2, 0) \in X_1({}^3/_4)$ there exists no element $y \in X_1(1)$ with the property that $y \geq {}^3/_4(0, {}^1/_2, {}^1/_2, 0)$. Hence it is clear the conditions of both theorem 9.4.2 and theorem 9.4.3 are violated. Moreover, it is not difficult to check that Λ_1-sets are all empty, and as such we can take $\delta_1 = 1$. Hence this family $\{X_1(\alpha)\}$ can be used to produce product solutions to arbitrary games of the form $J \otimes K$—solutions that theorem 9.4.2 or 9.4.3 cannot predict. For further results in this connection, refer to [8] and [9].

REFERENCES

[1] Gillies, D. B. (1959). Solutions to general non-zero-sum games. *Ann. Math. Studies* No. 40, pp. 47–85.
[2] Lucas, W. F. (1968). The Proof That a Game May *Not* Have a Solution, Rand Corporation, R.M. 5543, PR.
[3] McKinsey, J. C. C. (1952). "Introduction to the Theory of Games," McGraw-Hill Book Company, New York.
[4] Mills, W. H. (1959). The four-person game—finite solutions on the face of the cube. *Ann. Math. Studies* No. 40, pp. 125–145.
[5] Parthasarathy, T. (1966). A note on compound simple games. *Proc. Am. Math. Soc.* **17**, 1334–1340.
[6] Shapley, L. S. (1964). Solutions of compound simple games. *Ann. Math. Studies* No. 52, pp. 267–305.
[7] von Neumann, J., and Morgenstern, O. (1943). "Theory of Games and Economic Behavior," Princeton University Press, Princeton, New Jersey.
[8] Parthasarathy, T. (1969). Product Solutions for Simple Games II. *Proc. Am. Math. Soc.* **20**, 107–114.
[9] Parthasarathy, T. (1969). Product Solutions for Simple Games III. *Proc. Am. Math. Soc.* **23**, 412–420.

Chapter Ten

STOCHASTIC GAMES

In this chapter we prove that the stochastic game has a value if and only if certain dummy games have a value using the results on dynamic programming when the state space is countable. The problem remains open when the state space is uncountable. We shall also establish that for the limiting average effective payoff on perfect information or cyclic stochastic games there are optimal stationary strategies for the two players. We shall present an example to indicate that the result on limiting average effective payoff cannot be extended in general.

10.1 STATEMENT OF THE PROBLEM

Consider a system with a countable state space S, which we take to be the set of natural numbers. The states will be denoted by s or s'. Once a day players I and II observe the current state s of the system, and then player I chooses an action a from a finite set A of actions, and player II chooses an action b from a finite set B of actions. As a result of this: (1) player I receives an immediate income $r(s, a, b)$, depending on the current state s of the system and the actions a, b chosen, and (2) the system moves to a new state s' with probability $q(s'/s, a, b)$, which also depends on s, a, b. We assume that $|r(s, a, b,)| \leq M$ for all $s \in S$, $a \in A$, $b \in B$. Further, there is specified a discount factor β, $0 \leq \beta < 1$, so that the value of the unit income n days in the future is β^n. Payments accumulate throughout the course of the play.

Player I wants to maximize his accumulated income, whereas player II wants to minimize the same. The problem is to choose a strategy for player I that will maximize his total expected income and to choose another strategy for player II that will minimize the total expected income of player I.

The concept of stochastic games was introduced by Shapley [8] with the proviso that, with probability 1, play terminates. He tackled the problem and gave a complete solution when S is finite. Gillette [4] and Hoffman and Karp [5] have considered the case when the play never terminates. Takahashi [9] considered the problem when S is finite and A, B are infinite.

We consider the case when the state space S is countable and A, B are finite or countable. In essence we have established, in the next section, that both the players have optimal (or ε-optimal) stationary strategies. Also, the stochastic game has a value if and only if certain dummy games have a value. It is clear from Aumann's theorem (p. 639 [1]) that in a game of perfect recall players can restrict themselves to playing only behavior strategies. In fact, in our setup it is enough if the players employ only stationary strategies—this is stated in [8] without proof. (See theorem 10.3.2 of this chapter.)

10.2 DEFINITION AND NOTATION

Let $S = \{1, 2, 3, \cdots\}$, and A, B are finite. A behavior strategy Π for player I is a sequence $(\Pi_1, \Pi_2 \cdots)$, where $\Pi_n \in Q(A|H_n)$ and

$$H_n = S \times A \times B \times S \times A \times B \times \cdots \times S \times A \times B \; ((3n - 3) \text{ factors})$$

is the set of possible histories of the system when the nth act must be chosen. Let $Q(Y|X)$ denote the set of all conditional probabilities on Y given X. If each $\Pi_n \equiv \mu$, then Π is called stationary, and it will be denoted by $\mu^{(\infty)}$. Let $\lambda^{(\infty)}$ stand for stationary strategy for player II, and let Γ stand for behavior strategy for II. Let $\mu(s)$, $\lambda(s)$ stand for stochastic vectors on A, B, respectively.

With each pair (μ, λ) associate (1) the vector $r(\mu, \lambda)$ whose sth coordinate is $r(s, \mu(s), \lambda(s))$ in $l^{(\infty)}(S)$ (denoted by $l^{(\infty)}$ hereafter) and (2) the infinite stochastic matrix $Q(\mu, \lambda)$ whose (s, s')th element is given by $q(s'/s, \mu(s), \lambda(s))$. If the players use the policy $\mu^{(\infty)}$ and $\lambda^{(\infty)}$, the system is a Markov chain with n-step transition matrix $Q_n(\mu, \lambda) = [Q(\mu, \lambda)]^n$. Now

$$r(s, \mu(s), \lambda(s)) = \sum_i \sum_j r(s, a_i, b_j) \, \xi_i(s) \, \eta_j(s)$$

where

$$\mu(s) = (\xi_1(s), \xi_2(s), \cdots) \quad \text{and} \quad \lambda(s) = (\eta_1(s), \eta_2(s), \cdots).$$

Similarly $q(s' \mid s, \mu(s), \lambda(s))$ is defined.

The total expected payoff for player I from the strategies (Π, Γ) is an $l^{(\infty)}$ vector and is denoted by $W(\Pi, \Gamma)$; the sth coordinate of $W(\Pi, \Gamma)$ is the total expected payoff to player I if the system starts in state s.

Definition 10.2.1: A strategy Π^* is optimal for player I if

$$W(\Pi^*, \Gamma) \geq \inf_{\Gamma} \sup_{\Pi} W(\Pi, \Gamma)$$

for all Γ. A strategy Γ^* is optimal for player II if

$$W(\Pi, \Gamma^*) \leq \sup_{\Pi} \inf_{\Gamma} W(\Pi, \Gamma)$$

for all Π.

Definition 10.2.2: The stochastic game has a value if

$$\inf_{\Gamma} \sup_{\Pi} W(\Pi, \Gamma) = \sup_{\Pi} \inf_{\Gamma} W(\Pi, \Gamma)$$

It is not hard to see that

$$W(\mu^{(\infty)}, \lambda^{(\infty)}) = \sum_{n=0}^{\infty} \beta^n Q_n(\mu, \lambda) \, r(\mu, \lambda)$$

Let (μ_1, Π) stand for the strategy for player I when he plays μ_1 on the first day and from the second day onward he employs Π. A straightforward computation shows the following:

$$W((\mu_1, \mu^{\infty}), \lambda^{(\infty)}) = r(\mu_1, \lambda) + \beta Q(\mu_1, \lambda) \, W(\mu^{(\infty)}, \lambda^{(\infty)})$$

We associate with each μ the transformation L_μ which maps the $l^{(\infty)}$ vector w into

$$L_\mu w = r(\mu, \lambda) + \beta Q(\mu, \lambda) \, w$$

Clearly, $L_\mu w \in l^{(\infty)}$. Further, it is clear that

$$L_{\mu_1} L_{\mu_2} \cdots L_{\mu_n} W(\mu^{(\infty)}, \lambda^{(\infty)}) = W((\mu_1, \mu_2, \cdots, \mu_n, \mu^{\infty}), \lambda^{(\infty)})$$

Let $v(\lambda^{(\infty)}) = \sup_{\mu^{(\infty)}} W(\mu^{(\infty)}, \lambda^{(\infty)})$. If we fix $\lambda^{(\infty)}$—that is, if player II always chooses the strategy $\lambda^{(\infty)}$—then it becomes a problem in dynamic programming. Soon after the appearance of Wald's work in sequential analysis, Richard Bellman recognized the broad applicability of the methods of sequential analysis, named this body of methods of dynamic programming, and applied the methods to many problems. Our formulation of the dynamic programming problem is somewhat narrower than Bellman's. For us, a dynamic programming problem is specified by four objects—S, A, q, r—where S and A are any nonempty Borel sets; q associates with each pair $(s, a) \in S \times A$, a probability distribution $q(\cdot | s, a)$ on S; and r is a bounded Baire function on $S \times A$. Periodically, say once a day, you observe the current state, s, of the system, then choose an act $a \in A$. Then the system moves to a new state, s', selected according to $q(\cdot | s, a)$, and you receive a reward $r(s, a)$. Your problem is, given the initial state of the system, to maximize your total expected reward over the infinite future.

When S is countable, A and B are finite, and if we fix $\lambda^{(\infty)}$, a result due to Blackwell (see p. 227, [2]) or Maitra (see p. 244, [6]) asserts the following:

$$\max_{\mu} L_\mu v(\lambda^{(\infty)}) = \max_{\mu^{(\infty)}} W(\mu^{(\infty)}, \lambda^{(\infty)})$$

$$= v(\lambda^{(\infty)})$$

10.3 EXISTENCE OF OPTIMAL STRATEGIES

Before stating the main theorems we shall prove some lemmas. Also, we shall assume that the players use only stationary strategies. Later we shall remove this restriction. We assume S to be countable and A and B finite.

LEMMA 10.3.1

$$L_{\mu_1} L_{\mu_2} \cdots L_{\mu_N} W = \sum_{n=0}^{N-1} \beta^n Q_n r(\mu_{n+1}, \lambda) + \beta^N Q_N W$$

where $Q_n = \prod_{K=1}^{n} Q(\mu_K, \lambda)$ for $n \geq 1$, and Q_0 is identity matrix. Proof follows by induction.

LEMMA 10.3.2: Let $\Pi = \{\mu_n : n = 1, 2, \cdots\}$ and $W \in l^{(\infty)}$. Then

$$\lim_{N \to \infty} L_{\mu_1} L_{\mu_2} \cdots L_{\mu_N} W = W(\Pi, \lambda^{(\infty)})$$

PROOF: From lemma 10.3.1, we have

$$\left\| L_{\mu_1} L_{\mu_2} \cdots L_{\mu_N} W - \sum_{n=0}^{N-1} \beta^n Q_n r(\mu_{n+1}, \lambda) \right\| \leq \beta^N \| W \| \to 0$$

as $N \to \infty$, since $\beta < 1$. Hence lemma 10.3.2. (Note that for $x \in l^\infty$, $\|x\| = \sup_{1 \leq i \leq \infty} |x_i|$.)

Define
$$v^* = \inf_{\lambda^{(\infty)}} v(\lambda^{(\infty)}) \quad \text{and} \quad \lambda^{(\infty)}$$

$$Tv^* = \inf_{\lambda} \max_{\mu} (L_\mu v^*)$$

$$= \inf_{\lambda} \max_{\mu} [r(\mu, \lambda) + \beta Q(\mu, \lambda) v^*]$$

LEMMA 10.3.3: $\qquad\qquad Tv^* = v^*$

PROOF: From Blackwell's result, it follows that $\max_{\mu} L_\mu v(\lambda^{(\infty)}) = v(\lambda^{(\infty)})$.

It is clear that $v^* \leq v(\lambda^{(\infty)})$, and the L-operator is monotone. Hence we can conclude that

$$Tv^* \leq v^*$$

We shall now establish that equality must hold good. If possible, let

$$(Tv^*)_s < v_s^* \quad \text{for some} \quad s$$

Let λ' be any optimal strategy for the minimizing player for the dummy game with payoff

$$r(\mu, \lambda) + \beta Q(\mu, \lambda) v^*$$

Such a strategy exists by von Neumann's minimax theorem. Note that λ' will be of the form $\lambda' = \{\lambda'(s): s \in S\}$.

$$\{r(\mu, \lambda') + \beta Q(\mu, \lambda') v^*\}_s \leq (Tv^*)_s < v_s^*$$

Now once for all fix λ'.

Since $L_\mu^n v^* \leq L_\mu v^*$ for every n (where $L_\mu^n = L_\mu L_\mu \cdots L_\mu$),

$$\lim_{n \to \infty} L_\mu^n v^* \leq L_\mu v^*$$

where

$$L_\mu v^* = r(\mu, \lambda') + \beta Q(\mu, \lambda') v^*$$

That is, $W(\mu^{(\infty)}, \lambda'^{(\infty)}) \leq L_\mu v^*$ (from lemma 10.3.2). Hence it follows that

$$W(\mu^{(\infty)}, \lambda'^{(\infty)})_s < v_s^* \quad \text{for all} \quad \mu^{(\infty)}$$

This, together with Blackwell's result, yields $v_s^* < v_s^*$, which is impossible. Hence the lemma.

LEMMA 10.3.4: If μ' is optimal for the maximizing player in the dummy game with payoff

$$r(\mu, \lambda) + \beta Q(\mu, \lambda) v^*$$

then $\mu'^{(\infty)}$ is optimal for the stochastic game.

PROOF: Since μ' is optimal for the maximizing player, we have

$$r(\mu', \lambda) + \beta Q(\mu', \lambda) v^* \geq Tv^* = v^* \quad \text{for all} \quad \lambda$$

As before, using the monotone nature of the L-operator and using lemma 10.3.2, we obtain the following result:

$$W(\mu'^{(\infty)}, \lambda^{(\infty)}) \geq v^* \quad \text{for all} \quad \lambda^{(\infty)}$$

This proves the conclusion of lemma 10.3.4. Thus we have proved the following theorem.

Theorem 10.3.1

When S is countable, and A and B are finite, the stochastic game has a value that is

$$\min_{\lambda^{(\infty)}} \max_{\mu^{(\infty)}} W(\mu^{(\infty)}, \lambda^{(\infty)}) = \max_{\mu^{(\infty)}} \min_{\lambda^{(\infty)}} W(\mu^{(\infty)}, \lambda^{(\infty)})$$

Further, $\mu^{(\infty)}\{\lambda^\infty\}$ is optimal for the maximizing (minimizing) player for the stochastic game if and only if $\mu\{\lambda\}$ is optimal for the maximizing (minimizing) player for the dummy game. Now we shall prove the following theorem.

Theorem 10.3.2

$$\min_{\Gamma} \max_{\Pi} W(\Pi,\Gamma) = \max_{\Pi} \min_{\Gamma} W(\Pi, \Gamma)$$

$$= \min_{\lambda^{(\infty)}} \max_{\mu^{(\infty)}} W(\mu^{(\infty)}, \lambda^{(\infty)}) = \max_{\mu^{(\infty)}} \min_{\lambda^{(\infty)}} W(\mu^{(\infty)}, \lambda^{(\infty)})$$

We need some preliminaries for the proof of theorem 10.3.2.

Definition 10.3.1: A strategy $\Pi = \{\Pi_n\}$ is called a Markov strategy if each $\Pi_n \in Q(A|H_n)$ is a degenerate element of $Q(A|H_n)$; that is, $\Pi = (\mu_1, \mu_2, \cdots)$, where μ_n is the strategy chosen by the player on the nth day, and he chooses his action according to the distribution $\mu_n(s)$ if the current state of the system is in s. In other words, μ_n depends only on the current state of the system.

Let Π be any behavior strategy for player I. Then we can assume that Π is already Markov from some point on, say $n > N$, since any two strategies that agree to the first N days have for fixed $\lambda^{(\infty)}$

$$\| W(\Pi, \lambda^{(\infty)}) - W(\Pi', \lambda^{(\infty)})\| \le \beta^N M|(1 - \beta)$$

We can now show that, as Blackwell has shown (see pp. 229–230, [2]), if $\Pi = (\Pi_1, \Pi_2, \cdots, \Pi_N, \mu_{N+1}, \mu_{N+2}, \cdots)$ is Markov for $n > N$, then $\Pi_1, \Pi_2, \cdots, \Pi_N$ can be replaced by $\mu_1, \mu_2, \cdots, \mu_N$, where $\mu_1, \mu_2, \cdots, \mu_N$ will depend only on the current state. The following result is given in [2] for fixed $\lambda^{(\infty)}$:

$$\max_{\Pi} W(\Pi, \lambda^{(\infty)}) = \max_{\mu^{(\infty)}} W(\mu^{(\infty)}, \lambda^{(\infty)})$$

Similarly, for fixed $\mu^{(\infty)}$,

$$\min_{\Gamma} W(\mu^{(\infty)}, \Gamma) = \min_{\lambda^{(\infty)}} W(\mu^{(\infty)}, \lambda^{(\infty)})$$

PROOF OF THEOREM 10.3.2:

$$\sup_{\Pi} \inf_{\Gamma} W(\Pi, \Gamma) \ge \sup_{(\mu^\infty)} \inf_{\Gamma} W(\mu^{(\infty)}, \Gamma)$$

$$\ge \max_{\mu^{(\infty)}} \min_{\lambda^{(\infty)}} W(\mu^{(\infty)}, \lambda^{(\infty)})$$

Hence,

$$\sup_{\Pi} \inf_{\Gamma} W(\Pi, \Gamma) \ge \min_{(\mu^\infty)} \max_{(\lambda^\infty)} W(\mu^{(\infty)}, \lambda^{(\infty)})$$

But

$$\min_{\lambda^{(\infty)}} \max_{\mu^{(\infty)}} W(\mu^{(\infty)}, \lambda^{(\infty)}) \ge \min_{\lambda^{(\infty)}} \max_{\Pi} W(\Pi, \lambda^{(\infty)})$$

$$\ge \inf_{\Gamma} \sup_{\Pi} W(\Pi, \Gamma)$$

Hence

$$\sup_{\Pi} \inf_{\Gamma} W(\Pi, \Gamma) \geq \inf_{\Gamma} \sup_{\Pi} W(\Pi, \Gamma)$$

This proves the theorem.

Remark 10.3.1: It is easy to see that theorems 10.3.1 and 10.3.2 are true if instead of assuming, as we have done, that there is a uniform upperbound on the number of actions available in each state, we stipulate only that the number of actions available in each state be finite.

Remark 10.3.2: We can give the following interpretations to the operator: $L_\mu W \cdot L_\mu W$ is the expected payoff to player I, as a function of the initial state, if players I and II start using $\mu^{(\infty)}$ and $\lambda^{(\infty)}$, respectively, but the game is terminated at the beginning of the second day with a final reward, $W(s')$, where s' is the state at termination. One can give L_μ^n a similar interpretation, replacing "second" by $n + $ 1st.

We shall now assume that S, A, and B all are countable. We will also assume that players I and II use only stationary strategies. We shall now state and prove the following theorem [7].

Theorem 10.3.3

The stochastic game has a value if and only if the dummy games with the following payoff have a value

$$r(s, \mu(s), \lambda(s)) + \beta \sum_{s'=1}^{\infty} q(s'| s, \mu(s), \lambda(s)) v_{s'}$$

where

$$v_{s'} = \inf_{\lambda^{(\infty)}} \sup_{\mu^{(\infty)}} W(\mu^{(\infty)}, \lambda^{(\infty)}) (s')$$

Let $v = (v_1, v_2, \cdots)$.

PROOF: We shall assume that dummy games have a value. That is,

$$\inf_{M} \sup_{N} [r(s, \mu(s), \lambda(s) + \beta Q(\mu, \lambda) v]$$

$$= \sup_{N} \inf_{M} [r(s, \mu(s), \lambda(s)) + \beta Q(\mu, \lambda) v] \quad \text{for every} \quad s$$

where $M = N = $ the space of stochastic vectors defined over $A = B$.

The above fact enables us to conclude that there are ε-optimal strategies, μ_ε, λ_ε, for both players for the dummy game. That is, given any ε, there exists μ_ε, λ_ε such that

$$r(\mu_\varepsilon, \lambda) + \beta Q(\mu_\varepsilon, \lambda) v \geq \inf \sup [r(\mu, \lambda) + \beta Q(\mu, \lambda) v] - \varepsilon \quad \text{for all} \quad \lambda$$

and

$$r(\mu, \lambda_\varepsilon) + \beta Q(\mu, \lambda_\varepsilon)\, v \le \sup_\mu \inf_\lambda \left[r(\mu, \lambda) + \beta Q(\mu, \lambda)\, v \right] + \varepsilon \quad \text{for all} \quad \mu$$

Also, we know from [2] dynamic programming that

$$\sup \left[L_\mu v(\lambda^{(\infty)}) \right] = v(\lambda^{(\infty)})$$

where $v(\lambda^{(\infty)}) = \sup\limits_{\mu^{(\infty)}} W(\mu^{(\infty)}, \lambda^{(\infty)})$. These facts enable us to conclude that lemma 10.3.3 is valid. Also, it can be proved that, if μ_ε is ε-optimal for the dummy game, then $\mu_\varepsilon^{(\infty)}$ is $\varepsilon/(1 - \beta)$-optimal for the stochastic game. Hence the stochastic game has a value.

Conversely, suppose that the stochastic game has a value. That is,

$$\inf_{\lambda^{(\infty)}} \sup_{\mu^{(\infty)}} W(\mu^{(\infty)}, \lambda^{(\infty)}) = \sup_{\mu^{(\infty)}} \inf_{\lambda^{(\infty)}} W(\mu^{(\infty)}, \lambda^{(\infty)})$$

Now, consider the dummy games with payoff $r(\mu, \lambda) + \beta Q(\mu, \lambda)\, v$. We know from dynamic programming that

$$\sup_\mu \left[r(\mu, \lambda) + \beta Q(\mu, \lambda)\, v\, (\lambda^{(\infty)}) \right] = v(\lambda^\infty)$$

and

$$\inf \left[r(\mu, \lambda) + \beta Q(\mu, \lambda)\, u(\mu^\infty) \right] = u(\mu^\infty)$$

where

$$v(\lambda^{(\infty)}) = \sup_{\mu^{(\infty)}} W(\mu^{(\infty)}, \lambda^{(\infty)})$$

and

$$u(\mu^{(\infty)}) = \inf_{\lambda^{(\infty)}} W(\mu^{(\infty)}, \lambda^{(\infty)})$$

Let

$$v = \inf \sup W(\mu^{(\infty)}, \lambda^{(\infty)}) = \sup \inf W(\mu^{(\infty)}, \lambda^{(\infty)})$$

Hence one can conclude the following:

$$\inf_\lambda \sup_\mu \left[r(\mu, \lambda) + \beta Q(\mu, \lambda)\, v \right] \le \inf_{\lambda^{(\infty)}} \sup_{\mu^{(\infty)}} W(\mu^{(\infty)}, \lambda^{(\infty)})$$

$$\le \sup_{\mu^{(\infty)}} \inf_{\lambda^{(\infty)}} W(\mu^{(\infty)}, \lambda^{(\infty)})$$

$$\le \sup_\mu \inf_\lambda \left[r(\mu, \lambda) + \beta Q(\mu, \lambda)\, v \right]$$

That is, the dummy games have a value.

Remark 10.3.3: This theorem includes theorem 1 of Takahashi [9].

Remark 10.3.4: We do not face any measurability difficulties, since S, A, and B are countable. When S is uncountable, measure structure is necessary, and difficult problems concerning the measurability of the value as a function of the state arise. We have not yet tackled this problem.

Note added in Proof: We have proved in [11] that the stochastic game has a value and that both players have optimal stationary strategies, under certain conditions, where S, A, and B are compact metric spaces. A similar result is true also for positive stochastic games. For details refer to [12].

10.4 STOCHASTIC GAMES WITH ZERO STOP PROBABILITIES

In this section we shall approach the problem under consideration from a different angle. Shapley [8] has defined a stochastic game to be a game consisting of a finite collection of positions among which two players pass according to jointly controlled transition probabilities. Thus, if at the ith position the players choose pure strategies k and l, the probability of passing to position j is given by P_{kl}^{ij}, and the (finite) payoff to the first player is a_{kl}^i. Payments are to accumulate throughout the game. Shapley has, in addition, assumed that at each position there is a positive probability s_{kl}^i of stopping the game. We may include the stop probability within our framework by requiring the existence of a position j with $a_{kl}^j \equiv 0$ and with $P_{kl}^{ij} \geq \delta > 0$ and $P_{kl}^{ji} = 0$ for all i, k, l.

In this section we shall investigate the consequences of relaxing the requirement that the game shall certainly end; that is, we shall allow the stop probability to be zero at some or all positions. It follows, of course, that the accumulated payoff to player I may not be well defined. As such, Gillette [4] introduces effective payoffs, defined over the entire game, and succeeds in finding optimum strategies for the players for a certain class of stochastic games. The results contained in this section are due to Gillette.

We shall assume that there are M positions. A stationary strategy, μ, for player I can be written as $\mu = (\mu^1, \mu^2, \cdots, \mu^M)$ where each μ^i is a probability distribution on the set of available alternatives $[1, 2, \cdots, K_i]$ at the ith position. Let λ stand for a stationary strategy for player II.

The probability of passing from position i to position j and the payoff to player I at position i, given stationary strategies (μ, λ), is

$$P_{ij}^1(\mu, \lambda) = \sum_{K,l} P_{Kl}^{ij} \mu_K^i \lambda_l^i$$

$$A_i(\mu, \lambda) = \sum_{K,l} a_{Kl}^i \mu_K^i \lambda_l^i$$

Let P_{ij}^0 denote the identity matrix and, for $n \geq 1$, let

$$P_{ij}^{n+1}(\mu, \lambda) = \sum_{m=1}^{M} P_{im}^n(\mu, \lambda) \, P_{mj}^1(\mu, \lambda)$$

The payoff that player I has accumulated after traversing $(N + 1)$ positions, $N \geq 0$, starting with position i, given stationary strategies (μ, λ) is

$$H_i^N(\mu, \lambda) = \sum_{n=0}^{N} \sum_{j=1}^{M} [P_{ij}^n(\mu, \lambda)] \, [A_j(\mu, \lambda)]$$

For nonstationary strategies, this formulation of $H_i^N(\mu, \lambda)$ is not valid; however, we shall retain the notation for the accumulated payment independent of the type of strategy.

An *effective payoff* is some function of the payoff's accumulated position by position during the course of the game. If the payoff after $(n + 1)$ positions have been played is discounted by $(1 - s)^n$, $0 < s < 1$, we write the effective s-discounted payoff for strategies (μ, λ) as

$$D_i^S(\mu, \lambda) = \sum_{n=0}^{\infty} (1 - s)^n \, (H_i^n(\mu, \lambda) - H_i^{n-1}(\mu, \lambda))$$

where $H_i^{-1}(\mu, \lambda) = 0$.

We also define a *limiting average payoff*, $L_i(\mu, \lambda)$, by

$$L_i(\mu, \lambda) = \liminf_{N \to \infty} \left(\frac{H_i^N(\mu, \lambda)}{N} \right)$$

From theorem 10.3.2 it follows that for the s-discounted payoff on a stochastic game—possibly with stop probabilities—there are stationary strategies μ^*, λ^* such that for all strategies μ, λ and all $i = 1, 2, \cdots, M$,

$$D_i^s(\mu, \lambda^*) \leq D_i^s(\mu^*, \lambda^*) \leq D_i^s(\mu^*, \lambda)$$

Moreover, if the game is of perfect information, a solution exists in stationary pure strategies. To obtain an analogous result for the limiting average payoff, we need the following lemmas.

LEMMA 10.4.1: Suppose that f_n is a sequence of functions converging uniformly on $X \times Y$ to a function f. If for each n there exist $(x_n, y_n) \in X \times Y$ such that for all $(x, y) \in X \times Y$:

$$f_n(x, y_n) \leq f_n(x_n, y_n) \leq f_n(x_n, y) \cdots \circledast$$

Then

(1) $$\inf_Y \sup_X f(x, y) = \sup_X \inf_Y f(x, y)$$

Moreover, for each $\varepsilon > 0$ there is an n such that, if (x_n, y_n) satisfies ⊛, then for all $m \geq n$:

(2)
$$f_m(x_n, y_n) - \varepsilon < f_m(x_n, y)$$
$$f_m(x_n, y_n) + \varepsilon > f_m(x, y_n)$$

PROOF: Choosing n so that for all $m \geq n$ and for all x, y,

$$|f_m(x, y) - f(x, y)| < \varepsilon/4$$

one may easily verify (2) from which (1) follows.

LEMMA 10.4.2: If a_n is a sequence of nonnegative numbers, then

$$\limsup_{N \to \infty} \frac{1}{N} \sum_{i=1}^{N} a_i \leq \limsup_{s \to 0^+} s \sum_{n=1}^{\infty} (1 - s)^{n-1} a_n$$

$$\liminf_{N \to \infty} \frac{1}{N} \sum_{j=1}^{N} a_i \geq \liminf_{s \to 0^+} s \sum_{n=1}^{\infty} (1 - s)^{n-1} a_n$$

For a proof, see Titchmarsh (p. 227 [10]).

LEMMA 10.4.3: If $P = \|P_{ij}\|$ is an M-dimensional stochastic matrix, with P^n the nth power of P, then there is a stochastic matrix $Q = \|q_{ij}\|$ such that

$$\lim_{N \to \infty} \frac{1}{N} \sum_{i=1}^{n} P^i = Q$$

Moreover, if there is an integer m and a number $\delta > 0$ such that

$$\min_{i, j} P_{ij}^m \geq \delta$$

then the numbers q_{ij} are independent of i, and

$$\left| \frac{1}{N} \sum_{n=0}^{N} (P_{ji}^n - q_{ij}) \right| \leq \frac{1}{N} \frac{1}{(1 - M\delta) [1 - (1 - M\delta)^{1/m}]}$$

These results one can get from Doob (pp. 173–175 [3]).

We impose the condition that the payoff to player I be nonnegative. From the definition of the limiting average effective payoff we see that this results in no loss of generality. We first concentrate on stationary strategies. It is not hard to see from lemma 10.4.3 that

$$L_i(\mu, \lambda) = \lim_{N \to \infty} \frac{1}{N} \sum_{n=1}^{N} \sum_{j=1}^{M} P_{ij}^n(\mu, \lambda) A_j(\mu, \lambda)$$

It follows from lemma 10.4.2 that

$$L_i(\mu, \lambda) = \lim_{s \to 0^+} sD_i^s(\mu, \lambda) \cdots \cdots \cdots \cdots \cdots \cdots \textcircled{U}$$

There is always a solution for the s-discounted payoff; consequently, from lemma 10.4.1 there are at least ε best strategies in the limiting average effective payoff, provided the passage to the limit is uniform over a set of (μ, λ) simultaneously containing solutions of an infinite sequence of s-discounted payoffs with the discount approaching zero. We consider essentially two cases: (1) perfect information stochastic games, and (2) cyclic stochastic games.

Case 1. *Perfect Information*: The convergence in $\textcircled{U}$ is uniform over the (finite) set of all pure stationary strategies. Since this set also contains solutions for the s-discounted games for all s, we find that there are pure stationary strategies μ^*, λ^* so that for all pure stationary strategies (μ, λ) for all i,

$$L_i(\mu^*, \lambda) \geq L_i(\mu^*, \lambda^*) \geq L_i(\mu, \lambda^*)$$

Let (μ, λ) be any pair of strategies. From lemma 10.4.2 we have

$$L_i(\mu, \lambda^*) \leq \limsup_{N \to \infty} \left[\frac{1}{N} H_i^N(\mu, \lambda^*) \right]$$

$$\leq \limsup_{s \to 0^+} s \sum_1^\infty (1 - s)^{n-1} [H_i^n(\mu, \lambda^*) - H_i^{n-1}(\mu, \lambda^*)]$$

$$\leq \limsup_{s \to 0^+} [sD_i^s(\mu, \lambda^*)]$$

$$\leq \limsup_{s \to 0^+} [sD_i^s(\mu^*, \lambda^*)] = L_i(\mu^*, \lambda^*)$$

$$\leq \liminf_{s \to 0^+} [sD_i^s(\mu^*, \lambda^*)]$$

$$\leq \liminf_{s \to 0^+} [sD_i^s(\mu^*, \lambda)]$$

$$\leq \liminf_{s \to 0^+} s \sum_1^\infty (1 - s)^{n-1} [H_i^n(\mu^*, \lambda) - H_i^{n-1}\mu^*, \lambda)]$$

$$\leq \liminf_{N \to \infty} \frac{1}{N} H_i^N(\mu^*, \lambda) = L_i(\mu^*, \lambda)$$

Thus we have the following theorem.

Theorem 10.4.1

For the limiting average effective payoff on a stochastic game of perfect information there are pure stationary strategies (μ^*, λ^*) such that, for all strategies (μ, λ) and all $i = 1, 2, \cdots, M$,

$$L_i(\mu^*, \lambda) \geq L_i(\mu^*, \lambda^*) \geq L_i(\mu, \lambda^*)$$

We shall now consider cyclic stochastic games.

Definition 10.4.1: A cyclic stochastic game is one in which there exists a $\delta > 0$ and $N > 1$ such that

$$\min_{\substack{i,j,K_m,l_m \\ 1 \leq m \leq N}} \left\{ \sum_{j_1,\cdots,j_{N-1}} \left[p_{K_1,l_1}^{ij_1}, p_{K_1,l_2}^{j_1,j_2}, \cdots, p_{K_N,l_N}^{j_{N-1},j} \right] \right\} = \delta$$

Remark 10.4.1: Here, independent of the choice of strategies, there is a true probability of passing from position i to position j in exactly N moves. In particular, there can be no "absorbing subsets" of the positions of the stochastic game, nor is there a stop position.

It is clear that the transition probabilities $P_{ij}^1(\mu, \lambda)$ for a cyclic game satisfy the conditions of the second part of lemma 10.4.3 simultaneously in all stationary strategies. As a result of this it may be shown that, for stationary strategies, the passage to the limit ① is uniform over all stationary strategies. Consequently, from lemma 10.4.1 and theorem 10.3.1, for stationary strategies (μ, λ) and all i,

$$\inf_{\lambda} \sup_{\mu} L_i(\mu, \lambda) = \sup_{\mu} \inf_{\lambda} L_i(\mu, \lambda)$$

Using the norm,

$$\|x - \bar{x}\| = \sum_{i,K} |x_K^i - \bar{x}_K^i|$$

it may be shown that the functions $sD_i^s(\mu, \lambda)$ are continuous in μ, λ. From the uniform convergence, it follows that $L_i(\mu, \lambda)$ is continuous so that inf and sup may be replaced by min and max. Using this fact and using lemma 10.4.2 as in case 1, we have theorem 10.4.2.

Theorem 10.4.2

For the limiting average effective payoff on cyclic stochastic games there are stationary strategies μ^* and λ^* such that, for all i,

$$L_i(\mu, \lambda^*) \leq L_i(\mu^*, \lambda^*) \leq L_i(\mu^*, \lambda)$$

Remark 10.4.2: It is not in general true that the stochastic game will have a solution in stationary strategies for the limiting average effective payoff.

Consider the following probabilities and payoffs for a three-position game [4]:

$$P_{KI}^{11} = \begin{vmatrix} 1 & 0 \\ 1 & 0 \end{vmatrix}, \qquad P_{KI}^{12} = \begin{vmatrix} 0 & 1 \\ 0 & 0 \end{vmatrix}, \qquad P_{KI}^{13} = \begin{vmatrix} 0 & 0 \\ 0 & 1 \end{vmatrix}$$

$$P_{KI}^{21} = |0|, \qquad P_{KI}^{22} = |1|, \qquad P_{KI}^{23} = |0|$$

$$P_{KI}^{31} = |0|, \qquad P_{KI}^{32} = |0|, \qquad P_{KI}^{33} = |1|$$

$$a_{KI}^{1} = \begin{vmatrix} 1 & 0 \\ 0 & 1 \end{vmatrix}, \qquad a_{KI}^{2} = |0|, \qquad a_{KI}^{3} = |1|$$

It may be checked that

$$\min_{\lambda} \max_{\mu} L_1(\mu, \lambda) = 1$$

$$\max_{\mu} \min_{\lambda} L_1(\mu, \lambda) = \tfrac{1}{2}$$

Here we have a noncyclic game (positive 2 may be considered a stop position). Convergence in ⓤ is nonuniform, and the function $L_i(\mu, \lambda)$ is discontinuous in λ at $\lambda = (1, 0)$.

Remark 10.4.3: In general, the value of a stochastic game depends on the starting position. However, with the limiting average effective payoff for a cyclic game, the value is independent of the starting position (this is a consequence of lemma 10.4.3).

Remark 10.4.4: It is not hard to check that, if there are finite sets X and Y such that if, for all $s < s_0$, optimal strategies of the s-discounted payoff lie in (X, Y), then there is an $s_1 \leq s_0$ and a pair (x, y) in (X, Y) that is uniformly the solution for the s-discounted payoff, $s \leq s_1$, and for the limiting average payoff. However, such a result is not valid when one considers the cyclic stochastic games. For a counter example interested readers can refer to [4].

REFERENCES

[1] Aumann, R. J. (1964). Mixed and behavior strategies in infinite extensive games. *Ann. Math. Studies* No. 52, p. 639.

[2] Blackwell, D. (1965). Discounted dynamic programming. *Ann. Math. Stat.* **36**, 226–235.

[3] Doob, J. C. (1953). "Stochastic Processes," John Wiley & Sons, New York.

[4] Gillette, D. (1957). Stochastic games with zero-stop probabilities. In "Contributions to the Theory of Games," Vol. 3, pp. 179–187 (*Ann. Math. Studies*), edited by M. Dresher, A. W. Tucker, and P. Wolfe, Princeton University Press, Princeton, New Jersey.

[5] Hoffman, A. J., and Karp, R. M. (1966). On nonterminating stochastic games. *Management Sci.* **12**, 359–370.

[6] Maitra, A. (1965). Dynamic programming for countable state systems. *Sankhya* **27**, 244.

[7] Maitra, A., and Parthasarathy, T. (1967). "Dynamic Programming Approach to Stochastic Games with Countable State Space," Technical Report No. 1/67, Indian Statistical Institute, Calcutta.

[8] Shapley, L. S. (1953). Stochastic games. *Proc. Natl. Acad. Sci. U.S.* **39**, 1095–1100.
[9] Takahashi, M. (1962). Stochastic games with infinitely many strategies. *J. Sci. Hiroshima Univ.* **26**, 123–134.
[10] Titchmarsh, E. C. (1939). "The Theory of Functions," 2nd edition, Oxford University Press, London.
[11] Maitra, A., and Parthasarathy, T. (1970). On stochastic games. *J. Opt. Theory Appl.* **5**, 289–300.
[12] Maitra, A., and Parthasarathy, T. On stochastic games II. Submitted for publication.

Chapter Eleven

SOME OPEN PROBLEMS

The purpose of this chapter is to mention a few unsolved problems and make a few remarks about them.

1. Let $K(x, y)$ be a real-valued function on $X \times Y$, where X and Y are subsets of finite-dimensional Euclidean spaces. If X and Y are compact, $K(x, y)$ is jointly continuous in (x, y), and further, if every 2×2 matrix $\|K(x_i, y_j)\|$, $i, j = 1, 2$; $x_i \in X$, $y_j \in Y$, has a saddle point, then it is known that

$$\min_{Y} \max_{X} K(x, y) = \max_{X} \min_{Y} K(x, y)$$

This result is of little interest from the practical point of view. It will be of some interest if one can give a set of workable sufficient conditions without the convexity assumption for the existence of saddle points.

2. When the state space, S, and the actions A and B of the two players are countable, we know that the stochastic game has a value, provided that certain dummy games have a value. The problem seems to be difficult when S is uncountable, even if we assume that A and B are finite. To be precise, let S be a Borel subset of some complete separable metric space. Assume that A and B are finite. Further, suppose that the two players employ only measurable stationary strategies. That is a strategy Π for player I is of the form

$$\Pi = (f, f, f, \cdots)$$

where f is a measurable function from $S \rightarrow P_A$, where P_A stands for the set of all probability vectors on A. Our conjecture is that the value of the stochastic game as a function of the state space is absolutely measurable. One can also consider the same problem when S, A, and B are arbitrary Borel subsets of some complete separable metric space.

3. We know now that there exists a ten-person game which has no solution in the von Neumann–Morgenstern sense. Now we can raise the following questions in this connection: (1) What is the smallest n for which there exists no solution on the von Neumann–Morgenstern sense? (2) What sort of economic interpretations can one give for games with no solution?

For other problems, the interested reader can refer to the literature [1, 2, 3].

REFERENCES

[1] Isaacs, R. (1965). "Differential Games, A Mathematical Theory with Applications to Warfare and Pursuit, Control and Optimization." John Wiley & Sons, New York.

[2] Luce, R. D., and Raiffa, H. (1957). "Games and Decisions—Introduction and Critical Survey," John Wiley & Sons, New York.

[3] McKinsey, J. C. C. (1952). "Introduction to the Theory of Games," McGraw-Hill Book Company, New York.

AUTHOR INDEX

Numbers in parentheses indicate the numbers of the references when these are cited in the text without the names of the authors.

Numbers set in *italics* designate the page numbers on which the complete literature citation is given.

Addison, J. W., 150, *151*
Akers, S. B., Jr., 139(2), *151*
Arrow, K. J., *86*
Ascoli, – –, 141
Aumann, R. J., 239, *251*

Balinski, M L., 169(1), *183*
Banach, S., 10
Barankin, E. W., *86*
Bellman, R., 191, *211*
Berge, C., 19, *30, 56, 111,* 112, *137*
Berkovitz, L.D., 184, 187, 190, 204, *211, 212*
Blackwell, D., 38(2), *56, 86,* 150(4), 151, *151,* 240, 241, 242, 243, *251*
Bliss, G. A., *212*
Bohnenblust, H., 23, *86*
Brouwer, L. E. J., 27, 155, 156
Boltyanskii, V. G., *212*
Burger, E., *56*

Dalkey, N., *56*
Dantzig, G. B., *111*
Doob, J. L., 248, *251*
Douglas, R. G., 132, *137*
Dresher, M., *56, 86,* 151, 187, *212,* 251
Dubins, L.E., *212*
Dunford, N., 30, 137, *183*

Fan, Ky, 112, 119, 120, 128(6), *137*
Farkas, J., 22, 23, 108
Fleming, W.H., 184, 190, 207(13), *212*
Fomin, F. V., 1, *30, 212*
Friedman, A., *212*
Frobenius, G., 84

Gale, D., *56, 86, 111,* 138, 149, *151*
Gamkrelidze, R. V., *212*
Gantmacher, F., *86*

Ghouila-Houri, A., *56*
Gillette, D., 238, 246, 251(4), *251*
Gillies, D. B., 227, 229, *237*
Girshick, M. A., 38(2), *56, 86*
Glicksberg, I. L., 124, 136, *137,* 156, *183*
Grenander, U., *212*
Gross, D., 136, *137,* 138(6), *151*
Gross, D. -, 138(8), *151*

Hahn, 10
Helly, E., 129
Hoffman, A. J., 238, *251*
Howson, J. J. Jr., 159, 161(5), 166(5), *183*

Isaacs, R., 184, 203, *212,* 253(1), *254*
Isoda, K., 154, *183*

Kakutani, S., 28, *56,* 156
Kaplansky, I., *86,* 181
Karlin, S., 23, *56, 86,* 124, 130(3), *137, 212*
Karp, R. M., 238, *251*
Kelley, J. L., 3(3), *30,* 136(4), *137*
Knaster, B., 27
Kneser, H., 112, 114, 128(5), *137*
Kolmogorov, A. N., 1, *30, 212*
Krein, M. G., 14, 18, 59, 133, 145, 146, 147, *151*
Kuratowski, C., 27
Kuhn, H. W., 38(9), 56, *57,* 86, 87, 106, 159, 176, *183*

Lemke, C. E., 159, 161(5), 166(5), *183*
Loomis, L. H., *57*
Lucas, W.V., 213, 222, 227, *237*
Luce, R. D., *57, 111,* 159(6), *183,* 253(2), *254*
Lusin, N., 138, 139(10), *151*

255

SUBJECT INDEX